Advanced Manufacturing

Springer-Verlag London Ltd.

Other titles published in this series:

A Systems Approach to AMT Deployment
D.R. Towill and J. Cherrington (Eds)

Human-Intelligence-Based Manufacturing
Y. Ito (Ed.)

Intelligent Manufacturing: Programming Environments for CIM
W.A. Gruver and J.C. Boudreaux (Eds)

Automatic Supervision in Manufacturing
M. Szafarczyk (Ed.)

Modern Manufacturing
M.B. Zaremba and B. Prasad (Eds)

Advanced Fixture Design for FMS
A.Y.C. Nee, K. Whybrew and A. Senthil kumar

Intelligent Quality Systems
D.T. Pham and E. Oztemel

Computer-Assisted Management and Control of Manufacturing Systems
S.G. Tzafestas (Ed.)

The Organisation of Integrated Product Development
V. Paashuis

Advance Manufacturing: Decision, Control and Information Technology
S.G. Tzafestas (Ed.)

A.Y.C. Nee, S.K. Ong and Y.G. Wang (Eds)

Computer Applications in Near Net-Shape Operations

With 243 Figures

Springer

A.Y.C. Nee, PhD
S.K. Ong, PhD
Mechanical & Production Engineering Department,
National University of Singapore, 10 Kent Ridge Crescent, Singapore 119260

Y.G. Wang, PhD
Huazhong University of Science & Technology,
11-602, West Second #zone Wuhan, Hubei, 430074, PR China

Series Editor
Professor Duc Truong Pham, PhD, DEng, CEng, FIEE
University of Wales Cardiff, School of Engineering, Systems Division,
P.O. Box 917, Cardiff CF2 1XH, UK

ISBN 978-1-4471-1159-7

British Library Cataloguing in Publication Data
A catalogue record for this book is available from the British Library

Library of Congress Cataloging-in-Publication Data
Computer applications in near net-shape operations / A.Y.C. Nee, S.K. Ong, and Y.G. Wang (eds.).
p. cm. - (Advanced manufacturing series)
Includes bibliographical references and index.
ISBN 978-1-4471-1159-7 ISBN 978-1-4471-0547-3 (eBook)
DOI 10.1007/978-1-4471-0547-3
1. Near net shape (metalwork) – Automation. 2. CAD/CAM systems.
I. Nee, A.Y.C. (Andrew Yeh Chris), 1948- . II. Ong, S. K., 1969- . III. Wang, Y. G., 1938- . IV. Series: Advanced manufacturing series (Springer-Verlag)
TS213.C66 1999 99-35688
671.3—dc21 CIP

Typesetting: Camera ready by editors

69/3830-543210 Printed on acid-free paper SPIN 10520955

Foreword

Having edited "Journal of Materials Processing Technology" (previously entitled "Journal of Mechanical Working Technology") for close on 25 years, I have seen the many dramatic changes that have occurred in the materials processing field. Long gone are the days when the only "materials processing" carried out was virtually the forming of conventional metals and alloys, and when the development of a new product or process in a great number of cases called for several months of repetitive trial-and-error, with many (mostly intuition- or experience-based) expensive and time-consuming modifications being made to the dies, until success was achieved. Even when a 'successful' product was formed, its mechanical properties, in terms of springback and dimensional accuracy, thickness variations, residual stresses, surface finish, etc., remained to be determined. Bulk-forming operations usually required expensive machining to be carried out on the product to impart the required dimensional accuracy and surface finish.

Over the years, the experience-based craft of metal forming has given way to the science of materials processing. With the use of the computer, forming operations can be simulated with accuracy, to determine the best forming route and the associated forming loads and die stresses, and to predict the mechanical properties of the formed product, even down to its surface texture. The accuracy of the products has increased remarkably, with research now being undertaken into the allowances to be made in the manufacture of dies to compensate for the effect of their elastic distortion on the dimensions of the product. The blanking operation now extends down to such small size components as the lead frames of ICs, having sub-millimeter widths, requiring previously unheard of accuracy of the alignment of the punch and the die, and requiring the use of the fine-blanking process. Of tremendous interest internationally is near net-shape forming, where components can be produced of high dimensional accuracy, requiring no further operations to be performed upon them before being put into service.

It is with great pleasure therefore, to see the present new book written by A.Y.C. Nee, S.K. Ong and Y.G. Wang, covering as it does all the latest developments over the whole field of materials processing, under the generic title "Near Net-Shape Operations". That the authors are admirably qualified to write such a book is beyond question, judging from their great many published works in the areas concerned. The book is not overly academic, but considers the industrially-relevant aspects of the work, and provides many examples of application of the techniques and systems involved for the benefit of the reader. CAD/CAM/CAE/CNC/FMC/IPD/FEM, expert systems, etc are all introduced and explained, and examples given of their application in all of the relevant areas of near net-shape forming, such as: sheet metal forming; stamping, blanking and fine blanking; progressive stamping; bending; nibbling; punch and tool selection; die, punch and ejector design and construction; wire EDM and its programming; bulk metal forming; design of dies; calculation of forming loads; injection moulding; wire-frame, surface and solid modelling; the design of plastic injection moulds; flow simulation; cooling simulation; information transfer; tool path generation; etc. The chapters on FEM and CNC are particularly expansive, and should be required reading for all those, whether student, teacher or industrialist, working or researching in these areas.

I congratulate the authors on having produced this splendid new book.

F.W. Travis

Preface

The proposal to write this book originated when Professor Y.G. Wang from the Huazhong University of Science & Technology (HUST), PR China visited the Department of Mechanical & Production Engineering, National University of Singapore (NUS) in 1995. In search of a joint collaboration, Professor Wang and Professor Nee suggested that since both universities had done a good amount of work in the application of computer techniques to the design, simulation and analysis of near net-shape operations, a book to detail the research findings would be very appropriate.

We started with a thorough search on the availability of such a book and were soon convinced that such a book did not exist. Although there were many articles written on similar topics, it will be useful to group them under one cover. Since the book was planned to cover many topics, it would not have been possible for both of us to write all of them on our own. Therefore it was suggested that researchers from both universities would be approached as contributors to the different chapters of this book.

Planning continued for a year, but unfortunately Professor Nee was appointed Dean of Faculty of Engineering from 1995 to 1998 and this slowed down the progress of the book tremendously. Fortunately, a young faculty, Dr S.K. Ong joined the department and she was most dedicated in assisting the compilation and editorial work of this book. The authors must also thank Professor Duc Truong Pham, the Series Editor and Mr Nicholas Pinfield of Springer-Verlag for their kind patience in waiting for more than three years for this manuscript to be ready.

We intend this book to be of interest to researchers, graduate students and practising engineers involved in the design, simulation and analysis of near net-shape operations. We gratefully acknowledge the published information of many distinguished researchers worldwide who had laid the foundation of this book. We have included some of their contributions either in references or bibliography. We would also like to thank many graduate students of the two universities who had contributed to both the theoretical and experimental investigations of near net-shape operations over the last decade.

The authors are most grateful to Professor F.W. Travis, the Chief Editor of the Journal of Materials Processing Technology, for his kind and informative Foreword to this book.

A.Y.C. Nee
National University of Singapore
S.K. Ong
National University of Singapore
Y.G. Wang
Huazhong University of Science and Technology

Contents

List of Abbreviations

1-D	One-Dimensional
2-D	Two-Dimensional
3-D	Three-Dimensional
AC	Alternate Current
A/D	Analogue-to-Digital
AGFPO	Automatic Generation of Forming Process Outline
ALPID	Analysis of Large Plastic Incremental Deformation
APT	Automatically Programmed Tools
BEM	Boundary Element Method
B-Reps	Boundary Representation
CAD	Computer Aided Design
CADD	Computer Aided Design and Draughting
CAE	Computer Aided Engineering
CAM	Computer Aided Manufacturing
CAPP	Computer Aided Process Planning
CNC	Computer Numerical Control
CPU	Central Processing Unit
CRT	Cathode-Ray Tube
CSG	Constructive Solid Geometry
D/A	Digital to Analogue
DC	Direct Current
DDE	Dynamic Data Exchange
DXF	Drawing Exchange Format
EDM	Electro-Discharge Machining
EPROM	Erasable Programmable Read Only Memory
FEM	Finite Element Method
FMC	Flexible Manufacturing Cell
HUST	Huazhong University of Science and Technology
IGES	Initial Graphics Exchange Specification
IMOLD	Intelligent Mould Design System
IPD	Intelligent Progressive Dies
LED	Light emitting diode
MBR	Model-Based Reasoning
MFC	Microsoft Foundation Class

NC	Numerical Control
NUS	National University of Singapore
PC	Personal Computer
PLC	Programmable Logic Control
PRF	Part Representation File
PWM	Pulse Width Modulation
RAM	Random Access Memory
RXD	Received Data
STD	Standard
STEP	STandard Exchange of Product Data
SUMT	Sequential Unconstrained Minimisation Techniques
TXD	Transmitted Data
WCS	World Co-ordinate System

List of Authors

Chapter 1

Introduction to near net-shape operations

S.K. Ong and A.Y.C. Nee

1.1 Introduction

Traditionally, metal working industries have been relying on basic forming and fabrication techniques such as casting, forging, stamping, shearing and machining operations to achieve specified geometrical shapes of their products. In recent years, because of stiffer industrial competition, shorter product to market time, the development of new materials, shortage of certain metals, and the increase in energy costs, these traditional processing methods are being critically analysed and re-evaluated. In addition, manufacturers are always concerned with high material wastage in metal forming and machining operations. These concerns have become more acute in recent years as metal prices have risen and the availability of certain metals has become sporadic, while the demands for these scarce metals are growing rapidly. It is therefore highly desirable to produce parts in fewer processing steps and with as little wastage as possible. Several novel techniques for manufacturing components to "net-shapes" or to "near net-shapes," based on traditional processes, are being developed to meet these challenges of today and tomorrow.

The process of producing to final net-shapes is a desirable goal for many metal working industries. Casting, which is a very old and time-proven process, can produce a specified solid object directly from the molten state. Casting can thus be considered as a forerunner of the newer family of near net-shape processes. This book examines a number of metal working processes that have recently been developed to achieve net-shapes or near net-shapes and brings together related research topics, particularly the development of computer aids, at the National University of Singapore (NUS) and the Huazhong University of Science and Technology (HUST) in PR China.

1.2 Classification of near net-shape operations

Modern manufacturing industries are more concerned with the efficient utilisation of materials because of the pressing environmental issues and the need to conserve materials and energy. This has led to the development of near net-shape processes. Cost reduction is another important factor favouring any initial manufacturing processes that could result in configurations closer to final dimensions. Billions of dollars are being spent annually in removing materials to reach final part shapes and specifications. Producing parts closer to the final specifications in fewer processing steps is obviously more viable economically from several aspects such as reduction in energy consumption, material wastage, as well as lower environmental burden.

There are certain drawbacks in near net-shape operations, e.g., close tolerances and surface finish are difficult to maintain without secondary machining operations or processes. Several manufacturing processes, such as casting and forging, and the sheet metal working processes such as stamping, bending and rolling, are currently regarded as good near net-shape candidates. Injection moulded parts have reached a point where they can compete with machined counterparts from strength, tolerance and finish considerations.

In the following sections, several near net-shape operations are briefly described under three broad categories, *viz.*, sheet metal forming operations, massive (bulk) metal forming operations, and injection moulding processes. Reference is made to the contributions by researchers from NUS and HUST in the subsequent chapters of this book. The applications of computer tools in the development of novel machines for near net-shape operations are briefly mentioned.

1.2.1 Sheet metal forming

Sheet metal forming processes are among the most versatile of all metal forming operations. They are generally used to produce components with high ratios of surface area to thickness. Products that are made using the sheet metal forming processes can easily be found in our daily life, such as metal desks, filing cabinets, electrical and mechanical appliances, car bodies, aircraft fuselages, beverage cans, etc. Compared to the other near net-shape processes, sheet metal parts offer the advantages of good strength to weight ratio and the ability to assume relatively complex shapes. These near net-shape operations can be classified into two groups, namely, the shearing and forming operations. The shearing operations are used to remove blanks of suitable shapes and dimensions from larger sheets. They include punching, blanking, nibbling and fine blanking. Forming operations are used to shape sheet metal parts into intricate shapes, they include press brake bending, roll forming, stretch forming and deep drawing.

Chapter 2 of this book covers a number of common shearing and forming operations such as progressive stamping, fine blanking, nibbling, press brake forming, and wire EDM that is commonly used to make stamping dies. Computer-aided techniques such as feature modelling of stampings and the application of expert systems in design are introduced.

Chapter 8 reports an Intelligent Progressive Die (IPD) system developed at the NUS. This system uses AutoCAD as the modelling platform and the model-based reasoning approach for die design. It is able to generate a three-dimensional strip layout with very little input from the user. This system is undergoing substantial beta testing for preliminary design and verification by several industrial partners in Singapore.

1.2.2 Massive (bulk) metal forming processes

1.2.2.1 Forging

Forging is one of the traditional metal forming processes dating back several hundred, if not thousand of years. Open-die hammer forging is one of the oldest forms, followed by drop forging with closed impressions. Press forging employs a slower squeezing action in deforming the heated plastic metal, compared to rapid blows in drop forging. Upset forging is used to deform the heated end of a bar stock and causing this to conform to the shape of a closed die. Closed die forgings have much better dimensional and finish control and would require relatively few secondary operations.

In each case, the design of forging dies is complex and would require computer tools to simulate and solve material flow and heat transfer problems.

Chapter 3 describes several massive (bulk) forming processes such as cold upsetting and closed die forming. The use of CAD/CAM and CAE tools in arriving at optimal process parameters, design of die cavities and blocker impressions is reported.

1.2.2.2 Casting

Casting can take many forms with wide ranging dimensional tolerance and surface finishes. Permanent mould castings such as hot and cold chamber die-casting, gravity casting, centrifugal casting, etc. can produce parts with very good tolerances that would require few secondary machining operations. Precision investment casting is able to produce castings of very high qualities suitable for final applications.

Although casting has not been dealt with specifically in this book, the computer tools used to perform analysis and design of injection moulds can be applied to die-casting moulds. CAM tools can be used to fabricate casting moulds and inserts.

1.2.2.3 Extrusion

Extrusion is usually a hot-working process where metal is extruded, under pressure, to uniform, cross-sectional shapes through a die-opening. The advantages of extrusion include the ability to produce intricate shapes with good accuracy and surface finish at high production rates.

Chapter 5 describes the application of the finite element method to near net-shape operations in simulating metal flow, load requirements and the process variables such as the accumulated strain, strain rate and temperature. Specific examples include the material flow patterns in forward and backward extrusion processes, and the improvement of the design of extrusion dies.

1.2.3 Injection moulding

Plastic injection moulding is a process where polymeric granules are plasticised in a heated cylinder into a semi-liquid paste, and is squeezed out of the other end of the cylinder through a nozzle into a closed mould. It then hardens upon cooling, taking closely the shape of the mould cavity. After cooling, the mould opens and the finished part is ejected. Injection moulding is very economical for high volume production. The initial equipment cost is high and the moulds are costly to design and fabricate. The process, however, is very rapid upon set-up, with common cycle times ranging from several seconds to minutes.

Chapter 4 presents the CAD/CAE techniques in the detailed design of plastic injection moulds and their components such as mould bases, cores and cavities, runners and gates. Single-, two- and three-dimensional flow and cooling analyses are covered in detail. The use of CAM tools in tool path generation and die manufacturing is reported.

Chapter 7 describes the various functional modules of a commercialised intelligent mould design system IMOLD® developed at NUS. Currently, additional design modules are being developed.

1.2.4 Machines for near net-shape operations

Apart from process improvement through the use of CAE tools, the requirement for improved and novel machines for near net-shape operations is another important issue. Chapter 6 describes the design of machines for near net-shape operations. Aspects considered include kinetic and dynamic analysis, strength analysis, control system design, etc., using various modelling and simulation techniques. Universal CNC systems in machines for near net-shape operations are also reported.

1.3 Near net-shape operations: past, present and future

The demand for improved cost-effectiveness coupled with better material utilisation has brought about recent developments and advancements in near net-shape processes. Material savings from both flashless and other near net-shape processes can be substantial, as at least 30% saving in raw materials can be achieved. Machining time is also reduced, and in general, most near net-shape processes could reduce machining costs ranging from 30% to 100%, not to mention savings in energy costs.

Spurred by economic concerns, the aerospace industry is currently showing more than a passing interest in near net-shape operations and processes. Economic issues, however, are not the only reason for the pursuance of near net-shape processing technologies. Tighter performance standards, rising costs of super-alloys, more stringent mechanical property requirements have contributed to this technology shift.

Although near net-shape processes often require extra capital investment and/or special handling procedures, the overall cost per part can be substantially decreased in many cases. The elimination of trimming, machining, and some final finishing operations, the better use of critical materials and, in some cases, the substitution of materials, as well as the possible decrease in total energy consumption provide the justification for these techniques.

The implications for near net-shape processes both now and in the future are staggering. The new technologies coming on stream open up avenues for promising breakthroughs in metallurgy and materials engineering applications.

Rapid prototyping, which is a radically new approach for making components to their final shapes by adding material instead of material deformation and/or removal, is a promising process although at this stage, the parts produced using this technique are mainly semi-functional. Rapid prototyping processes are not covered in this book as there are already a number of texts and reference books on this subject matter.

Bibliography

Alexander J M, Brewer R C 1963 *Manufacturing Properties of Materials*. Van Nostrand, London.

Altan T (consulting ed) 1998 *Metal Forming Handbook*. Schuler Company, Germany. Springer-Verlag

Altan T 1982 Computer Aided Design and Manufacturing (CAD/CAM) of Hot Forging Dies. *Journal of Applied Metalworking* January 1982:77-85

Altan T, Oh I S, Gegel H 1983 *Metal Forming, Fundamentals and Applications*. American Society for Metals, USA

Bittence J C 1979 Edging closer to 'net-shape'. *Materials Engineering* May 1979:47-50

Blazynski T Z (ed) 1989 *Plasticity and Modern Metal-Forming Technology*. Elsevier Applied Science

Blazynski T Z 1976 *Metal Forming Tool Profiles and Flow*. The Macmillan Press Ltd

Blazynski T Z 1986 *Design of Tools for Deformation Processes*. Elsevier Applied Science Publishers

Bradley E F 1979 Near Net Shape Processing for Gas Turbine Components. *Journal of Applied Metalworking* July 1979:73-79

Chandler H E 1978 Appliance makers Evaluate Net-Shape Process. *Metal Progress* December 1978:22-25

Chenot J L, Onate E 1988 *Modelling of Metal Forming Processes*. Kluwer Acadmic Publishers

Demeri M Y (eds) 1990 *Expert System Applications in Materials Processing and Manufacturing*. The Minerals, Metals & Materials Society, Pennsylvania

Dwivedi S N, Paul A J, Dax F R (eds) 1992 *Concurrent Engineering Approach to Materials Processing*. The Minerals, Metals & Materials Society

Gold R 1978 Forging technologies of the twenty-first century. *Precision Metal* November 1987:40-43

Gunasekera J S, Fischer C E, Anbajagane R 1993 A Three Stage Approach to the Design of Manufacturing Process. In: *Proceedings of the Near-Net-Shape Manufacturing: Examining Competitive Processes Conference*, Pittsburgh, Pennsylvania, 1993. pp 65-70

Harvey R E 1979 Cutting Metal Loss Tied to Near Net Shapes. *Iron Age* November 1979:57-63

Hoffman R 1993 Automation of the Design of Forming Processes. In: *Proceedings of the Near-Net-Shape Manufacturing: Examining Competitive Processes Conference*, Pittsburgh, Pennsylvania, 1993. pp 61-64

Johnson W, Mellor P B 1962 *Plasticity for Mechanical Engineers*. Van Nostrand, London

Kobayashi S, Oh S I, Altan T 1989 *Metal Forming and the Finite Element Method*. Oxford University Press

Lee P W, Fergusan B L 1993 *Near-Net-Shape Manufacturing*. The Materials Information Society, ASM International Materials Park, Ohio

Ong S K, de Vin L J, Nee A Y C, Kals H J J 1997 Fuzzy set theory applied to bend sequencing for sheet metal bending. *Journal of Materials Processing Technology* 69:29-36

Voller V R, Marsh S P, El-kaddah N 1994 *Materials Processing in the Computer Age II*. The Minerals, Metals & Materials Society, Pennsylvania

Chapter 2

CAD/CAM for sheet metal forming and related processes

Z.G. Li, N.F. Choong, K.H. See Toh, H.T. Loh and A.Y.C. Nee

2.1 Introduction and basic techniques

2.1.1 Feature modelling of stampings

Stamping processes such as punching, bending, drawing, etc., have been widely used to manufacture sheet metal parts or stampings. The design and manufacture of stamping dies plays a very important role in the production of stampings.

In the past two decades, CAD and CAM of stamping dies has been under continuous development. Several developed systems can be used in practice to design and manufacture dies [1], [2]. However, the majority of these CAD/CAM systems cannot cover all the stages in the design and manufacture of stamping dies. Due to insufficient information in CAD product models, these systems cannot solve many tasks such as stage layout and die construction design which are highly dependent on the knowledge and experience of the designers. This is particularly true in progressive die design. Therefore, it is most appropriate to use knowledge-based methods to establish CAD/CAM systems for the design of stamping dies.

Feature modelling techniques provide a good approach. Although feature modelling of mechanical parts has been studied in detail in the past, it is still necessary to address feature modelling of stampings. This is because features of stampings are quite different from those features found on mechanical parts made from machining, casting and forging [3].

2.1.1.1 Features of stampings

From an examination of a large number of stampings, it can be seen that most stampings have the following features:

1. Stampings usually consist of sections that are formed by drawing, bending, flanging or other stamping operations. Additional shapes that can be found in stampings include holes, notches and tabs.
2. Drawn, flanged and bulged areas of stampings are usually of regular shapes. In most cases, there are connections between these areas.
3. Bent areas are usually made by simple bending, with the bent area having a cylindrical shape with planar shapes connected to it.
4. There exists certain constraining relations between two adjacent areas of a stamping. For example, the bend radius should be larger than a certain value.
5. Blanks can be made of different materials and have a variety of shapes.
6. Stampings should meet all the technical requirements such as dimensions, tolerances and burr height.

Stamping features listed above fall into three groups, *viz.*, form features, precision features and material features. A stamping can be properly modelled using these features.

2.1.1.2 Representation of form features

The operations required to manufacture a stamping can be mainly determined from its form features. Hence, the representation of form features has an important bearing on stamping process planning.

A stamping can be regarded as a shell. The shape of a stamping can be represented by a number of primitives called feature shells. Feature shells are divided into four groups according to the stamping operations used to make them, *viz.*, shape-shells, plane-shells, bend-shells and punch-shells. Shape-shells are used to represent drawing, flanging and bulging form features. Bend-shells describe bending form features. Plane-shells define planar areas in stampings. Punch-shells deal with the holes, notches and tab-like projections on a stamping. A potentiometer part shown in Figure 2.1 consists of six form features. Shells 1, 2 and 3 are plane-shells, while shells 4 and 5 are bend-shells. Shape-shell 6 represents the cylindrical area formed by the drawing operation.

2.1.1.3 Definition of feature shells

Feature shells can be defined by characteristic lines. There are two types of characteristic lines. The lines that determine the positions of feature shells are called position characteristic lines. The other type of characteristic lines is the boundary characteristic lines, which are used to define the boundaries of feature shells. Since the shape definition of stampings is based on part drawings, characteristic lines of a feature shell can be easily determined from these part drawings.

Position characteristic lines of a plane-shell are two parallel lines that determine the position, direction and thickness of the shell (Figure 2.2). The contour of a plane-shell is formed by its boundary characteristic lines.

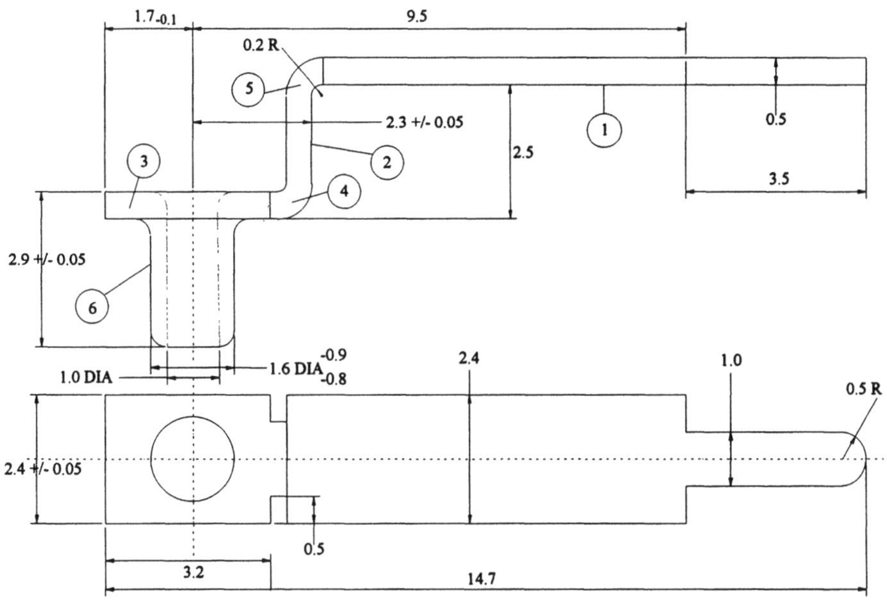

Figure 2.1: A potentiometer part.

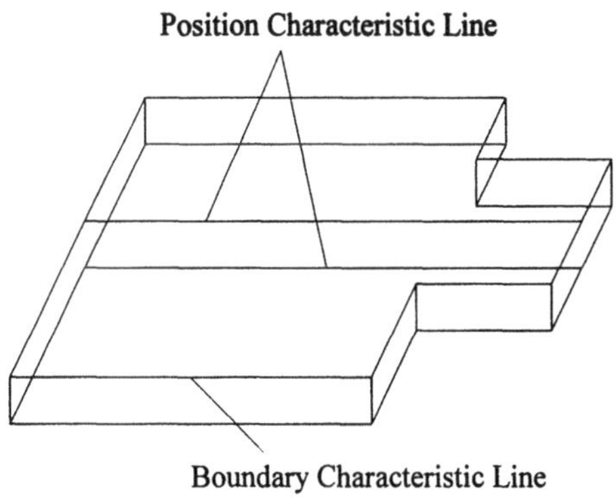

Figure 2.2: Characteristic lines of a plane-shell.

A shape-shell is defined by the sweeping method. Rotational sweeping is used when an axi-symmetric shape-shell is defined, as shown in Figure 2.3a. In this case, the position characteristic line is the rotational axis, and the boundary characteristic line is the generatrix. A non-axi-symmetric shape-shell can be defined by general sweeping, and the trajectory is the position characteristic line, as illustrated in Figure 2.3b.

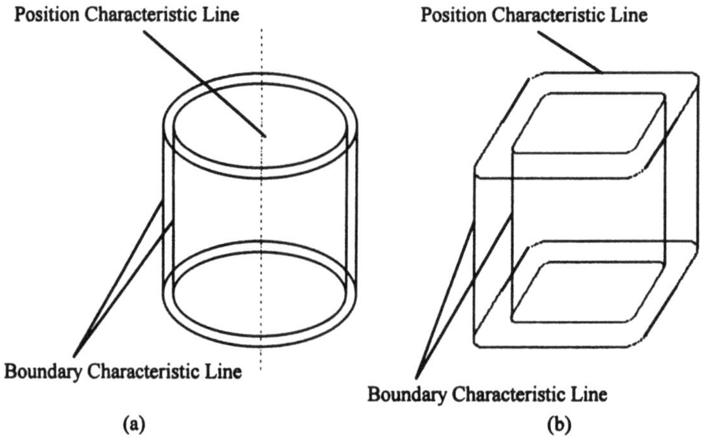

Figure 2.3: Shape-shell defined by sweeping.

The position characteristic lines of a bend-shell are two arcs (Figure 2.4). Its boundary characteristic lines are generated by the system based on different bending parameters.

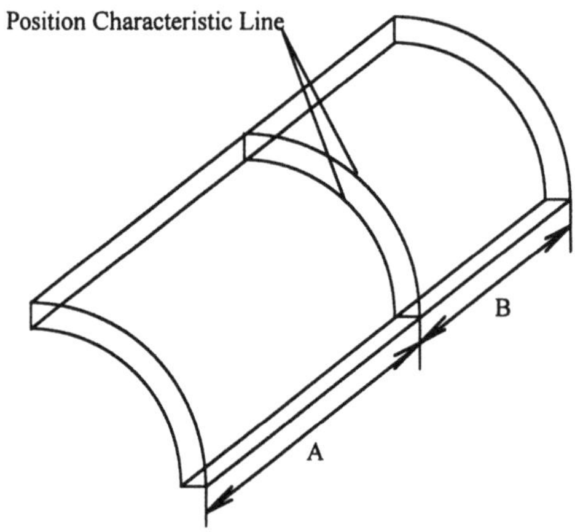

Figure 2.4: Definition of a bend-shell.

The contour of a punch-shell is its boundary characteristic line. Its position characteristic line is the same as the punched shell.

Using the geometry editor of the system, characteristic lines of feature shells can be input interactively. After the characteristic lines have been specified, the geometric shape of the feature shell and the corresponding topological relations are automatically generated.

Figure 2.5 is a wire-frame display of the potentiometer part shown in Figure 2.1 that has been defined using the feature modelling system for stampings.

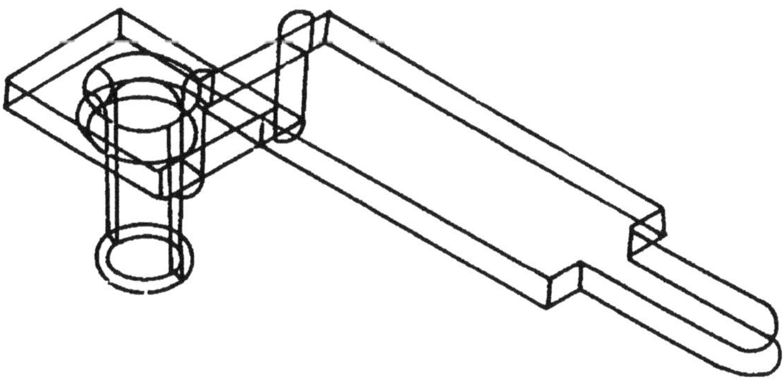

Figure 2.5: Wire-frame display of a stamping.

2.1.1.4 Structure of the feature modelling system

The structure of the feature modelling system for stampings is shown in Figure 2.6. The designer can use the geometry editor to define feature shells. The functions of the system include creating, modifying, deleting, moving, copying and displaying controls. To ensure the consistency and uniqueness of feature definitions, the volumetric boundary representation is used to represent stampings. Material features, precision features and form features are stored in the system library.

Based on feature shells, the system creates a relation model that records the relations between the shells. Shell relations are stored in a tree structure as shown in Figure 2.7. In this figure, the nodes represent feature shells and the arrowhead lines indicate the connective relations between shells.

Besides the relation model, the system creates the geometric model of a stamping based on boundary representations (Figure 2.8). The data structure adopted is the vertex-edge-face type, where faces are the centres of this structure. The information of the definition datum and additional shapes is also stored in the geometric model. Additional shapes refer to the outer rings, inner rings, notches, tabs, etc. As shown in Figure 2.8, each additional shape has an edge ring. Each edge has five pointers that point to end-points, adjacent faces and the definition datum.

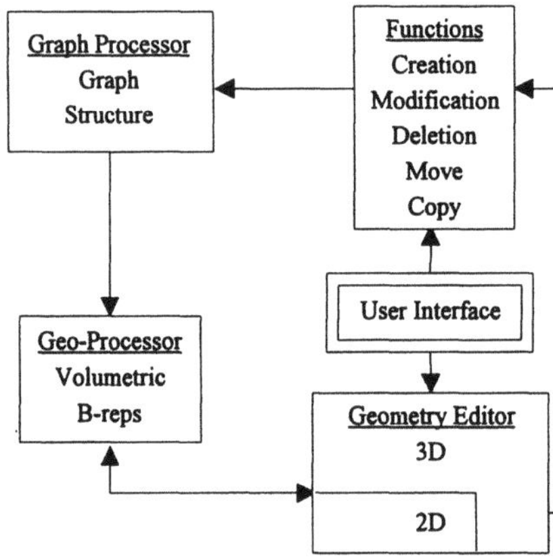

Figure 2.6: Structure of the system.

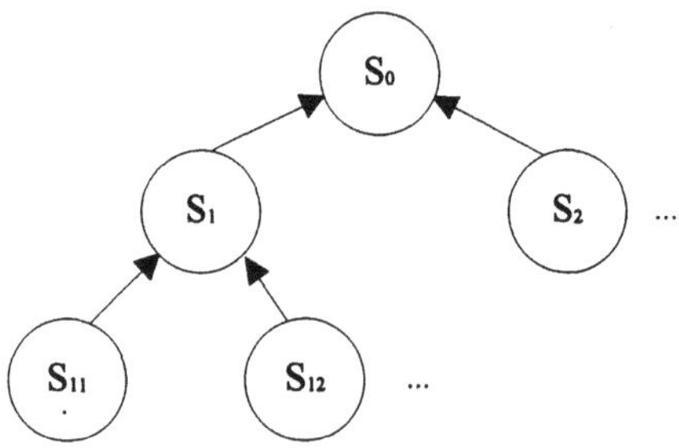

Figure 2.7: Relation model.

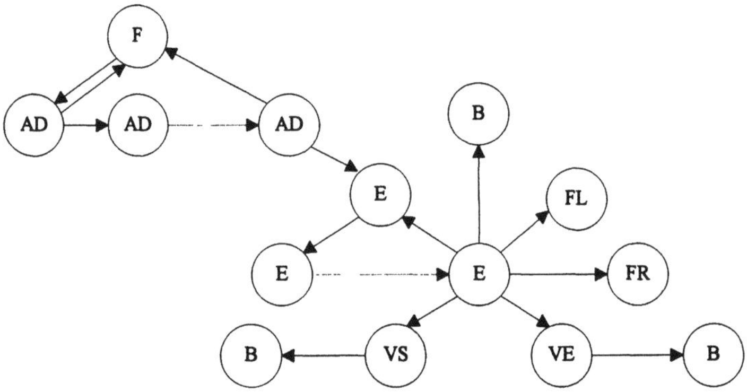

Figure 2.8: Boundary representation of the geometric model.

The product model created by the feature modelling system contains enough information to meet the requirements for the design and manufacture of stamping dies. For example, stamping process planning for progressive die design, which is called strip layout, is very difficult to complete based on the model created using a conventional geometric modelling system. However, it can be easily done based on the product model generated using the feature modelling system.

2.1.2 Application of expert systems to stamping die design

Stamping die design is an "art." It depends greatly on the experience of the designer, which can only be obtained through years of practice. This is particularly true for progressive die design where the experience and skill of the designer are the major deciding factors.

At present, many CAD/CAM systems have been used to design and manufacture stamping dies. Applications of these systems are mainly confined to drafting and NC programming. There are many tasks in stamping die design that cannot be handled using these existing systems. More intelligent CAD/CAM systems for stamping die design have to be developed.

Several expert systems have been developed for stamping process planing. The AGFPO (Automatic Generation of Forming Process Outline) system can automatically generate process plans for axi-symmetric deep-drawn parts [4]. In this system, concatenation of volumetric shape elements is used to represent the shape of a stamping. The problem solving strategy of the system is generate-test-and-rectify. In the experimental expert system developed by Reha [5], the operation sequence for a rectangular box-shape part with bending and round holes can be determined. Rules are used for knowledge representation in this system. The product shape is input using an interactive text dialogue method. Round holes and bent edges are regarded as features. The advantages of using an expert system include the utilisation of the knowledge of domain experts, high efficiency and flexibility. Development of expert systems for stamping process planning can also promote systematisation and standardisation of the knowledge in sheet metal forming.

Xiao et al [6] have developed an expert system for the stamping process to improve the CAD and CAM of progressive dies. Feature modelling is used to define products, and the object-oriented method is employed to represent knowledge and features of the products.

2.1.2.1 Description of sheet metal parts

The process plan of a sheet metal part is determined by its shape, accuracy and other technical requirements. Conventional geometric modelling techniques describe sheet metal parts from the geometric viewpoint. The shapes of these parts are usually represented using B-Reps or CSG. Non-geometric information, such as the accuracy and functions of the parts, is not included in these geometric models. However, expert systems use symbols for knowledge inferencing. It is insufficient to make inference merely from the geometric information of the parts.

The feature modelling method is adopted in the developed expert system to solve the problems present in ordinary geometric modelling systems. Process features of sheet metal parts, such as punching, bending, drawing and local forming, can be extracted from the models. Both geometric and non-geometric information, which is needed for the expert system to infer, can be provided by the incorporated feature modelling system.

In this feature modelling system, sheet metal parts can be defined using features and their parameters, as discussed in Section 2.1.1. Process features extracted from the feature models by the expert system include planar features, bending features, drawing features, hole features and local forming features (Figure 2.9). The features of sheet metal parts are treated as objects in the system. The operation sequence to form a sheet metal part can be determined based on its process features.

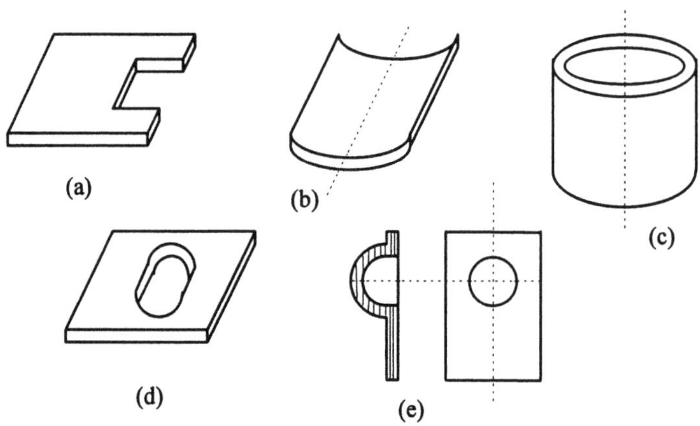

Figure 2.9: Process features of sheet metal parts, (a) plane features, (b) bending features, (c) drawing features, (d) hole features, (e) local forming features.

2.1.2.2 *Representation of stamping process planning knowledge*

Knowledge base and inference engine are the two major elements in an expert system. The knowledge base consists of facts, rules, etc. It is difficult to collect domain knowledge to create the knowledge base. The stamping process planning knowledge stored in the system comes from die design handbooks and the industry. Comprehensive investigation and collection of domain knowledge are conducted during system development. After the knowledge has been formalised in the standard format, it is stored in the knowledge base using a text editor.

The knowledge of stamping process planning includes descriptions of the stampings, process calculation and design experience. It is difficult to represent this knowledge using a single knowledge representation method. The knowledge representation method used in the system is a hybrid representation scheme. Frames are used as the backbone of the system, supporting both rule-based and process-based representations. Objects are described by frames, and feature attributes and design knowledge are encapsulated in the structures of these objects.

There are four types of slots in a frame:

1. Relation slot: This represents the relations between objects. It is used to link the frames with each other.
2. Attribute slot: The slot contains static information of an object.
3. Method slot: This is used to represent process data, and control procedures of the objects.
4. Rule slot: Knowledge in rule form is stored in this slot.

In this knowledge representation scheme, process features are regarded as objects. Different kinds of knowledge can be included in an object. For example, the attributes of a bending feature include the bend angle, bend radius, relations with adjacent areas and springback. Bending features are treated as a class of objects, and represented by the bend frame below.

```
Bend-frame:
  Radius-slot: R
  Angle-slot: Angle
  Rule-slot: rule1:
           if R<1
           then the bend radius is too small
           rule 2:
           ... ...
  Method-slot: Spring-back (design,ang)
          var
          ang: real;
          begin
          ... ...
          end;
Endframe.
```

2.1.2.3 *Inference process of the expert system*

The architecture of the expert system is shown in Figure 2.10. A knowledge acquisition module is used to create the knowledge base. There is an explanation module to answer questions from the users, and provide explanation of the results of the inference process. External functions, which are written in C-language, are used as supporting functions of the system. Using these functions, stamping process planning results can be displayed on the screen, and the necessary calculations can be performed.

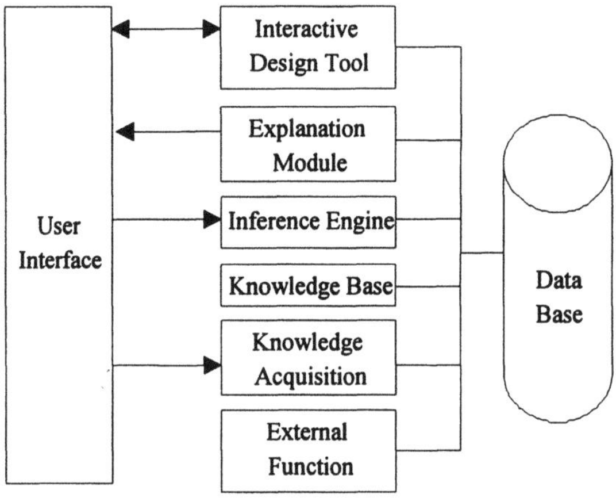

Figure 2.10: Architecture of the expert system.

After the required product has been defined by the feature modelling system, the expert system can be initialised by instancing different classes of objects. Instance frames are generated dynamically. Through the relation slots of the frames, a frame tree that describes the product is formed, as shown in Figure 2.11. Each node in the figure represents an instance frame, and the lines between the nodes represent the relations of the frames.

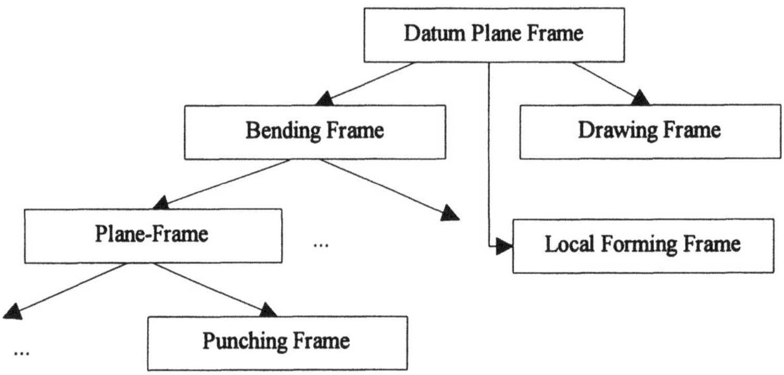

Figure 2.11: Frame tree to describe a product.

Different inference methods correspond to different representations of knowledge. Frame inference engine, rule inference engine and process inference engine are used in this system. A meta inference engine, which can decide the inference order and control the different inference engines, is developed to co-ordinate these inference engines.

Strip layout for progressive die design is carried out in two steps. In the first step, the inference engine generates the operations by searching through the process features and the knowledge base. Next, the operations are assigned to the proper stations of the strip.

An interactive tool package is incorporated into the system to combine automatic design with interactive design to enhance the practicality of the system.

A strip layout of a sheet metal part is shown in Figure 2.12. The required punching and bending operations and their sequences for this part are designed by the expert system. In this example, the pilot-hole location and scrap shapes are designed using an interactive tool package.

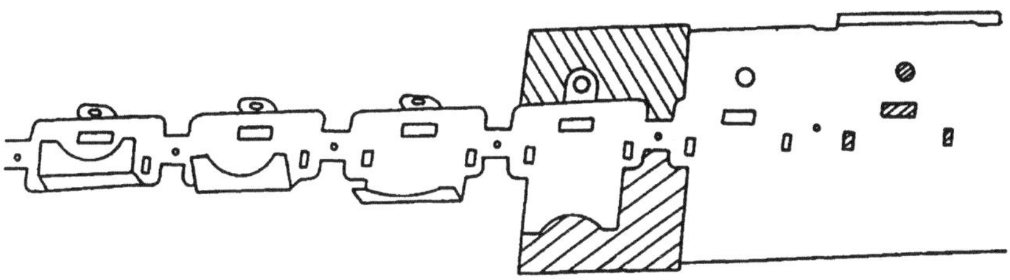

Figure 2.12: An example of strip layout.

2.2 Optimisation of blank layout

Material cost makes up about 60% of the total cost of stamped parts. Economic utilisation of materials is therefore of paramount importance in the production of metal stampings.

The purpose of blank layout is to search for an optimal arrangement of stamped parts, i.e., highest material utilisation ratio, among all the possible arrangements. It is difficult to find an optimal blank layout manually because the shapes of the stamped parts can be quite complex and varied, and there are too many possible arrangements. Computer-aided blank layout has superiority over the manual method as it can increase material utilisation ratio significantly. An average increase of 3% to 7% in material utilisation can be achieved with computer-aided blank layout.

2.2.1 Mathematical description of blank layout

Blank layout schemes commonly used in stamping production are illustrated in Figure 2.13, where (a) and (b) are single-row layouts, (c) and (d) are double-row layouts, and (b) and (d) are turn-round or pair-wise layouts.

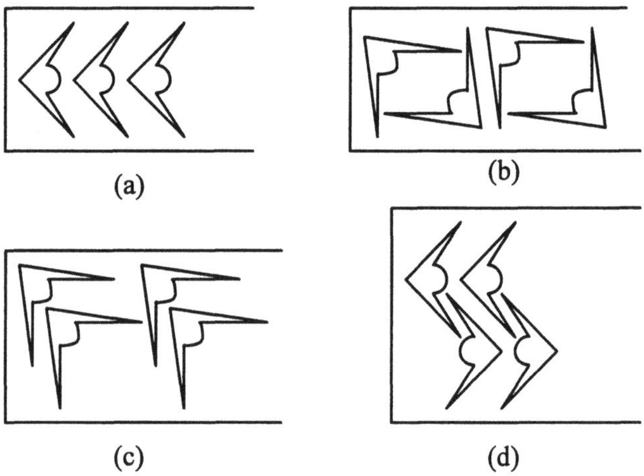

Figure 2.13: Commonly used blank layout schemes.

Blank layout is a non-linear programming problem with an objective function of finding the highest possible material utilisation ratio. For the case of blanking coil material, blank layout can be evaluated by the material utilisation ratio expressed as the feeding pitch. It can be calculated using the following expression.

$$\eta = S/BH \tag{2.1}$$

S is the total area of the parts arranged in a pitch, B is the coil width, and H is the feeding pitch.

For the case of blanking on strip material, the material utilisation ratio is as follows:

$$\eta = nS_1/BL \tag{2.2}$$

S_1 is the area of a part, n is the number of the parts arranged in the strip, and B and L are the width and length of the strip respectively.

For the case of blanking on plate material, the material utilisation ratio is expressed as:

$$\eta = {NS_1}/{A_1B_1} \tag{2.3}$$

N is the number of parts blanked from the plate, and A_1 and B_1 are the length and width of the plate respectively.

In general, blank layout is determined by parameters ϕ and λ, as shown in Figure 2.14. The domain of the parameters is as follows.

$$G\{0 \le \phi \le \pi, -\beta(\phi) \le \lambda \le \beta(\phi)\} \tag{2.4}$$

$\beta(\phi)$ is a single-valued function of ϕ. This function reflects the relation between the width of a part in the y-direction with the angle ϕ.

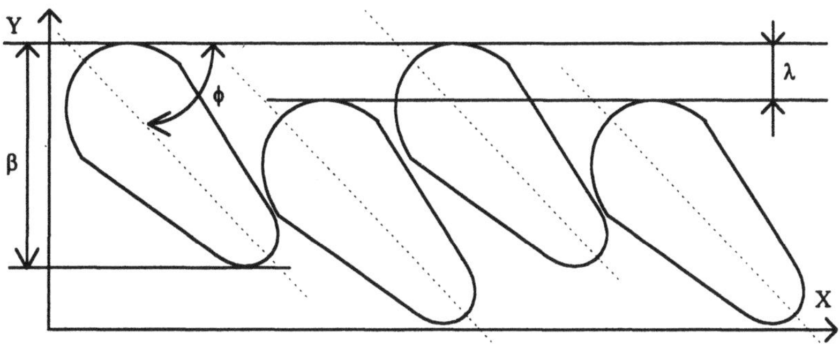

Figure 2.14: Blank layout parameters.

The objective of blank layout optimisation is to find the optimal values of ϕ and λ from the following objective functions, to obtain the maximum values in the domain of G.

$$\eta(\phi,\lambda) = \frac{S}{B(\phi,\lambda)H(\phi,\lambda)} \quad \text{(for coil)} \tag{2.5}$$

or

$$\eta(\phi,\lambda) = \frac{N(\phi,\lambda)S_1}{A_1B_1} \quad \text{(for plate)} \tag{2.6}$$

Due to the shape complexity of sheet metal parts, it is impossible to express the objective function of blank layout with a unified formula. Although many methods have been used in computer-aided blank layout, the basic way to solve the problem is still the same, i.e., to select the optimal arrangement for the blanked part from all the possible schemes. The generation of layout schemes, calculation of material utilisation ratio, and comparison and selection of layout schemes are performed automatically in computer-aided optimisation of blank layout. Two methods adopted by Li et al [7] and Yu et al [8] in a CAD/CAM system for designing blanking dies are discussed in this section.

2.2.2 The polygon method

The main feature of the polygon method is the approximation of the shapes of the blanked parts by polygons. The layout schemes are generated by rotations and movements of these polygons [7]. The procedure of the polygon method is described next.

A. Polygonisation of blanked part
The polygonisation of a blanked part is to substitute straight lines for arcs, in order to approximate a part as a polygon. Figure 2.15 illustrates the polygonisation of a part.

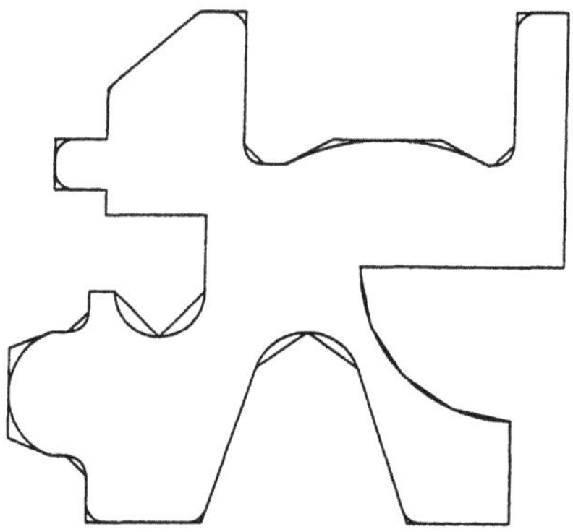

Figure 2.15: Polygonisation of a part.

B. Equi-distant enlargement
The region between two blanked parts on a strip is called the web or bridge. The polygon manipulated in computer-aided blank layout is an equi-distantly enlarged polygon of the part. The enlargement on the part is equal to half of the web width, i.e., $\Delta/2$. Consequently, when two equi-distantly enlarged polygons are tangent to each other, the distance between the parts is equal to the web width Δ.

C. Rotation and movement of the polygons
The equi-distantly enlarged polygons are rotated and moved until they are tangent to each other to generate a layout scheme.

D. Comparison and selection of schemes
The newly generated scheme is compared with the scheme that has been previously stored, and the layout scheme that has a higher material utilisation ratio is selected and kept. If all the permissible layout schemes are exhausted, the system goes to step E. Otherwise, step C is taken.

E. Output results of blank layout

The polygon method is suitable for all situations. However, it is a time-consuming optimal blank layout procedure.

2.2.3 The height function method

From a blank layout drawing, it can be observed that the axes of the parts are always parallel to one another (Figure 2.16). Based on the parallelism of the axes, the height function method takes the height difference h_{ij} of the parts as parameters for layout optimisation [8]. In the case of a double-row layout as shown in Figure 2.17, h_{12} and h_{23} are the layout parameters.

The following relations exist between parameters ϕ, λ and h_{12}, h_{23},

$$\phi = \pi/2 - \tan^{-1}(h/r) \tag{2.7}$$

$$\lambda = h_{12}\sin\phi - r_{12}\cos\phi \tag{2.8}$$

As shown in Figure 2.16, the parts are placed in their own co-ordinate systems. The co-ordinate systems of the three parts are $x_1o_1y_1$, $x_2o_2y_2$, and $x_3o_3y_3$ respectively. r_{12} and h_{12} are the co-ordinates of the origin of the second co-ordinate system in the first co-ordinate system. r_{23} and h_{23} are the co-ordinates of the origin of the third co-ordinate system in the second co-ordinate system. In equation (2.7), h is the algebraic sum of h_{12} and h_{23}. It is equal to the distance from o_3 to axis x_1. r is the distance between y_3 and y_1, given by the following equation.

$$r = \max\{[r_{12}(h_{12}) + r_{23}(h_{23})], r_{13}(h)\} \tag{2.8}$$

2.2.3.1 *Domain and tangential condition*

If the height of a part in its local co-ordinate system is t, as shown in Figure 2.17, the domain for optimal blank layout determination is as follows:

$$G\{-t \le h_{12} \le t, -t \le h_{23} \le t\} \tag{2.9}$$

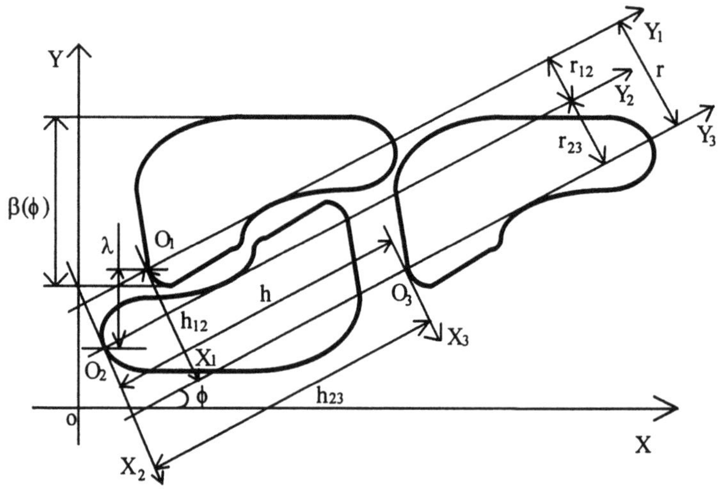

Figure 2.16: Relation between parts in blank layout.

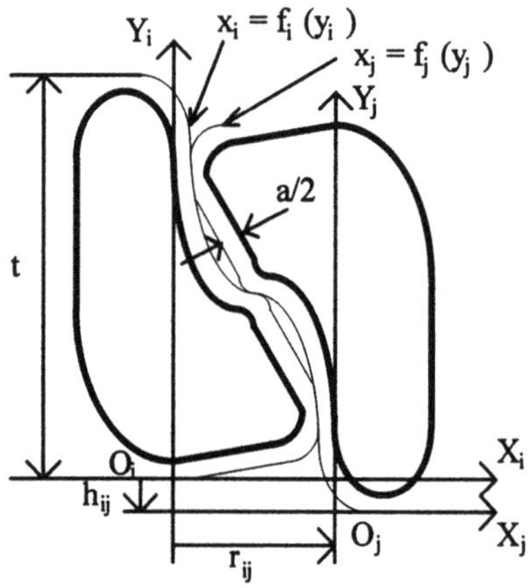

Figure 2.17: Relation between tangential figures.

Since the axes of the parts are parallel to each other, and the heights of the parts are equal, G is a square domain. If the grid of the domain is $2m \times 2m$, the increment of h_{12} and h_{23} during blank layout determination is $\Delta t = t/m$.

The optimisation process is carried out in two stages. In the first stage, primary optimisation is performed. In the second stage, fine searching is achieved based on the optimal solution obtained in the first stage. If the optimal parameters achieved in the primary optimisation process are h_{12} and h_{23}, the optimisation domain in the second stage is as follows:

$$G'\{h_{12} - \Delta t \le h'_{12} \le h_{12} + \Delta t, h_{23} - \Delta t \le h'_{23} \le h_{23} + \Delta t\} \tag{2.10}$$

As in the polygon method, the parts are equi-distantly enlarged by half of the web width, and made tangent to each other in the layout. The algorithm to determine if the parts are tangent greatly affects the optimisation time. In Figure 2.17, the parts are divided into two sections by their top-points and bottom-points. If the right section of part i is expressed as $x_i = f_i(y_i)$, and the left section of part j is $x_j = f_j(y_j)$, when the two parts are tangent to each other, r_{ij} can be obtained using equation (2.11).

$$r_{ij} = \begin{cases} \max\left[f_i(y_i) - f_j(y_i - h_{ij})\right], t \ge h_{ij} \ge 0, t \ge y_i \ge h_{ij} \\ \max\left[f_i(y_i) - f_j(y_i - h_{ij})\right], -t \le h_{ij} \le 0, t + h_{ij} \ge y_i \end{cases} \tag{2.11}$$

r_{ij} is the abscissa of o_j in the co-ordinate system $x_i o_j y_j$. This algorithm to determine the tangential condition has higher efficiency as it avoids the iterations used in other optimisation algorithms.

2.2.3.2 *Calculation of strip width, pitch and utilisation ratio*

As shown in Figure 2.18, a part is placed in a polar co-ordinate system such that the origin and the x-axis of the polar co-ordinate system coincide with the origin and the x-axis of the local co-ordinate system of the part. Let the equation of the contour curve be $\rho(\theta)$, and the polar angle of axis l be φ. The maximum distance from the contour curve to axis l is as follows:

$$T(\varphi) = \max[\rho(\theta)\sin(\theta - \varphi)], 0 \le \theta \le 2\pi \tag{2.12}$$

Here, $T(\varphi)$ is defined as the height function of the part.

The angle ϕ in Figure 2.16 can be determined using equation (2.7). The included angle φ between the strip direction and axis x_i of the local co-ordinate system of the figure is as follows:

$$\varphi = \pi/2 - \phi \tag{2.13}$$

Hence, the strip width B can be determined using equation (2.14).

$$B = \max\left[T_1(\pi/2 - \varphi), T_2(\pi/2 - \varphi) + \lambda\right] - \max\left[T_1(3\pi/2 - \varphi), T_2(3\pi/2 - \varphi) + \lambda\right] + a \tag{2.14}$$

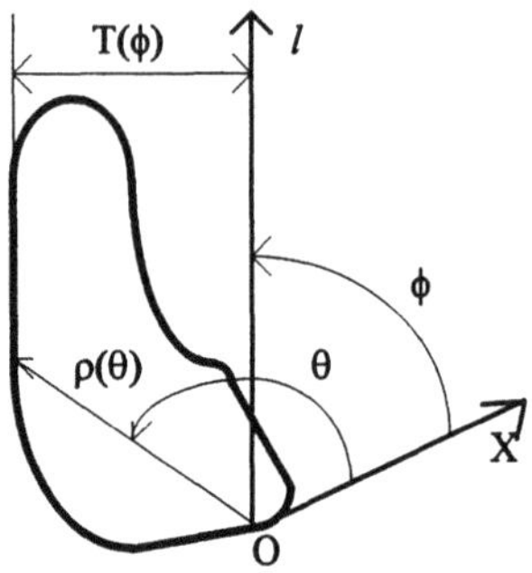

Figure 2.18: Definition of height function.

T_1 and T_2 are the height functions of the first and second parts. a is the web width. λ can be determined using equation (2.8). The value λ is the shift of part 2 relative to part 1, which can be positive or negative.

Referring to Figure 2.16, the feeding pitch H can be found using equation (2.15).

$$\begin{aligned} H &= \left(r^2 + h^2\right)^{1/2} \\ &= \left\{\left[\max\left(r_{12} + r_{23}, r_{13}\right)\right]^2 + \left(h_{12} + h_{23}\right)^2\right\}^{1/2} \end{aligned} \tag{2.15}$$

The material utilisation ratio of the strip η can be calculated as follows:

$$\eta = \frac{nS_1}{BH} = \frac{nS_1}{B(h_{12}, h_{23})H(h_{12}, h_{23})} \tag{2.16}$$

S_1 is the area of a part. n is the number of rows in the layout. Strip width B and feeding pitch H are functions of h_{12} and h_{23}. They can be determined using equations (2.14) and (2.15).

2.2.3.3 *Layout optimisation process*

The flowchart of the height function method is shown in Figure 2.19. Before optimisation, the data processing module converts the graphical information of the blanked part into a format that is convenient for layout optimisation. Pre-processing of the data includes the input of part drawing and blank layout scheme, area calculation of the part, selection of web width, and equi-distant enlargement of the parts. Next, the parts are polygonised, and their local co-ordinate systems are created. The values of the height functions of the parts are calculated. A data table, which can be used in layout optimisation, is generated according to equation (2.12). The range of the variable φ, which is equally divided by k, is $[0, 2\pi]$.

The searching grid is made on the domain of the layout. r_{ij} values are calculated using equation (2.11) to form a table. During optimisation, B and H can be rapidly determined by looking up this table. Blank layout optimisation is carried out in two stages, i.e., two rounds of searching. The optimal layout can be obtained within minutes on a PC.

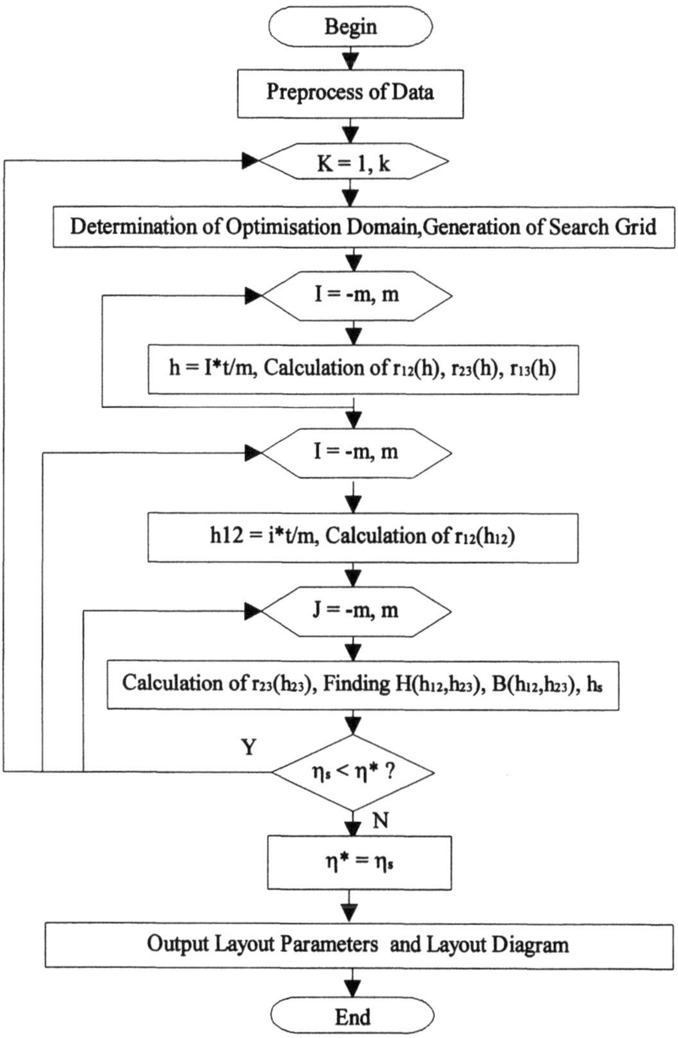

Figure 2.19: Flowchart of the height function method.

2.3 Fine blanking

2.3.1 Characteristics of the fine blanking process

The extent of plastic deformation that can be achieved in a workpiece during metal forming is greatly influenced by the stress state developed in the workpiece. The triaxial compressive stress state is favourable for plastic deformation because this stress state inhibits necking and fracture.

During blanking, plastic deformation occurs mainly near the edges of the punch and die. Since clearance exists between the punch and die, relatively high tensile stresses are induced. Hence, micro-cracks are liable to occur around the edges, and coarse angular fracture may result under these tensile stresses.

Fine blanking is used when the edge of the blanked part has important functionality and the blanked edge must be near to a full square surface without typical roll-over and coarse angular fractures. The clearance used in fine blanking dies is very small, usually around 0.1% of the sheet thickness. Special triple action hydraulic presses are usually used to operate the fine blanking die. High squeezing pressures and a vee ring are used so that triaxial compressive stresses are developed in the area near the edges. In fine blanking, a full square burnished blank edge can be produced. Parts that are made of thicker sheets and finished by broaching or machining are suitable for fine blanking. Most fine blanking dies are compound dies so that holes on the parts may be punched simultaneously. Figure 2.20 shows a typical die used in fine blanking.

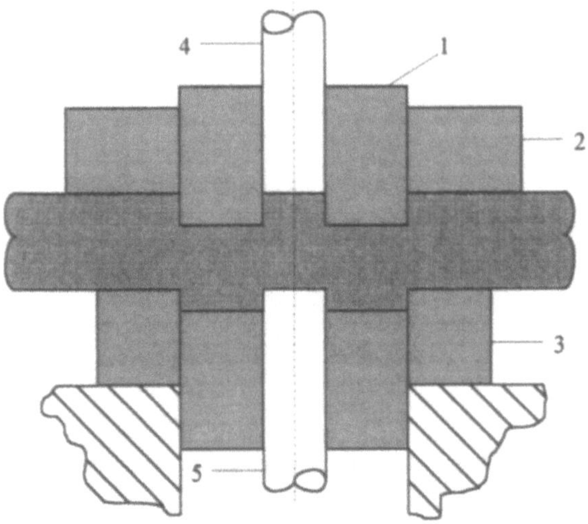

Figure 2.20: Fine blanking die: (1) punch-die, (2) blank holder, (3) die, (4) ejector, (5) punch.

Materials used in fine blanking should have good plasticity to prevent tearing during the blanking process. Steels that have a carbon content of less than 0.35% produce satisfactory results in fine blanking. For low alloy steels and steels with a carbon content of 0.35-0.70%, annealing should be carried out before fine blanking. This is because the metallurgical structures of the materials have significant influence on the quality of the blanked edge. It should be noted that many brittle materials, such as the high-carbon steel, can produce good results in fine blanking after they have been annealed to obtain a globular structure.

2.3.2 Application of the fine blanking process

Pressure in the fine blanking process is much higher than ordinary blanking because the clearance between the die and punch used in fine blanking is very small. Consequently, the requirements for accuracy, rigidity and strength of the dies are more stringent. In fine blanking, the required pressures are the punching pressure, blank holder pressure and the counter pressure. Expensive special triple action hydraulic presses are usually used to operate fine blanking dies.

A new type of die set has been developed at HUST that can be mounted on ordinary presses to perform fine blanking [9]. The constructions of the die set and the die are shown in Figure 2.21 and Figure 2.22 respectively.

A cylinder attached under the press bed provides the blank holder pressure. A small cylinder mounted in the upper holder of the die set supplies the counter pressure, which can be regulated by a valve. Different parts may be produced by changing the core die that is shown in Figure 2.22. The core die consists of the punch, die, pressure plate, ejector plate, etc. It can be conveniently positioned and changed in the die set.

Die sets of different sizes have been used in a factory to produce components of instruments and typewriters. For example, a clutch pawl made of cold-rolled steel was originally produced using twelve operations. After adopting the fine blanking process, seven operations and five fixtures were removed. The quality of the part was greatly improved.

2.3.3 CAD/CAM of fine blanking dies

A CAD/CAM system for fine blanking dies, the HJC system, has been developed at HUST [11]. The system consists of five parts and twelve modules (Figure 2.23).

The first part of the system deals with graphics input. Its functions include shape model generation and geometric information processing. The second part of the system performs punchability checks, blank layout, pressure calculation and selection of die sets. Die construction is designed in the third part of the system that performs core die design, determination of edge dimensions, blank holder design and ejector pin layout. The fourth and fifth parts of the system are draughting and NC programming respectively.

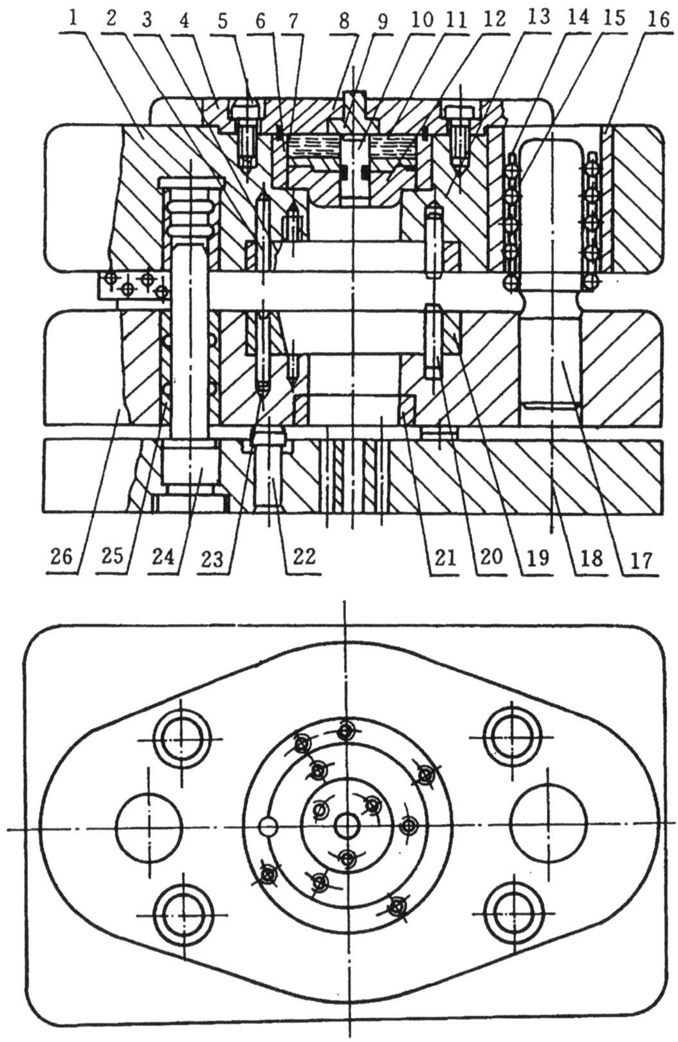

Figure 2.21: A die set for fine blanking: (1) upper holder, (2) pin, (3) upper tapered ring, (4) fixing plate, (5) bolt, (6) cylinder, (7) oil, (8) lid, (9) back plate, (10) supporting rod, (11) seal, (12) piston, (13) position pin, (14) retainer, (15) ball, (16) guide bush, (17) big guide pillar, (18) bed plate, (19) lower tapered ring, (20) position pin, (21) ring, (22) ejector rod, (23) pin, (24) small guide pillar, (25) small guide bush, (26) movable holder.

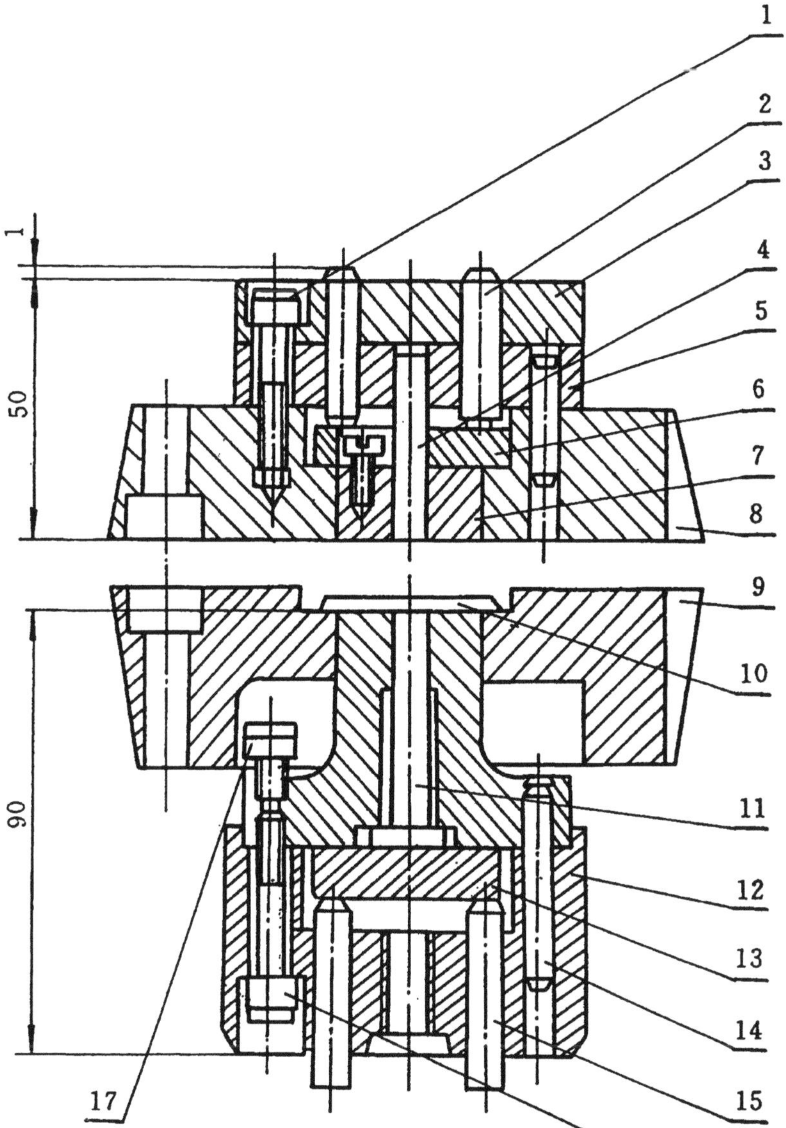

Figure 2.22: A fine blanking die assembly: (1) screw, (2) ejector rod, (3) back plate, (4) punch, (5) retaining plate, (6) link plate, (7) ejector plate, (8) die, (9) pressure plate, (10) punch-die, (11) ejector rod, (12) supporter, (13) plate, (14) pin, (15) ejector pin, (16) screw, (17) screw.

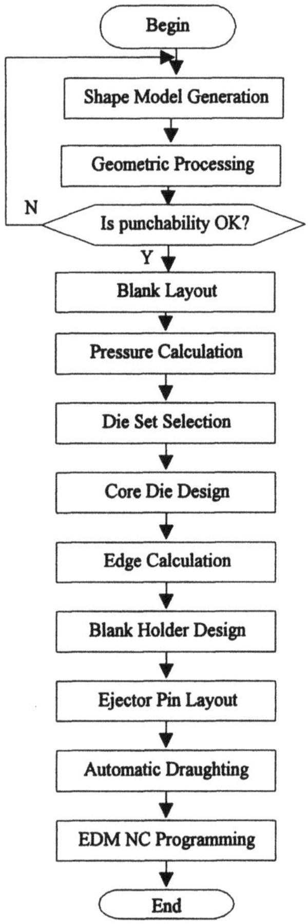

Figure 2.23: Structure of HJC system.

2.3.3.1 Graphic input of blanked parts

The graphic input of blanked parts is based on the wire-frame method [11]. The shape of a blanked part can be regarded as a union of three sets.

$$S = (A \cup B) \cup C \tag{2.17}$$

A, B and C are the sets of major elements, sub-elements and neutral elements respectively. Sub-elements are derived from major elements through transformation. Neutral elements refer to the elements that are not involved in any transformation. Major and neutral elements are described one by one according to the dimensional chain. Each of the major and neutral elements can be defined by one or two base elements that have been described earlier. If an element is described by two base elements, ten codes are needed to define it, as shown in Table 2.1.

Table 2.1: Coding format.

elem. num.	cont. num.	elem. type.	diam. num.	basis one	rel. form	dim. num.	basis two	rel. num.	dim. num.

Before coding, a co-ordinate system is created where the drawing of the blanked part is put in the first quadrant. The contours of the part are numbered. The number of the outermost contour is 1, and the inner contours are 2, 3 and so forth. The elements of the outermost contour and inner contours are numbered in counter clockwise and clockwise directions respectively. Next, the elements can be described according to the coding scheme. For the part shown in Figure 2.24, the data defining its shape are listed below.

19	11	13	16	19					
1	0	1	0	0	0	0	0	0	0
2	0	1	0	0	0	0	0	0	0
101	0	1	0	1	20	1	0	0	0
102	0	1	0	2	20	-2	0	0	0
3	1	3	0	101	40	0	102	40	0
4	1	3	0	101	40	-3	102	40	0
5	1	3	0	101	40	-3	102	40	-4
6	1	3	0	101	40	-5	102	40	-4
7	1	3	0	101	40	-5	102	40	-6
103	0	1	0	102	20	-7	0	0	0
9	1	3	0	101	40	-8	103	40	9
104	0	1	0	101	30	-10	9	40	0
8	1	3	0	103	40	0	104	40	0
11	1	-2	-11	101	20	-12	-102	20	-13
105	0	1	0	102	20	-13	0	0	0
106	0	1	0	105	20	-11	0	0	0
10	1	3	0	101	40	-8	106	40	-11
12	1	3	0	101	40	0	102	40	-13
13	2	-2	14	101	40	-15	102	40	-16

95.00000	0.00000	0.00000
40.00000	0.00000	0.00000
38.00000	0.00000	-0.62000
9.00000	0.18000	-0.18000
64.00000	0.00000	-0.74000
19.00000	0.00000	-0.52000
40.00000	0.00000	-0.62000
9.00000	0.18000	-0.18000
5.00000	0.15000	-0.15000
45.00000	0.00000	0.00000
4.00000	0.30000	0.00000
16.00000	0.21500	-0.21500
17.00000	0.00000	-0.43000
14.00000	0.43000	0.00000
25.00000	0.00000	-0.52000
11.00000	0.00000	-0.34000

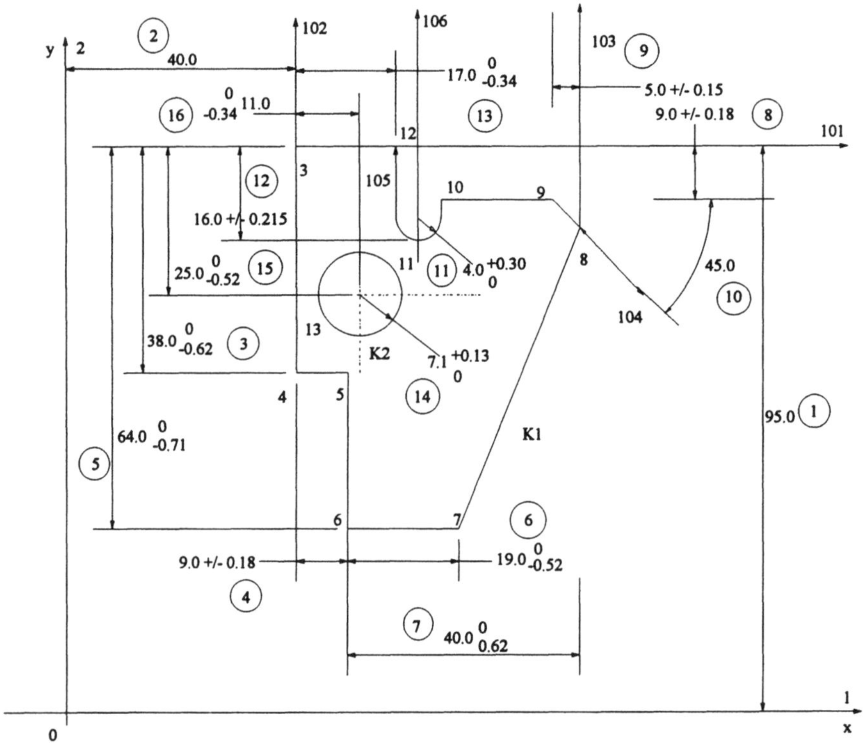

Figure 2.24: A blanked part.

2.3.3.2 Process analysis and die construction design

After the blanked part has been input, its punchability is analysed. The algorithm incorporated in the module can recognise features on a part, such as sharp corners, small holes and narrow slots, and compare them with the critical values of these features. If the required part is not suitable for fine blanking, the user is prompted to modify the design.

Optimisation of blank layout is carried out after the punchability check. The polygon method is adopted in this system. For most cases, the material utilisation ratio can be increased by more than 5%.

Since standard die sets and changeable core dies are used in this system, only parts that are related to the shape of the blanked part, such as the die opening position, die edge dimensions, vee ring on the blank holder, and the ejector pin layout, need to be designed.

The position of the die opening is determined based on the principle that the shear force should coincide with the centre of the core die. The dimensions of the die edge can be divided into three groups according to their wearing tendency, i.e., increasing, decreasing and constant dimensions. These three types of edge dimensions are recognised based on the geometric information of the shape model. They can be designed automatically.

One of the most noticeable differences between fine blanking and ordinary blanking is the presence of a vee ring on the blank holder. In general, the contour of the vee ring is parallel to that of the die opening. In some cases, the contour of the vee ring needs to be changed for strengthening or manufacturing purposes. Figure 2.25 shows a few examples of vee ring contours. Figure 2.26 gives the vee shape and its parameters used in the system.

It is necessary to find the depressions on a part in order to automatically design and modify the contour of the vee ring. Equi-distant enlargement is performed for this purpose. As shown in Figure 2.27, the contour of a part is equi-distantly enlarged. If the enlarged contour intersects itself, modifications on the contour of the vee ring should be made to keep away from the narrow region.

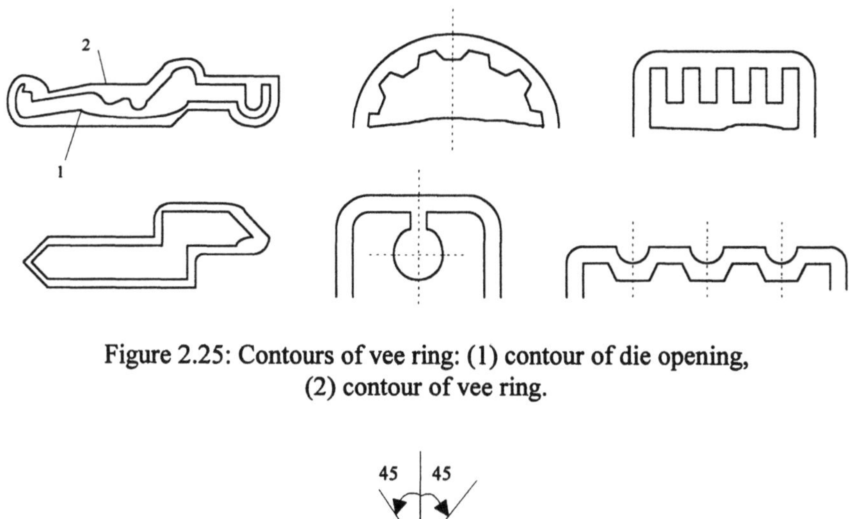

Figure 2.25: Contours of vee ring: (1) contour of die opening, (2) contour of vee ring.

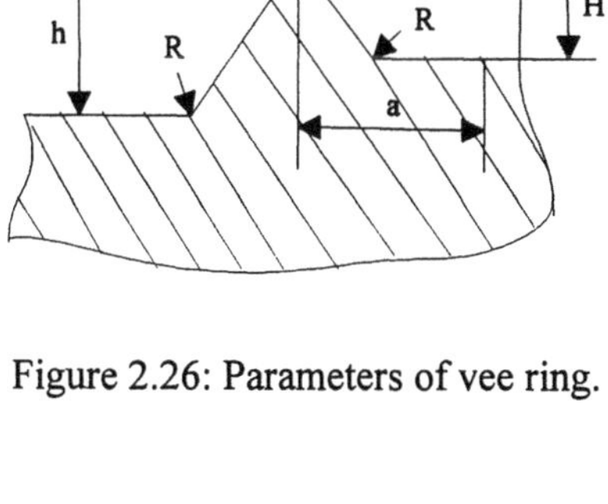

Figure 2.26: Parameters of vee ring.

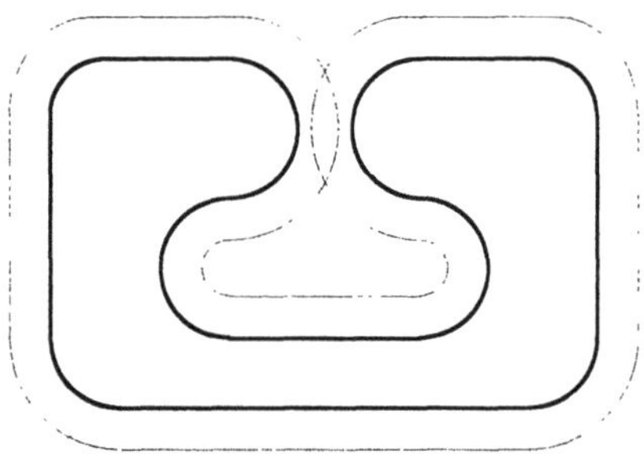

Figure 2.27: Modification of vee ring contour.

In fine blanking dies, ejector pins are needed to push the workpiece out of the die and/or provide counter pressure. Based on the requirements of the die construction and the strength restriction of the pin material, the system can determine the number and sizes of ejector pins needed. Objective functions are established in the system to optimise the layout of the ejector pins. This multi-objective optimisation problem is converted into a single objective optimisation problem using an evaluating function method. The constraint conditions can be removed by searching feasible meshes, and the positions of the pins needed can be layout interactively.

Besides the information for draughting, the die construction design module of the system provides the data required for wire EDM NC programming. The position of wire-through-hole, diameter of wire, discharge gap and cutting sequence can be selected interactively. The wire path and NC code are generated by the NC programming module.

The optimisation algorithm for ejector pin layout and NC programming for wire EDM will be elaborated later in this chapter.

2.4 Progressive stamping

Progressive stamping is a process where operations on different stations of the die are performed in one stroke. Progressive die is an important type of stamping tools widely used in the production of small and medium-sized sheet metal parts. It is characterised by its precision, and distinguished for its reliability and endurance.

Progressive die design consists of strip layout, die and punch design, pressure calculation, locating device design, die set and press selection, etc. It often takes several months to design and manufacture a progressive die. Li et al [12] have developed a CAD/CAM system for progressive dies design to improve product quality and speed up product renewal.

2.4.1 Software architecture of the system

The CAD/CAM system for progressive dies was implemented on a VAX workstation. The system has five modules, *viz.*, drawing input and feature modelling, strip layout, lower die design, upper die design and NC programming.

The architecture of the system is shown in Figure 2.28. In the system, the IGL interactive graphics library is used as the basic graphic support while the DOGS graphic package provides the draughting functions. The system runs in the VMS environment. As shown in Figure 2.28, there is an interface between the IGL graphics library and the modules of the system, which has the same format as GKS.

Part input and geometric modelling are essential in a CAD/CAM system for progressive die design. Presently, majority of the CAD/CAM systems for die design cannot cover all the stages in the design and manufacture of stamping dies because of the lack of sufficient information. Feature modelling is adopted in the system that has been discussed in Section 2.1.

The constructions of progressive dies are complex. They often consist of several hundred parts. Besides forming functions, progressive dies usually have various mechanisms to provide auxiliary functions, such as guiding, stripping, lifting and positioning strips, detecting error feeding, etc. Progressive dies used in high-speed presses require high accuracy and reliability.

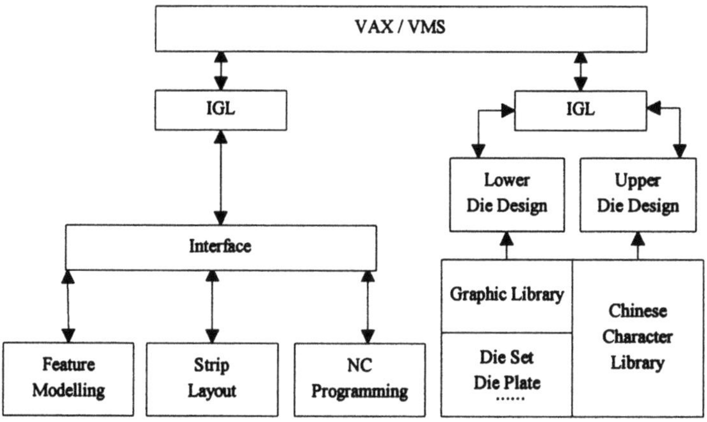

Figure 2.28: Software architecture of the system.

Die construction and die parts have been standardised during the development of the die construction design module. Four types of die constructions are formalised. Die parts for each type of construction can be divided into two groups, namely the variable and the invariable groups. Standard parts are stored in a graphics library that can be retrieved during die construction design. After a die construction type has been selected, the system can determine the basic dimensions of die plates. Results of strip layout are copied onto the die plates to facilitate the design of the split die and stripper.

The system can assemble die plates to generate the assembly drawing. The user may retrieve bolts, pins, ejectors, benders and other standard parts or mechanisms from the graphics library, and place them at proper locations. About 2000 Chinese characters are kept in a library for writing technical requirements and part list. Consideration was given to both efficiency and flexibility during the development of the system. For a typical die construction with standard parts, the automatic method is used. In situations when the die parts are variable, or the construction is too complicated to be dealt with automatically, an interactive method is adopted. Thus, the system is enhanced as the application range is extended. The structure of the die construction design module is illustrated in Figure 2.29.

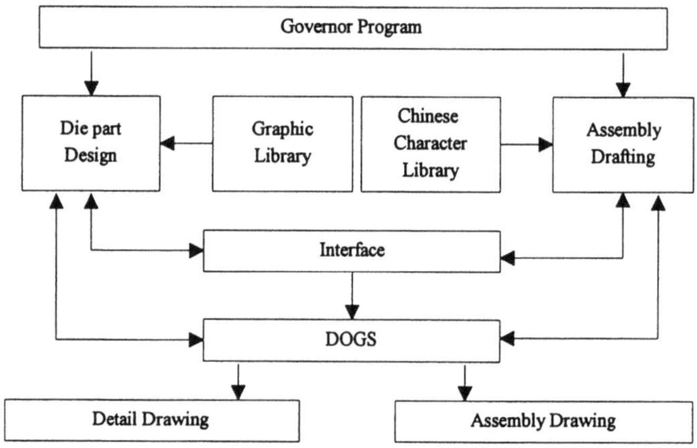

Figure 2.29: Die construction design module.

2.4.2 Strip layout

Stamping process planning, which is called strip layout in progressive die design, is of extreme importance as it determines the stamping operations required and affects the die construction. In conventional progressive dies design, strip layout is performed manually, and the quality of the die design depends heavily on the experience of the designer.

In the developed system, both automatic and interactive methods are used in strip layout. For processes that are well-proven in practice, the system can automatically determine the process parameters. For processes that are highly dependent on the skill of the designer, the system provides interactive functions during strip layout. The user can interactively alter the orientation of the part, adjust pitch, change strip width, and specify positions of the side cutter and pilot pins. The design of the scrap shape, and the adding or removing of a station on the strip can also be done this way.

Design rules and relevant mathematical model must be established to automatically carry out strip layout. In this chapter, punching and blanking process planning is used as an example to discuss the methods adopted in the system [13].

Many factors, such as the strength of the die and the accuracy of the stamping, should be considered in strip layout. Several rules must be followed.

1. Distances between the holes punched on one station must be larger than a certain value to ensure the die strength.
2. Holes with high position accuracy requirement should be punched in one station.
3. Narrow slots and projections are not allowed in a die as fracture may occur.
4. Smaller holes are usually punched first, and the blanking operation is arranged on the last station. The pressure centre and the die centre should be as close to each other as possible.
5. Empty stations can be arranged for strengthening purposes or for the ease of fixing punches.
6. Proper locating devices, such as pitch punch, stop pin and pilot pin, should be designed to ensure feeding accuracy.

Strip layout is determined by the shape of a part and its technical requirements. Holes on the part that have position accuracy requirements must be recognised so that they can be punched in one station. Therefore, the following issues should be determined for each dimension: (1) Does the dimension have any tolerance requirement? and (2) Which contours are related by the dimension?

The information needed for determining these issues is contained in the shape model of the part. The contours of a part consist of elements, and the shape model contains the information on the relations and dimensions between these elements. Hence, the contours that are related by a dimension can be found by searching through the elements related by this dimension.

A weighted graph is used to describe relations between the elements to facilitate searching. A part and its weighted graph are shown in Figure 2.30. The nodes in the graph represent the elements in the shape model. When two elements have direct dimensional relation, they are linked by an edge. The dimension number is taken as the weight of the edge. The graph shown in Figure 2.30(b) is also called a position relation model. Since it is derived from the shape model, indirect dimensional relations may exist. As shown in Figure 2.30(a), there is no edge between nodes 16 and 17 although a dimensional relation exists between them. Consequently, geometric information contained in the shape model is used together with the weighted graph to recognise the dimensional relations between elements. Contours with position accuracy requirements can therefore be detected.

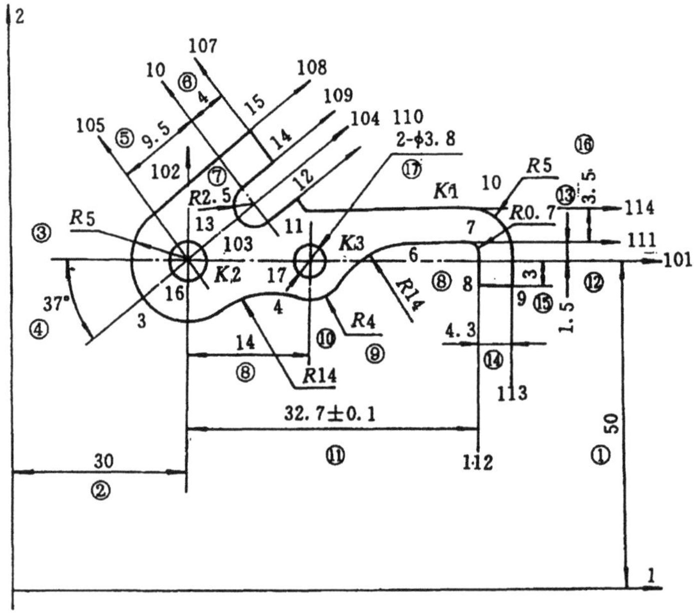

(a)

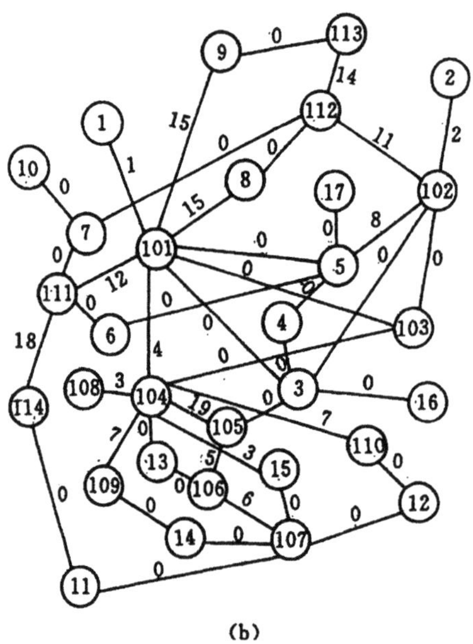

(b)

Figure 2.30: (a) A blanked part, (b) a weighted graph.

A blanked part can be regarded as a set of contours that is expressed as:

$$A = \{k_1, k_2, \ldots\ldots\ldots\ldots, k_i, \ldots\ldots, k_n\} \tag{2.18}$$

k_i (i = 1, 2, ..., n) represents a contour. Every contour that has position accuracy requirements forms a relation ρ_1 in the set A. If there exists a position accuracy requirement between contours k_i and k_j, the ordered pair of contours belongs to the relation, i.e., $(k_i,k_j) \in \rho_1$.

$$\rho_1 = \{\ldots, (k_i, k_j), \ldots\} \qquad k_i, k_j \in A \tag{2.19}$$

Thus, a position relation matrix M_1 can be obtained from ρ_1.

$$M_1 = \begin{bmatrix} a_{11} & a_{12} & \cdots & \cdots & a_{1n} \\ a_{21} & a_{22} & \cdots & \cdots & a_{2n} \\ \vdots & \vdots & & & \vdots \\ \vdots & \vdots & & & \vdots \\ a_{n1} & a_{n2} & \cdots & \cdots & a_{nn} \end{bmatrix} \tag{2.20}$$

$1 \leq i, j \leq n$, and n is the number of contours.

$$a_{ij} = \begin{cases} 1 \text{ if } (k_i, k_j) \in \rho_1 \\ \\ 0 \text{ if } (k_i, k_j) \notin \rho_1 \end{cases} \tag{2.21}$$

The distance between holes must be larger than a certain value, i.e., $D \geq D_{min}$ to ensure the die strength. The equi-distant enlargement method is used in the program to check for holes interference. If two holes interfere with each other, they should be arranged on different stations in strip layout.

An interference relation of the holes is defined as a relation ρ_2 in set A. The ordered pair $(k_i, k_j) \in \rho_2$ if contour k_i interferes with contour k_j. The interference relation matrix M_2 is derived in a similar way to matrix M_1.

$$M_2 = \begin{bmatrix} b_{11} & b_{12} & \cdots & \cdots & b_{1n} \\ b_{21} & b_{22} & \cdots & \cdots & b_{2n} \\ \vdots & \vdots & & & \vdots \\ \vdots & \vdots & & & \vdots \\ b_{n1} & b_{n2} & \cdots & \cdots & b_{nn} \end{bmatrix} \tag{2.22}$$

$1 \le i, j \le n$, and n is the number of contours.

$$b_{ij} = \begin{cases} 1 \text{ if } (k_i, k_j) \in \rho_2 \\ 0 \text{ if } (k_i, k_j) \notin \rho_2 \end{cases} \tag{2.23}$$

The position relation matrix M_1 and the interference relation matrix M_2 can be used as references during strip layout to meet the requirements of rules A, B and E that have been discussed earlier.

When there are narrow slots or thin projections on a part, it is difficult to blank the part in one station because of the weakened punch and die. In this case, it is a common practice to cut the material around the part in a piece-wise manner. In this way, the layout is called "scrap layout" since the material is cut off as scrap.

There are three types of scrap layouts, as shown in Figure 2.31. Local scrap layout (Figure 2.31a) is used for parts with narrow slots or projections. The scrap shown in Figure 2.31b is suitable for double-row layout of small parts. The third type of scrap layout (Figure 2.31c) is used for thin parts that are made by cutting scraps.

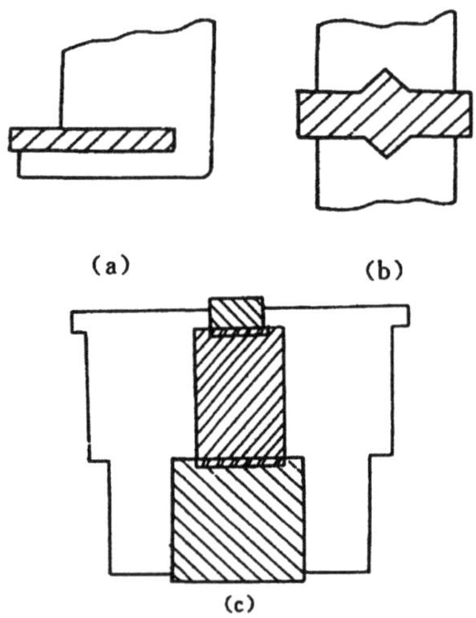

Figure 2.31: Three types of scrap layout.

Scrap design is carried out using the following steps.

1. The part drawing is displayed on the screen for the designer to determine the need to design scrap. If no scrap is needed, the program jumps to the step for station arrangement.

2. In the case when scrap is needed, the position and orientation of the part, which have been determined in blank layout, can be modified by transformations such as translation and rotation.
3. The scrap shape is interactively designed at appropriate places. The user may move the cursor to design scrap or position it with reference to the blank layout displayed on the screen. Scrap shape can be copied to different places so that scraps of the same shape can be efficiently designed. Each scrap is considered as a contour in strip layout.

After these stages, the program begins to arrange the stations, i.e., group the contours, including scrap contours, on different stations, so that the part can be cut progressively.

During station arrangement, contours interfering with each other should not be put in one station. Contours with position accuracy requirements are arranged in one station if possible. In essence, the station arrangement process divides the contour set A into subsets.

$$\bigcup_{i=1}^{m} B_i = A \qquad (2.24)$$

Subset B_i contains all the contours punched on station i, and m is the number of subsets.

Stations arrangement is made with reference to matrices M_1 and M_2. Contours with a_{ij} values of 1 are grouped on the same station, and contours with b_{ij} values of 1 are put on different stations. Next, the program sequences the stations according to the perimeters of the contours, and places the blanking operation at the last station.

Locating devices for material feeding are essential for achieving the accuracy of the required part. The system is able to design the following locating devices.

1. Pitch punch,
2. Pitch punch and pilot pin, and
3. Starting stop set and fixed stop pin.

Pitch punches and pilot pins are used in most cases. There should be holes on the part when pilot pins are used, otherwise auxiliary holes are needed.

The selection of locating devices depends not only on the shape and accuracy of a part, but also the production conditions. Therefore, it is interactively selected in the system. The dimension and position of the pitch punch can be determined automatically. The user may place the stop pin in an approximate position by moving the cursor. Then, an accurate position of the stop pin is determined by the program.

A strip layout is obtained from the above stages and displayed on the screen. The user can modify the result interactively if he is not satisfied with it. Figure 2.32 is the strip layout for the part shown in Figure 2.30.

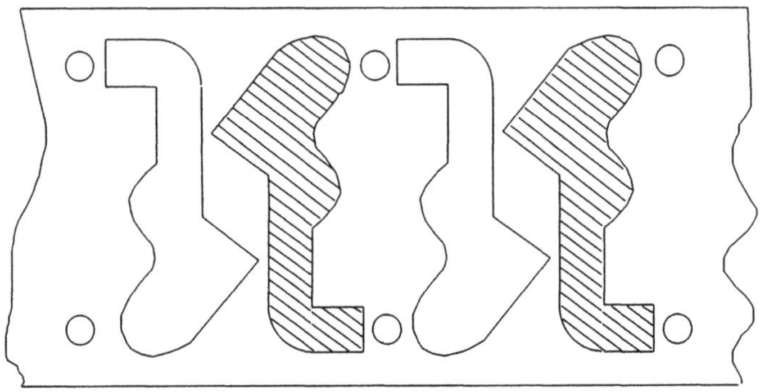

Figure 2.32: Strip layout of a part.

2.4.3 Construction design of progressive dies

With the rapid development in the manufacturing industry, progressive dies are required to have higher accuracy and longer working life. Rational design of die construction is essential for meeting these requirements.

Generally, progressive die constructions fall into two types, namely, the whole and split types. In the whole type, all the cavities are machined on one die plate. This type of die construction is only suitable for cases when the shape of the manufactured part is simple, or when the die has few stations. A split die construction consists of a number of segments and inserts, and complex die cavities are split into pieces. These dies are easy to machine, adjust and repair. They are advantageous in terms of improved accuracy and longer working life. This type of die construction is favourable for situations when the shapes of the die cavities are complex, or when more stations are needed.

Little research has been done in computer-aided design of split die construction and automatic division of die cavities. In the system developed by Li et al [14], the principles of graph theory and optimisation are utilised to incorporate the related design rules. Interactive functions are provided to facilitate the modification of the design to obtain satisfactory results. The flowchart for die construction design is shown in Figure 2.33.

The results of strip layout are used in die construction design. Cavities at each station are designed according to the forming operations and the intermediate shapes of the workpiece, which are determined from strip layout. Only the overall dimensions of the die need to be determined for the whole type die construction. Other details, such as screw and pin holes, are automatically designed according to the standards incorporated in the system. For split die construction, where the die is segmented, die sections and inserts may need to be designed.

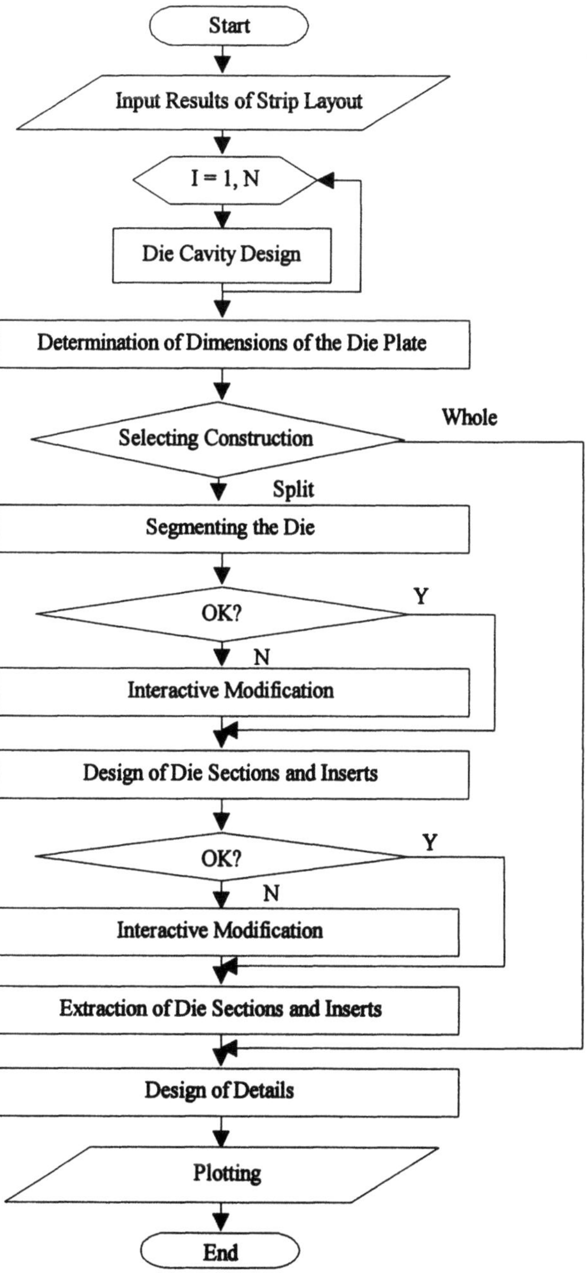

Figure 2.33: Block diagram of die construction design.

Interactive functions are provided to cope with various conditions. The user is permitted to modify design results. When split die construction design is completed, die sections and inserts are extracted to form detailed drawings of individual die parts. Finally, die drawings, including assembly and detailed drawings, are output to the plotter.

Based on the geometry of the required part, and the machining and assembling requirements, the die is segmented into a number of pieces in split die construction design. Segments may be fitted with inserts to form cavities of complex shapes, or to ease the replacement of parts that are liable to damage. Segments and inserts are assembled in a die plate and fastened by two clamps, as shown in Figure 2.34.

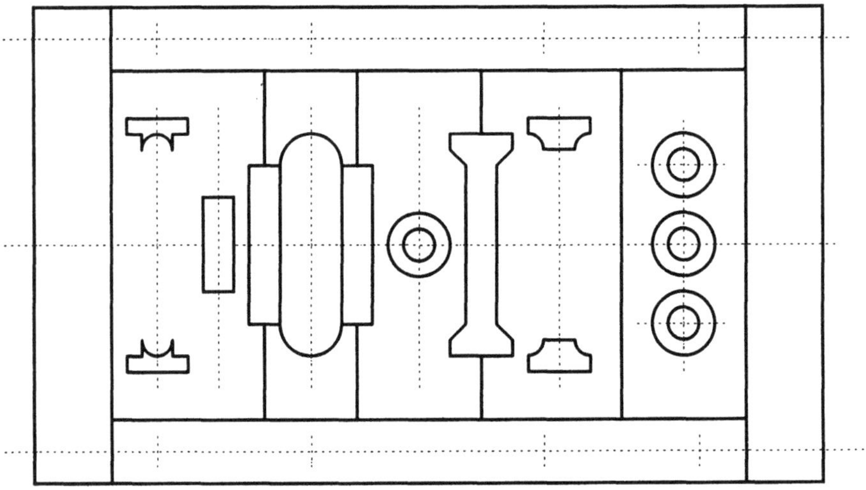

Figure 2.34: Split die construction.

In designing progressive dies of split construction, the rules listed below are usually followed.

1. The die segments should be assembled along the longitudinal direction if possible, i.e., the dividing lines should be perpendicular to the feeding direction.
2. The machining errors of the die sections should be compensated to facilitate adjustment and improve the accuracy of the assembly.
3. It is a common practice to segment dies according to the forming operations for ease of repair and replacement. When this is not feasible, either the punching edges or the cavities of other forming operations are made as inserts.
4. Die sections should be designed such that complex inner shapes are changed into outer shapes for machining.

5. Cavities with sharp corners should be divided at these corners to facilitate machining and avoid breakage.
6. Adjacent lines on the contours of die sections should make right angles or obtuse angles with each other; acute angles should be avoided.
7. Projections and depressions may be formed as die inserts for ease of replacement since they are easily worn out.
8. Dividing lines should not be tangent to arcs of the cavity contours; they usually intersect arcs at their mid points.
9. If the contour lines of a cavity are in the radial direction, divisions can be made in the same direction for ease of grinding.
10. Symmetrical cavities may be divided by their symmetry lines.
11. Contour lines on a die section should be parallel or perpendicular to each other if possible.
12. Shapes of inserts are usually simple, such as circular, rectangular or regular polygonal.

The formulation of mathematical models based on the above design rules is a difficult problem. Automatic design of split die construction is difficult because of the intractability involved.

In designing a die section, a cavity plan, i.e., the cavity contours on the upper plane of the die, is first generated from the results of the cavity design. These contours are then grouped and the die is segmented.

There are three types of segments, namely, segments with single punching edge, segments with punching edges arranged in the longitudinal or lateral direction, and segments with other forming cavities. For the third type of segments, some cavities are difficult to divide automatically owing to the complicated arrangement. Hence, this type of segments is divided interactively. For the second type of segments, each cavity contour is first divided. The dividing lines are then connected to form die sections. It can be seen that the division of the segments with a single edge is the key to the design of die sections.

Dividing schemes for eight typical contours as shown in Figure 2.35, are stored in the system. If a typical contour is found, segmentation will be performed by executing the corresponding routine.

Cavities with non-typical contours are divided using the following procedures.

A. Contour simplification

Arcs on the contours that have central angles less than or equal to 180° are replaced by straight lines (Figure 2.36).

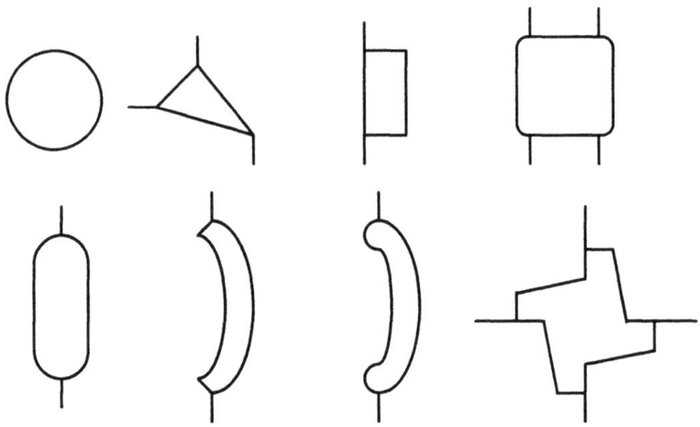

Figure 2.35: Dividing schemes for typical contours.

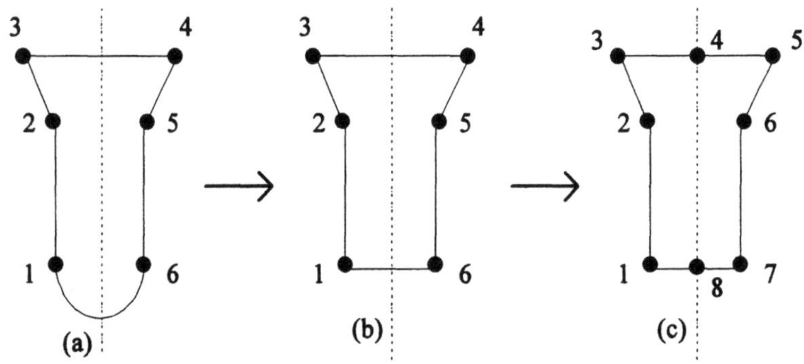

Figure 2.36: Contour simplification and the set of possible dividing points.

B. Projections and depressions determination

Three adjacent lines on a simplified contour can be expressed as vectors a_1, a_2 and a_3, as shown in Figure 2.37.

$$S_1 = a_1 \times a_2 = s_1 k \tag{2.25}$$

$$S_2 = a_2 \times a_3 = s_2 k \tag{2.26}$$

S_1 and S_2 are cross product and k is a unit vector. For a clockwise contour ring, these three adjacent lines form a projection when $s_1 > 0$ and $s_2 > 0$. A depression is formed when $s_1 < 0$ and $s_2 < 0$.

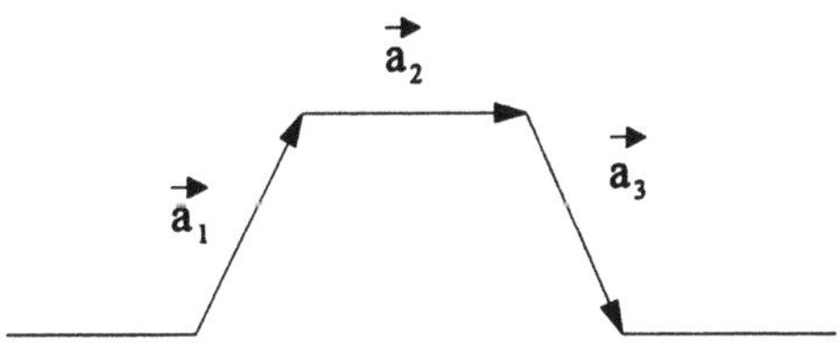

Figure 2.37: Projections and depressions determination.

C. Generation of a set of possible dividing points

The set of possible dividing points includes all the nodes. Tangent points between arcs, intersection points of symmetry axis and the contour, and mid points of depressed edges, (Figure 2.36c) and end-points of projecting edges are excluded from this set.

D. Directed and weighted graph composition

For every pair of possible dividing points, a check is performed to determine a path between them, i.e., whether the contour can be divided by these two points. If a path can be found, these two points are connected to generate a directed edge.

The weight of a directed edge is determined by the shape complexity of the contour section between the points. Contour sections of die cavities may be of I-shape, L-shape or U-shape (Figure 2.38).

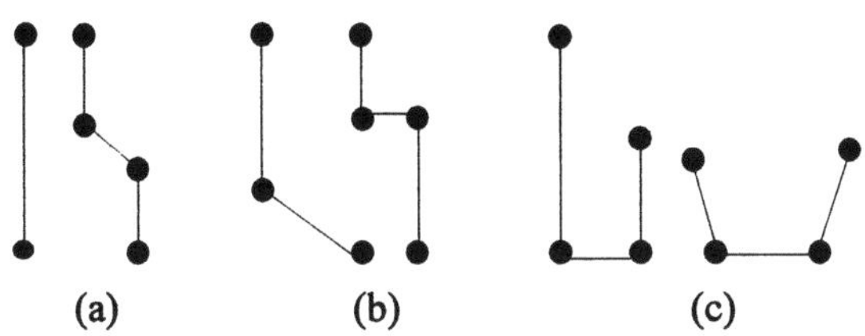

Figure 2.38: Three types of contour sections.

The weight W for the I-shape section is as follows, where L is the distance between the two dividing points.

$$W = \frac{1}{L} \tag{2.27}$$

The weight for the L-shape section is calculated as follows, where H_1 and H_2 are the lengths of the shorter and longer sides respectively, and a_1 is a coefficient.

$$W = \frac{a_1 H_1}{H_2} \tag{2.28}$$

The weight for the U-shape section is calculated using equation (2.29). H and B are the depth and width of the U-shape. a_2 is a coefficient, and M is a large number, for example 10000.

$$W = \begin{cases} \dfrac{a_2 H}{B} & (H \le 1.5B) \\ M & (H > 1.5B) \end{cases} \tag{2.29}$$

Thus, a directed and weighted graph can be composed, as shown in Figure 2.39.

E. Searching for the shortest path

There may exist a number of paths in a directed and weighted graph, i.e., a number of dividing schemes. An optimum scheme can be obtained by searching for the shortest path in the graph.

F. Design of dividing lines

After the dividing points have been determined, dividing lines are designed such that they make right or obtuse angles with adjacent contour lines.

If a user is not satisfied with the design results displayed on the screen, changes to the design can be made interactively.

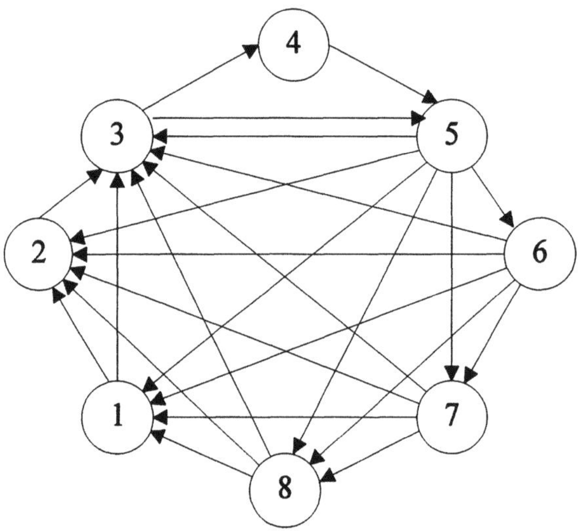

Figure 2.39: A directed and weighted graph.

2.5 An overview of a flat patterning and bending simulation system

In sheet metalworking, many hours of drawing and calculation efforts are spent to determine blank shapes and sizes, and feasible and effective bending sequences. Manual planning is experience-based, highly subjective, repetitive and error-prone. For parts with complex geometry, calculation bottlenecks for flat patterning and trial-and-error programming on the press brakes are often unavoidable.

To reduce the turn-around time of the production cycle, a PC-based system for sheet metal bending has been developed. The AutoCAD-based system consists of a feature-based design input module, a flat pattern development module and a bending simulation module. The output of the system is the blank profiles, the selected bending tools and the feasible bending sequences. Figure 2.40 depicts a schematic representation of the developed system.

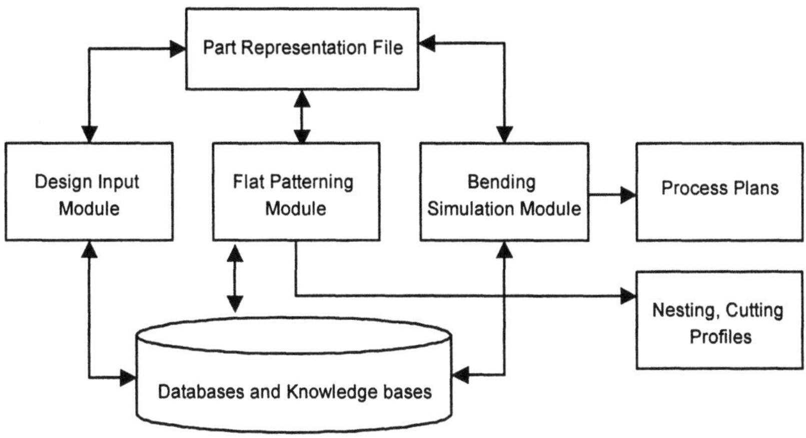

Figure 2.40: Schematic representation of the developed system.

2.5.1 Design input module

A feature-based approach for the design input is adopted. A 2-D or 3-D design can be input in the AutoCAD drawing editor in two ways. Firstly, the user selects standard features that constitute the part, such as walls, bends, holes, etc., from a feature library and parametrically specify the geometrical information, such as lengths, angles, origins, etc. The system will display the geometry of the selected features at the specified locations. Secondly, the user constructs the part using standard AutoCAD drawing aids, such as lines, arcs, circles, etc., or imports the drawing *via* the DXF or IGES file, and then interactively defines the features. To shorten the computing time, the part is graphically represented as a wire-frame model with zero thickness and bend radius. Design and manufacture rules have been embedded in the program to assist the user during design. A design can be modified by changing the created features or the entities of the features. Internally, a part representation file (PRF) is created to store the design information in a frame-like structure. The program automatically updates the on-line design information as design and modification proceeds.

2.5.2 Flat pattern development module

The flat pattern development module adapts the information in the PRF and performs unfolding and folding. During the operations, the plates are transformed to the corresponding locations and orientations, and the internal features such as holes, cut-outs, etc., are mapped onto the plates accordingly. In addition, bend allowances resulting from plastic deformation at bends can be accommodated.

In the interactive mode, unfolding and folding are performed progressively, i.e., bend-by-bend. Once a bend is specified, the program automatically identifies feature entities to be unfolded/folded and calculates unfolding/folding angles. The bend allowance and bend compensation are automatically computed based on material specifications, bend parameters and measurement methods. The user can override the calculated bend allowances and bend compensations, and specify the target plane onto which the parts are to be unfolded/folded. After unfolding, mould lines are displayed for manual dimensioning.

In the automatic mode, all plates are automatically unfolded/folded without any user-intervention. The program-calculated bend compensations are used for the operations.

The flat pattern information such as bend compensations, mould lines, etc., is written to the PRF so that they can be extracted when needed. According to the user's preferences, the program can further process the flat patterns to generate blanks that are required for nesting, nibbling, laser cutting, etc.

2.5.3 Bending simulation module

The bending simulation module simulates press brake operations to determine feasible bending sequences. To reduce unnecessary search for feasible bending sequences, the program adopts the 'reverse of unbending' approach. In this approach, a feasible unbending sequence is generated starting from the final product and reasoned towards the flat pattern state. The reverse of the unbending sequence represents a feasible bending sequence. Before unbending a bend, suitable tools can be selected from a tool database. The program validates the tool selections based on the tool configurations and the design information in the PRF, and positions the tools at the bend. Tool-work interference is checked before and after the bend is unfolded. A feasible unbending is the one that causes no interference during the operation. When an infeasible unbending is detected, the program can back-track the simulated operations. The output of the program is the flat pattern drawings with feasible bending sequence labels, which can be used to program CNC press brakes.

2.5.4 System details

2.5.4.1 Interactive unfolding

Figure 2.41 shows the flowchart of the interactive unfolding algorithm.

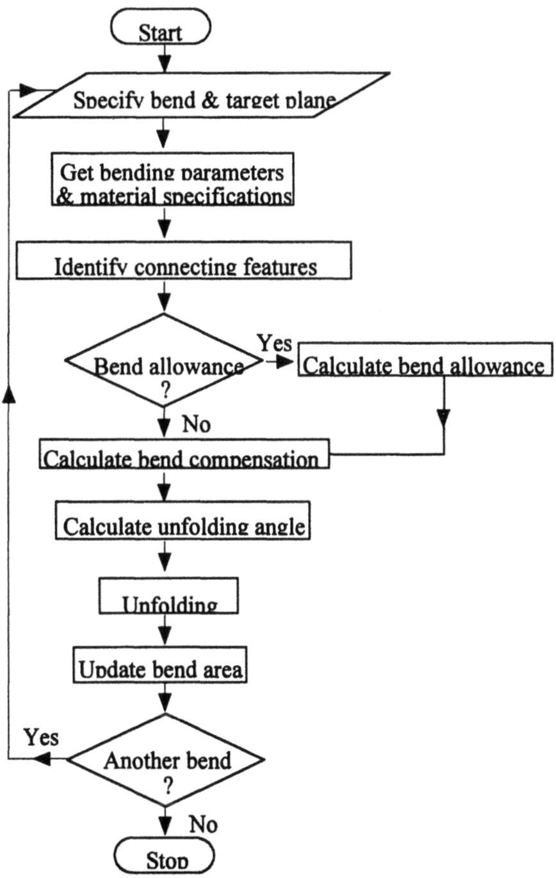

Figure 2.41: Flowchart of the interactive unfolding algorithm.

The interactive unfolding program allows a part to be unfolded bend-by-bend. After a bend is specified, the user can select a reference plane onto which the connected features will be unfolded or let the program perform the selection. For example, when unfolding bend "B1" in Figure 2.42, either wall "W1" or "W2" can be chosen as the reference plane.

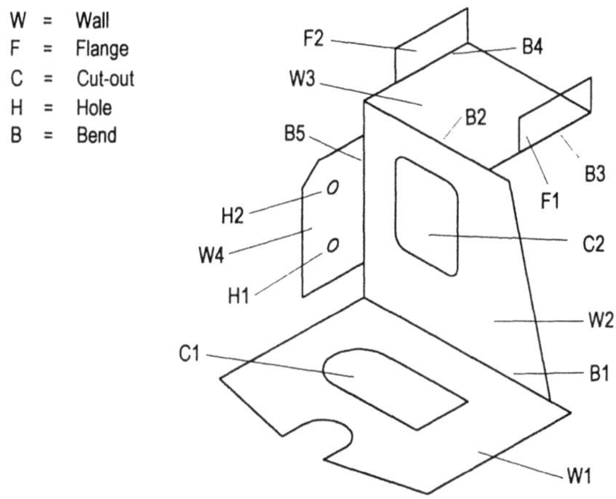

Figure 2.42: Features considered when unfolding bend "B1".

When unfolding a bend, the program performs a search for the connected features based on the feature relations that have been captured in the PRF. Figure 2.43 depicts the features considered when unfolding bend "B1". Since the slot on wall "W1" and the chamfer on wall "W4" form part of the outer profiles of the parent features, they are not considered during the connected features search. The geometric entities of the connected features are read from the feature sections in the PRF for unfolding.

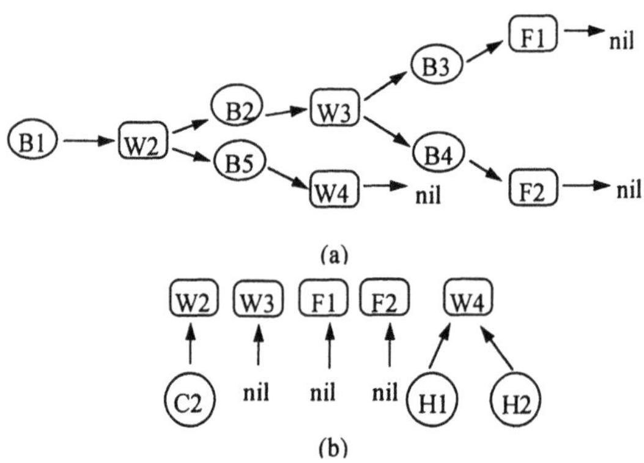

Figure 2.43: (a) Connected macro feature tree for bend "B1"; (b) the related co-feature tree.

Basically, unfolding involves two spatial transformations on the geometric entities: rotation and translation. Rotation will transform the connected features onto the reference plane while translation will re-size the unfolded parts to accommodate the bend compensation. After unfolding, the program updates the bend zone by generating the mould lines for the bending tool placements. Figure 2.44 shows the unfolding steps of the sample part.

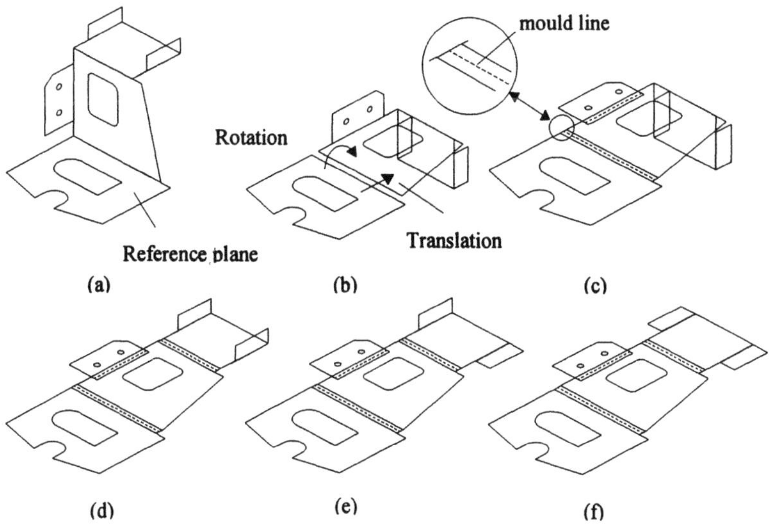

Figure 2.44: Unfolding steps for the sample part.

2.5.4.2 Interactive folding

The interactive folding program can be used to fold up an unfolded part or a 2-D part. Figure 2.45 shows the flowchart of the interactive folding algorithm.

When folding an unfolded bend, the program deletes the mould lines and searches the features to be folded. Folding is performed by translating the connected features based on bend compensation and rotating the features based on the specified bend angle.

The interactive folding program also folds up a 2-D design. Since nominal dimensions are used in the design, the folding will not involve feature translation which caters for bend compensation; only rotation of features is performed.

2.5.4.3 Automatic unfolding

Figure 2.46 depicts the flowchart of the automatic unfolding algorithm.

The automatic unfolding program performs unfolding without any user-intervention. The program first identifies all the bends to be unfolded and then unfolds the bends using the program-selected reference planes and the program-calculated bend compensation.

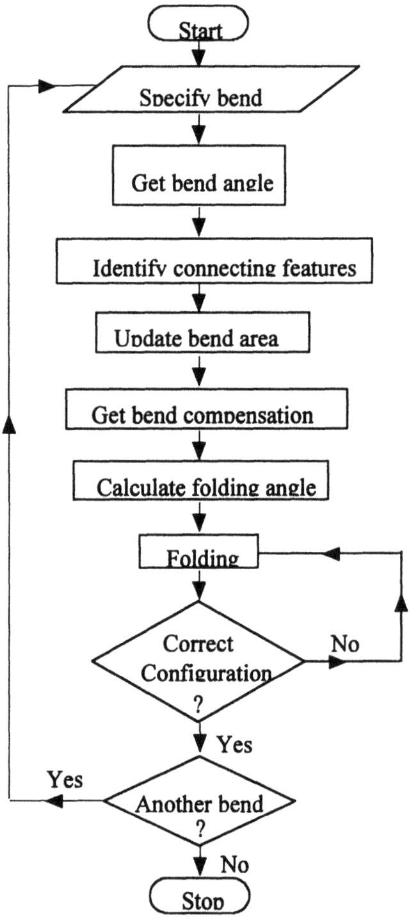

Figure 2.45: Flowchart of the interactive folding algorithm.

2.5.4.4 Automatic folding

Figure 2.47 depicts the flowchart of the automatic folding algorithm. Similar to the interactive folding program, the automatic folding program folds up an unfolded part or a 2-D design. No user-intervention is required as the program will check for all the bends to be folded and perform folding automatically.

2.5.4.5 Flat patterning information in the part representation file

Figure 2.48 depicts the part configuration and the flat patterning information captured in the PRF after unfolding bend "B1". The flat patterning data, which have been recorded in a "BZ" (bend zone) section and a "FPH" (flat patterning history) section, are required when performing automatic folding/unfolding and bend zone updating.

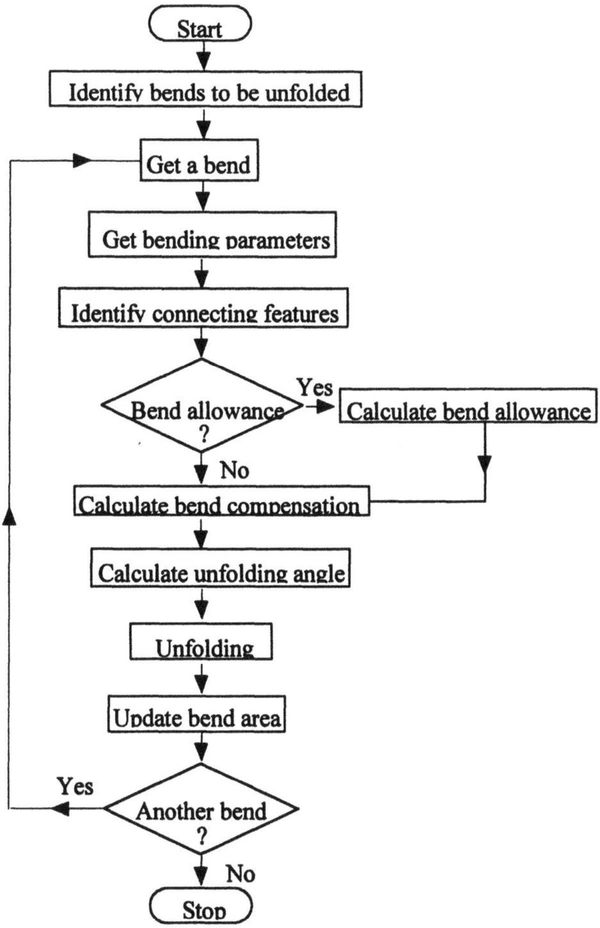

Figure 2.46: The flowchart of the automatic unfolding algorithm.

In a "BZ" section, the bend identity and the entity handle of the generated mould line are recorded. In a "FPH" section, the bend identity, the entity handles of the two bend lines, the bend compensation and the normal vector (before unfolding) of the adjacent connected feature are recorded. In the above example, since "W1" is used as the reference plane for the unfolding, the original normal vector of wall "W2" is recorded. The above data can be used to keep track of the following: the flat patterning history, for example, the identities of the bends that have been unfolded; the mould lines to be deleted during folding and; the original orientations of the features.

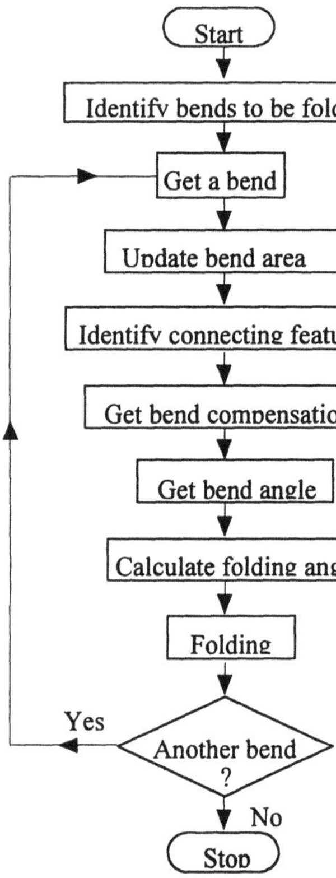

Figure 2.47: The flowchart of the automatic folding algorithm.

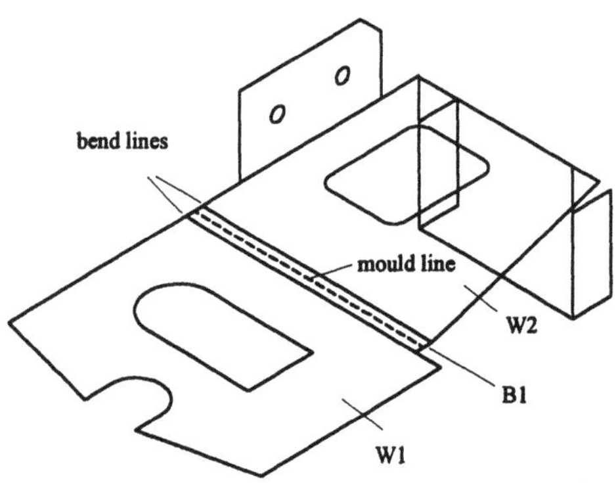

FPH	BZ
BEND	BEND
1	1
Normal	MouldLine
(0 1 0)	EA
BendLine	ENDBZ
3	******
BendLine	
22	
BendComp	
-6.7	
ENDFPH	
******	

Figure 2.48: The corresponding flat patterning data of the unfolded bend.

2.5.4.6 Blank generation

The generated flat patterns can be further processed by the program to obtain blank profiles which are required by the nesting, punching and bending computer software. The profiles can be exported in the forms of DXF and IGES file that are usually acceptable by the programs. Figure 2.49 shows the blank configurations that are possibly needed by the three operations.

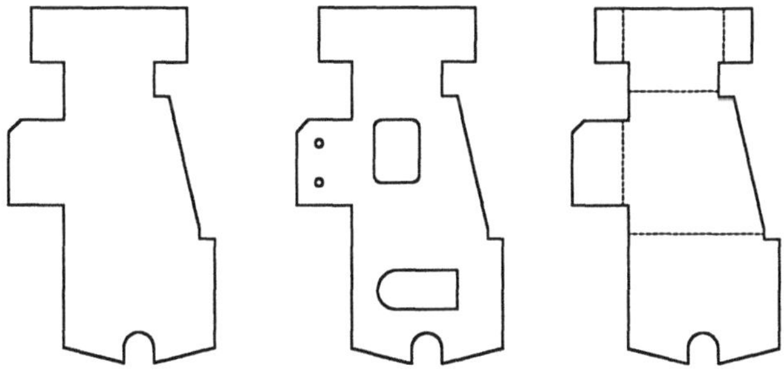

(a) (b) (c)

Figure 2.49: Blank configurations that are possibly required by (a) nesting, (b) punching and (c) bending programs.

2.5.5 Implementation of bending simulation module

The bending simulation module performs bending tool selection and feasible bending sequence generation. This section presents the implementation details for the program.

2.5.5.1 Feature-based bending simulation module

In bending simulation, both the part and the bending tools are represented in 3-D wire-frame configurations. The program is integrated with the design input and the flat pattern development modules. The feature-based part representation data are adapted for hardware interference checks before and after bending, and the folding/unfolding algorithms are used to simulate the bending operations.

2.5.5.2 Generation of feasible bending sequence

The present method adopts the 'reverse of unbending' approach for the generation of feasible bending sequences. A feasible unbending is one in which no hardware collisions would occur during the operation. A feasible unbending sequence is generated if all the unbending from the final product to the flat pattern state is feasible. By reversing the feasible unbending sequence, a feasible bending sequence is obtained. As the approach handles the more difficult bends at the early state, it can significantly reduce the computing time of the program. To further reduce the search space of feasible bending sequences, the program has been incorporated with tool selection heuristics and design modification suggestions based on expert knowledge.

Figure 2.50 depicts the flowchart for the bending simulation algorithm. The steps of the program are as follows:

1. The user selects a bend and the suitable bending tools from a tool database. If required, the bend angle can be modified to simulate overbending.
2. The program validates the selected tools before it positions the tools at the bend.
3. The program performs tool-work interference checks. If interference occurs, the user can select new tools or another bend to proceed.
4. If no interference occurs, the program will unfold the bend and check for tool-work interference at the unfolded stage. Where interference occurs, the program can back-track the performed operations. The back-tracking involves folding back the bends in the reverse sequence of unfolding.
5. The above steps are repeated until all bends are unfolded.

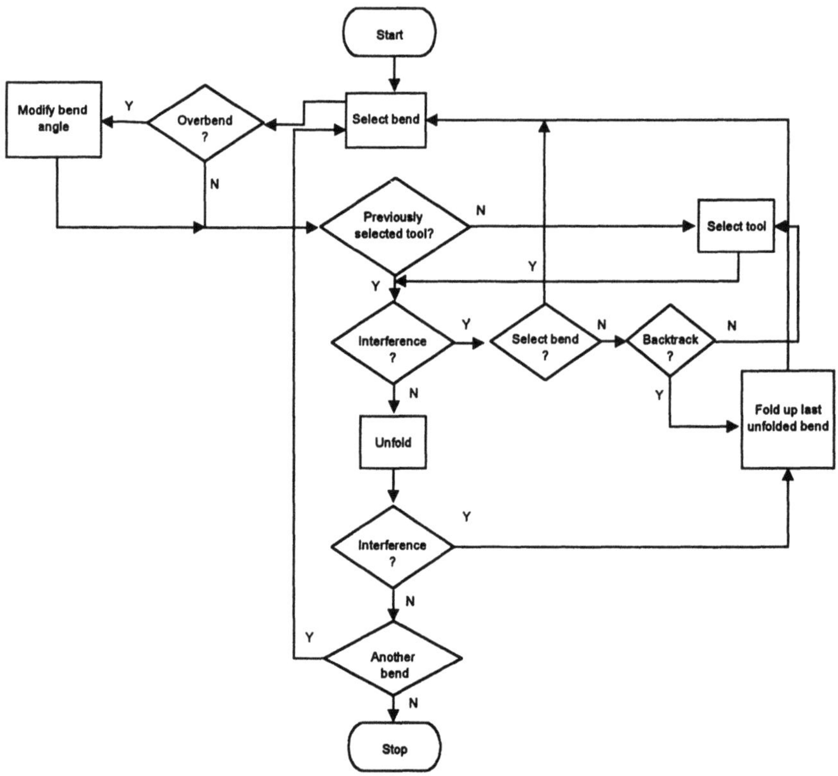

Figure 2.50: The algorithm of the bending simulation program.

Figure 2.51 depicts the bending simulation performed on the sample part. During the simulation, the configurations can be displayed in two views: a 3-D view and the side view.

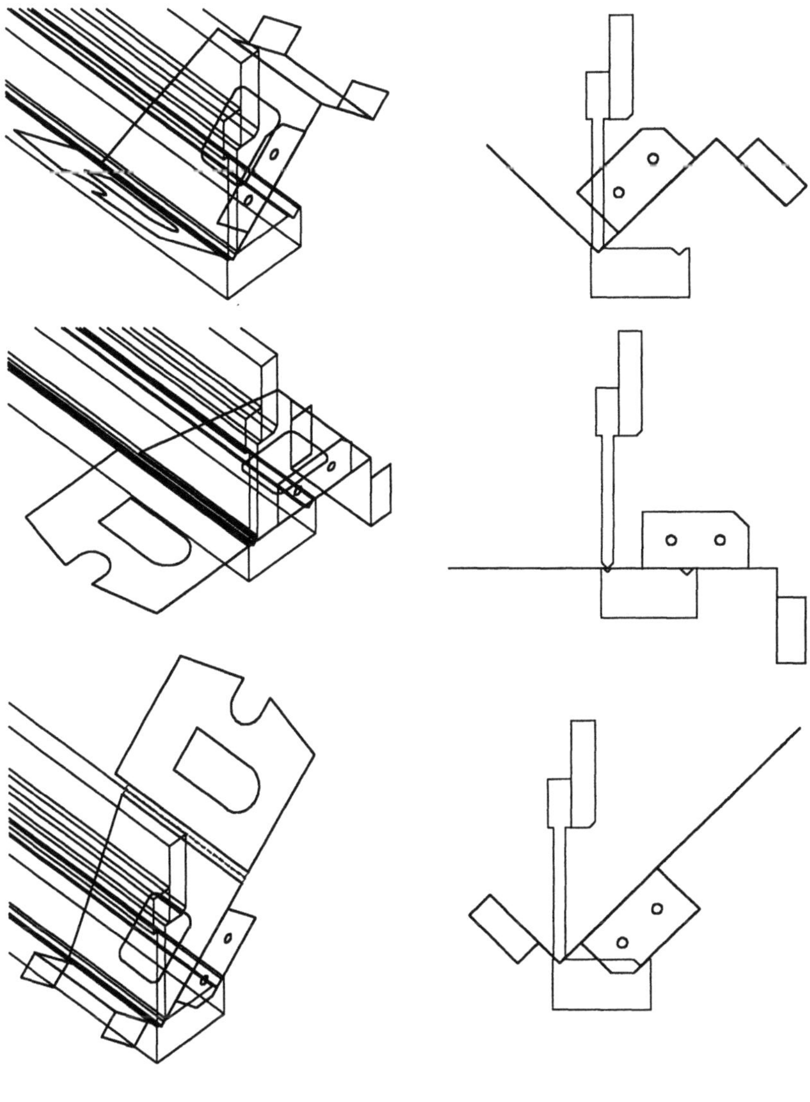

Figure 2.51: Bending simulation of the sample part.

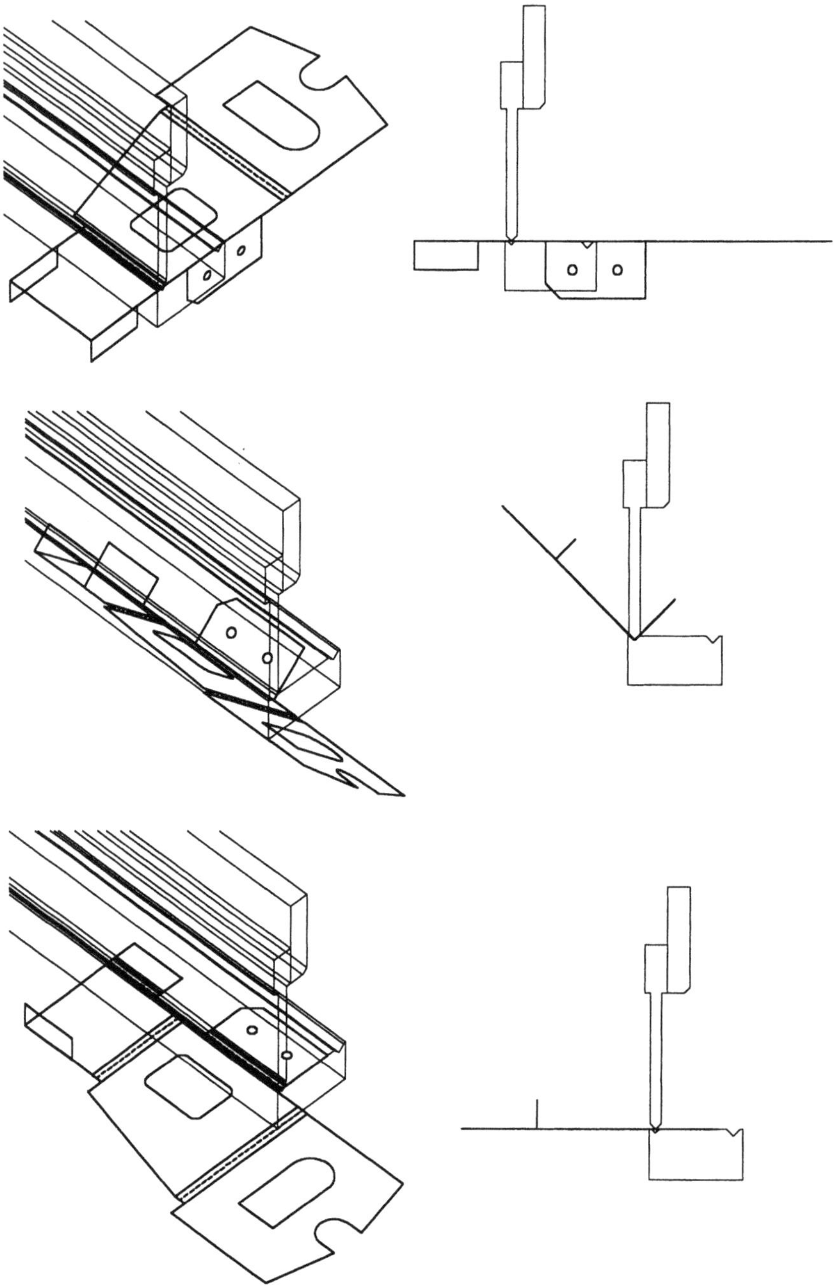

Figure 2.51: (*continue*).

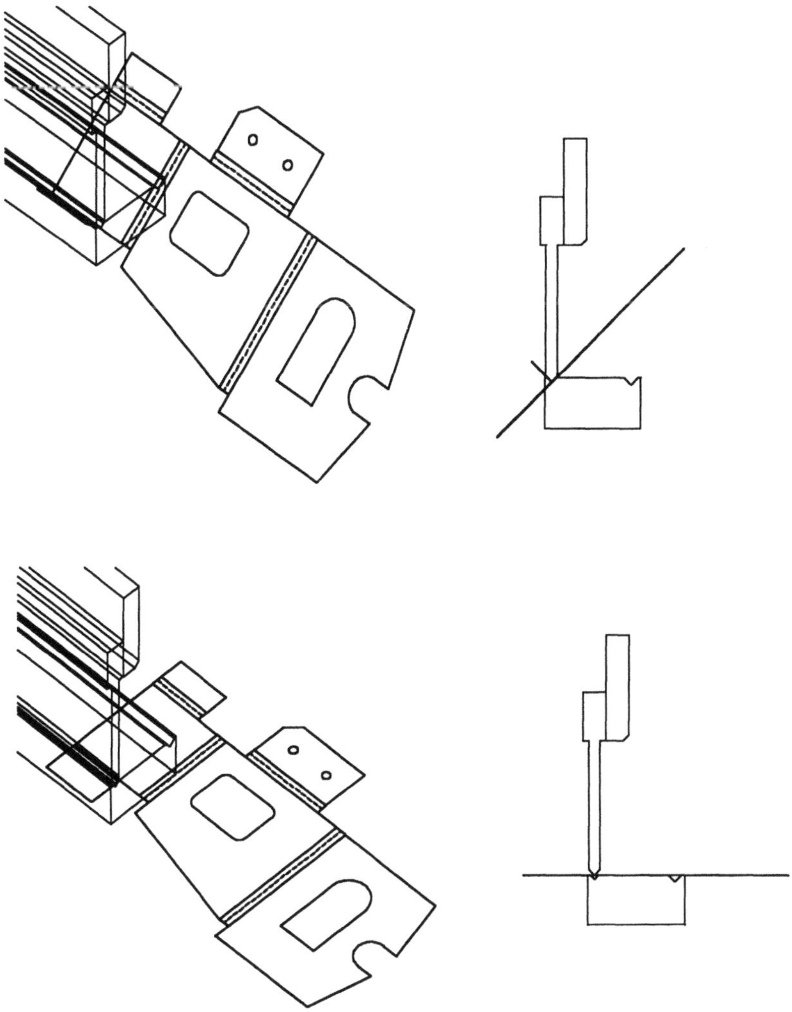

Figure 2.51: (*continue*).

2.5.5.3 Tool library and tool selection

The tool library consists of 3-D wire-frame drawings that define the tool geometry, and textual data files that store the tool data such as tool number, punch radius, die width, etc. The present tool library stores AMADA press brake punches and dies. However, the user can modify the existing tools and add new tools. Moreover, the tool geometry is not limited to punches and dies only. Machine casings and other hardware accessories can be included for the interference checks during bending simulation, where necessary.

After the tools are selected, the program checks for the suitability of the tools based on the tool geometry and the design parameters of the bend in-process. For example, a 90° straight punch will be unsuitable for an obtuse bend angle. To avoid time-consuming tool change, the program will first check for the suitability of the previously selected tools. If the tools are found to be unsuitable, the program will recommend other tools or suggest design modifications. For example, a bend radius can be modified to suit the selected punch and die. Where necessary, design modifications can be performed without abandoning the bending simulation program.

2.5.5.4 Tool-work interference check

Tool-work interference is checked before and after the unbending. The routine first forms two groups of plane surfaces using the constitutive geometric entities of the part features and the selected tools. The intersections between the two groups of surfaces are then checked. Where intersection is found, the program will highlight the corresponding surfaces. Figure 2.52 depicts an example where a tool surface interferes with a part surface.

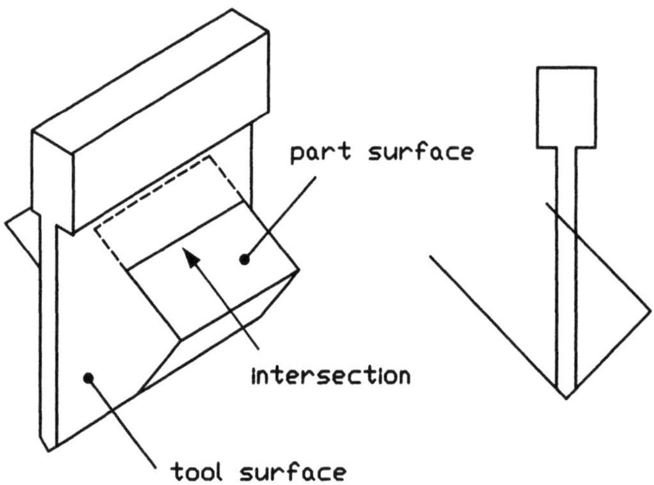

Figure 2.52: An intersection occurs between a tool surface and a part surface.

2.5.5.5 Bending information in the part representation file

After a feasible unbending operation has been determined, the program writes the bend identity and the corresponding tool information to the PRF. When a back-tracking is performed, the corresponding information will be removed from the PRF. Figure 2.53 shows the bending information generated in the PRF after unfolding bend "B1" of the sample part.

2.5.5.6 Output of bending simulation program

Based on the bending information in the PRF, the program generates a flat pattern drawing with the bending sequence labels, and creates a text file that stores the selected tool identities. Figure 2.54 shows the labelled flat pattern drawing and the corresponding tool information.

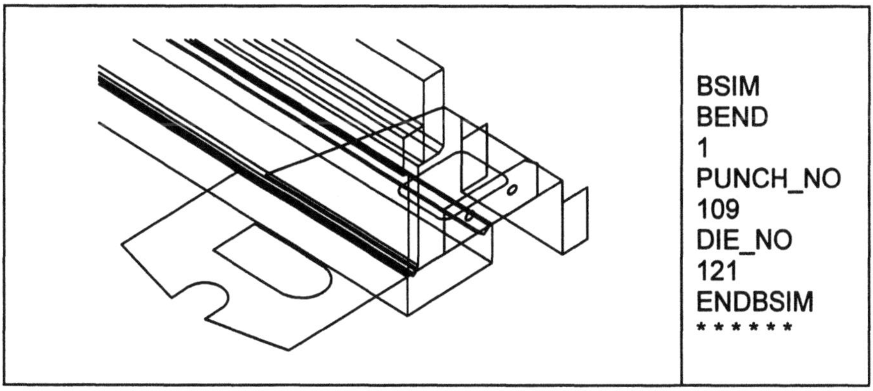

Figure 2.53: The corresponding bending data of the unfolded bend.

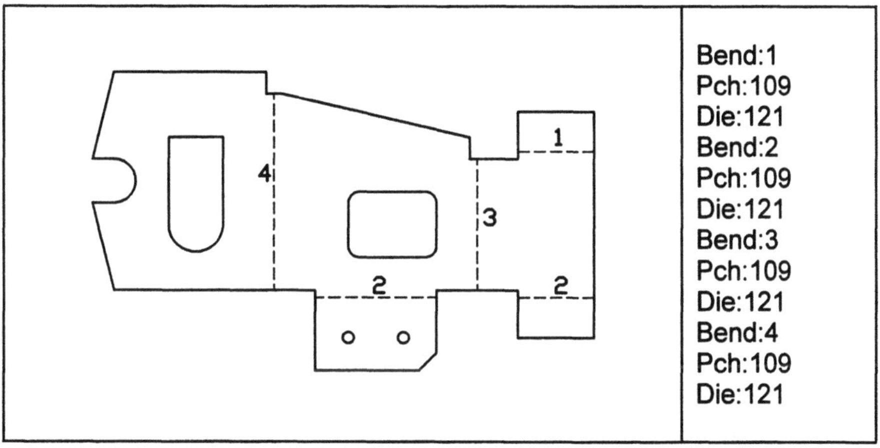

Figure 2.54: Flat pattern drawing with the bending sequence labels and the corresponding bending tool table.

2.6 CNC punching and nibbling

Nibbling is a progressive punching process to create sheet metal blanks of complex shapes. It uses a combination of a set of standard punches which may be square, rectangular, triangular or round in shape to produce a 2-D workpiece. Therefore, the construction of a mating punch and die set is not required. This reduces the lead time and investment cost for constructing new sets of toolings for manufacturing different parts.

During the nibbling process, a workpiece is moved stepwisely in the desired feed direction accompanied by constant up and down (continuous stroke rate) movements of punches. A special feed control assures that the workpiece is always stopped just before the punch dips into the workpiece surface. The next feed step begins as soon as the punch is again above the workpiece. By continuous repetition of this cycle - feed release, feed interruption during constant stroke movement of the tool - cutting out of the desired shape of the workpiece is sub-divided into a multitude of small individual cuts. Due to the division of the cutting process into a quick succession of partial punchings with a simple tool, the nibbling operation is especially advantageous for the following jobs :

1. Production of cut-outs and contours of any sizes and shapes which cannot be achieved in one down-stroke (single stroke) of the ram due to limitation in the punching capacity of the machine.
2. Manufacture of irregularly-shaped cut-outs where using special tools in one down-stroke of the ram would be too time-consuming and expensive.

CNC nibbling, which requires only the combination of several sets of standard punches for producing sheet metal parts, is particularly suitable for low to medium volume production of large and complex workpieces. However, workpiece tolerance and accuracy are difficult to control compared to stamping operations with matching tools and dies. Hence, nibbling is not suitable for high precision components.

2.6.1 Computer-based system for CNC nibbling

A sample part suitable for CNC nibbling is shown in Figure 2.55. The part can be produced by a combination of square, triangular or round punches and does not require the construction of mating punch and die sets. Therefore, it eliminates the need to make mating punch and die sets which are expensive and time-consuming.

In the punch selection process, the planner is required to search through a list of available punches to select those that would minimise the total number of strokes and the number of tool changes so as to minimise the total cycle time. If circular punches are selected, they must give the required surface roughness without violating the minimum allowable feed criterion of the machine. The generated surface roughness depends on the punch diameter, the geometry and the feed length. Smaller feed per stroke generates better surface finish. However, machine limitation restricts the feed to recommended minimum values to ensure high quality products. In the case of a TRUMPF nibbling machine, the minimum allowable feed is one half of the sheet metal thickness. After the generation of the NC program, a test run has to be carried out to check that the planned sequence could produce the object within tolerances specified.

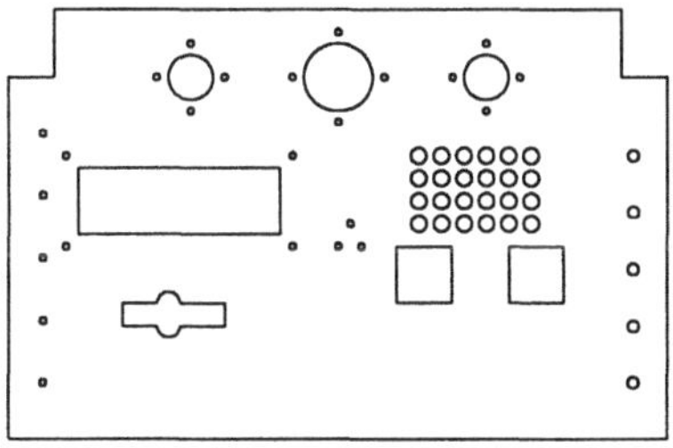

Figure 2.55: A sheet metal component typically suitable for punching and nibbling.

The above-mentioned processes involved are tedious and time-consuming. This is further aggravated when dealing with complex workpieces. Moreover, optimal solutions in terms of shortest cycle time are difficult to achieve. In addition, nibbling machines are expensive, labour cost of the planner is high and it would not justify long hours to iteratively select suitable punches to manufacture a small volume of parts.

To overcome all these problems, it is desirable to develop a computer-based system to select suitable punches, generate the tool paths graphically on screen, and provide information such as the number and type of punches required, total number of strokes and cycle time, and the indexing sequences. This will eliminate the tedious work involved and speed up the processes. Besides, several solutions can be generated easily and quickly for comparison and selection. Moreover, the graphical output of the tool path allows a planner to check for errors on screen without having to perform any test runs on the machine.

2.6.2 Punch libraries

The punch libraries are organised into square, rectangular, circular, triangular, fillet and special punches library. The special punch library stores punches that do not fall into any one of above-mentioned five types of standard punch libraries. Forming, engraving and louvering punches are some examples of this category. Special punches can only be created, updated or retrieved during manual tool selection. This is because special punches are not considered during automatic tool selection and are selected only through manual punch selection. The cross-sections of the five types of standard punches are shown in Figure 2.56.

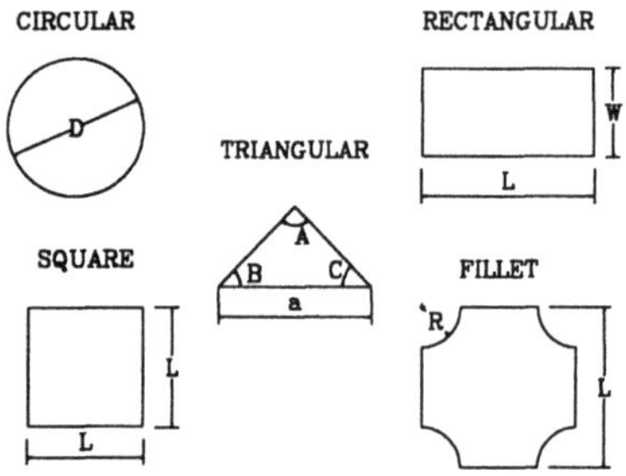

Figure 2.56: Cross-sections of the standard punches considered.

The required information that describes a punch is the punch dimensions and the quantity available. These are the basic punch data that are stored in the standard punch libraries. Special punch dimensions are generalised into overall length and width. A data field is provided for the descriptions of such punches.

2.6.3 Profile classification

Profiles of a feature can be categorised into general external, general internal and standard profiles. The general external and internal profiles are geometric entities made up of connected lines and arcs only. The standard profiles are square, rectangular, triangular, circular, oblong and arc-slot holes. In general, profiles usually belong to one of the followings (Figure 2.57):

1. general external profile
2. general internal profile
3. square hole
4. rectangular hole
5. triangular hole
6. oblong hole
7. arc-slot hole
8. circular hole

The test used to separate internal and external profile data is actually a modification of the even-odd method of determining whether a point lies in the interior or exterior of a polygon. In this method, to test whether a given point is inside or outside a polygon, a line segment is constructed between the point in question and a point known to be outside the polygon; and the number of intersections of this line segment with the polygon boundary is counted. An odd number of intersections indicate a point is inside the polygon and vice-versa. As mentioned above, the user is required to indicate three points inside the workpiece. Therefore, these points will be inside the polygon representing the external profile of the workpiece but outside the polygons representing the internal profiles of the workpiece. By applying the above test to each enclosed profile, the number of counts of line segment intersections with the boundary of the profile will indicate whether the profile in question is internal or external. An even-numbered count would indicate internal profile while odd-numbered count, external profile. Hence, all the extracted profile data can be separated into internal and external profile data. Three interior points are chosen instead of one because in some degenerate instances, the even-odd method can fail. By having three points, it is extremely unlikely that all three points will fall under the degenerate cases.

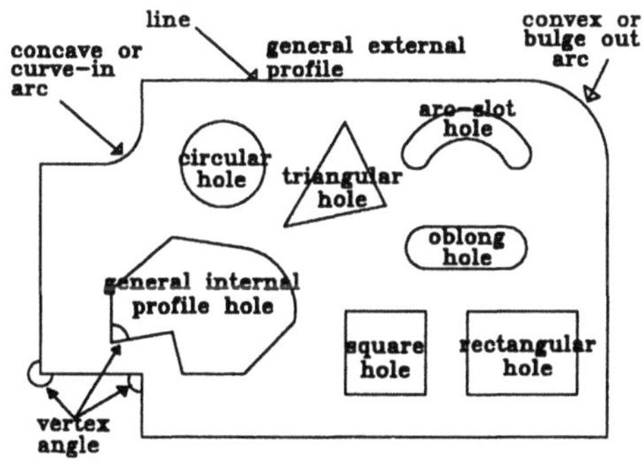

Figure 2.57: Classification of profiles.

After the test, the straight lines approximating the arcs are removed and the arcs are restored. The topological chain of a profile is then traced and recognised as a combination of straight lines and arcs, or just as circular holes. The external profile is arranged in the counter clockwise direction while all the internal profiles are arranged in the clockwise direction. However, in AutoCAD, the geometrical information of all arcs, i.e., the start and end points of arcs, are always stored in the counter-clockwise direction. In addition, arcs have to be distinguished as convex or concave so that suitable type of punches can be selected and directed to stamp on the unwanted material. Referring to Figure 2.58, convex and concave arcs can be distinguished by comparing the directions of the arcs with the direction of the topological chain of the profile. Convex arcs point along the same direction as the rest of the profile while concave arcs point in the opposite direction. The processed information is stored in three separate data files as either external profile, internal profile or circular hole.

The internal profile data are further processed into general profiles, square holes, rectangular holes, triangular holes, oblong holes and arc-slot holes. These data are stored into six separate data files.

2.6.4 Punch selection and optimisation

The slot and feature profiles often limit tool selection to only one of several types of standard punches. The geometry, in turn, restricts the size of the punches that can be used to obtain the required dimension without violating surface roughness and the allowable feed per stroke of the machine. Therefore, based on the shape and geometry of each profile, a set of possible tools can be selected. The total number of combinations of tools selected for each profile may be more than the total number of tool stations available in the machine. This would mean that more than one tool set-up may be required to produce the complete workpiece. In practice, one tool set-up is much preferred in order to achieve good tolerance and accuracy, and to minimise production cycle time.

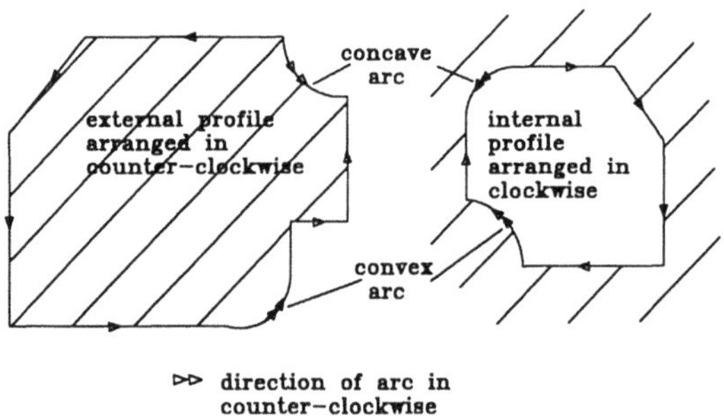

Figure 2.58: General profiles, concave and convex arcs.

In addition, similar tools with different orientations may also be selected (Figure 2.59). If a nibbling machine has auto-indexing stations or a rotary punching head, tools can be rotated at many orientations during nibbling. This can reduce the number of punches needed. However, if auto-indexing stations or rotary punches are not available, the total number of similar tools required is equal to the number of different orientations.

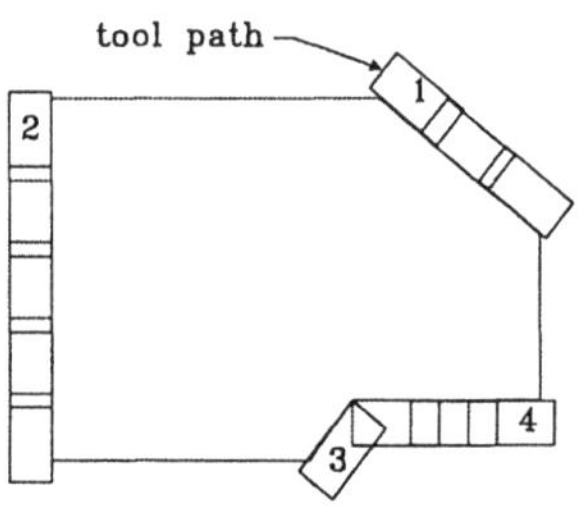

Figure 2.59: A rectangular punch used in four different orientations.

2.6.5 An approach to automatic tool selection

Based on the requirements of tool selection described earlier, the current algorithm divides this process into two stages. In stage one, or the primary tool selection stage, tools are selected for each profile based on its shape and geometry. In stage two, or the secondary tool selection stage, it reconsiders those independently selected tools in the primary stage and plans for a single tool set-up to produce the complete part. This is essential as a single tool set-up minimises the production cycle time, and is able to achieve better accuracy. Secondary tool selection is a necessary sequence following primary tool selection because tool selection in the earlier stage is considered individually for each profile of the part, and the total number of tools required is not known until all the profiles have been considered.

The automatic tool selection algorithm is implemented using C language. Correct tool selection requires lengthy computations to detect over-cutting and calculate surface finish requirements. The next two sections will discuss the primary and secondary tool selection processes in greater detail.

2.6.5.1 Primary tool selection

In primary tool selection, each profile is considered independently to select a set of suitable tools. Profiles can be categorised into eight different types. Each type of profile has a set of technological and heuristic rules governing the types of punches that can be chosen. Several alternative solutions are usually generated as a single solution may fail to be selected in the secondary selection stage.

The availability of tools, the required surface roughness and the minimum and maximum allowable feed lengths are the governing rules used for punch selection. Surface roughness depends on the geometry of a feature to be nibbled, the size of the punch and the amount of feed per stroke. However, the amount of feed per stroke for each type of standard nibbling punch is usually recommended within a range of values by the machine manufacturer to ensure high quality products. For example, the Trumatic Laserpress 240 recommends the following technological and heuristic rules for feed per stroke [15]:

1. *Circular punch*
   - minimum allowable feed = 0.5 x sheet thickness [mm]
2. *Square and rectangular punches with length, L*
   - minimum allowable feed = 1/3 x L [mm]
   - maximum allowable feed = L - 2 [mm]

Therefore, suitable punches are chosen by compromising these rules and the user's requirements.

As a rule-of-thumb, the largest punch that can satisfy all the above-mentioned constraints will be selected as the first choice. This is because a larger tool requires fewer strokes and would leave fewer nibbling marks, hence minimising the production cycle time and giving better surface finish.

Square, rectangular, triangular, circular, oblong and arc-slot holes are classified as standard internal profiles. Each profile is recognised as an object (hole) rather than a profile that consists of line and arc entities. Tool selection will therefore be based on the characteristics of these objects.

Standard internal profiles are best produced in a single stroke. Other alternatives to produce square holes include profiling them using rectangular or square punches of smaller sizes. For a rectangular hole, the next best choice is either a rectangular or square punch which matches the width of the hole and can nibble two or three sides of the profile in one stroke. Triangular and circular holes can be produced by nibbling with smaller punches with shapes similar to the holes. The best choice for an arc-slot hole is a circular punch with diameter equal to the width of the slot. The other alternative is to use a smaller circular punch to profile around each of the four arcs. An oblong hole requires the combination of a matching circular punch and a square or rectangular punch to create the profile with a minimum of three strokes. The two alternatives are either profiling the contour with a smaller circular punch or a combination of smaller circular and square or rectangular punches. The above-mentioned rules are implemented for standard internal profiles as follows:

SQUARE / RECTANGULAR HOLE

1. *select a sq/rect punch that fit hole-1 stroke*
2. *select a rect & sq punch to profile*
   (for rectangular hole only) select a rect/sq punch to nibble 2 or 3 sides in one stroke

TRIANGULAR HOLE

1. *select a tri punch that fit hole-1 stroke*
2. *select a smaller tri punch that similar shape to hole to profile without reorienting punch*
3. *select a smaller tri punch that 2 angles <= 2 smallest angles of hole to profile with reorienting punch*

CIRCULAR HOLE

1. *select a cir punch that fit hole-1 stroke*
2. *select a smaller cir punch to profile*

ARC-SLOT HOLE

1. *select a cir punch = width of hole to nibble along arc*
2. *select a smaller cir punch to profile the 4 arcs*

OBLONG HOLE

1. *select a cir punch that fit arc radius-1 stroke & select a sq/rect punch that fit sq/rect hole-1 stroke*
2. *select a smaller cir punch to profile arcs & sq/rect punch to profile sq/rect hole*
3. *select a smaller cir punch to profile arcs & sq/rect hole*

General profiles are usually made up of combinations of straight line segments, and convex and concave arcs. Each entity is characterised by its shape, dimension and vertex angles formed by the entity itself and the adjacent entities. Tool selection for general profiles is based on the selection of a suitable punch for each side of the profile.

Rectangular and square punches are preferred for nibbling lines with vertex angles greater than or equal to 90° as these tools produce the best surface finish. However, circular punches can also be selected as an alternative if rectangular or square punches cannot be used at a particular slant angle. Triangular punches are used for lines with sharp vertex angles (angles $< 90^\circ$). Convex arcs with 90° subtended angles can be produced in one stroke by fillet punches with matching radii. Concave arcs can also be produced in one stroke by circular punches with matching radii. The alternative way for producing both types of curves is profiling around the arcs using circular punches. Rules implemented for the general profiles are as follows:

LINES

if v_angle1 < 90 & v_angle2 < 90

select tri punch with angles <= v_angle1 & v_angle2

if v_angle1 < 90 or v_angle2 < 90

select tri punch with any one angle <= the smaller v_angle

if both v_angle1 >= 90 & v_angle2 >= 90

1. *select largest rect punch with length <= length of line, determine orientation of punch*
2. *select largest sq punch with length <= length of line, determine orientation of punch*
3. *select largest circular punch with diameter <= length of line & satisfies the required roughness & min. allowable feed rule*

ARCS

if convex arc with subtended angle=90

1. *select a fillet punch with radius fit to arc radius*
2. *select largest cir punch that satisfies the required roughness & min. allowable feed rule*

if other convex arc

1. *select largest cir punch that satisfies the required roughness & min. allowable feed rule*

if concave arc

1. *select a cir punch with radius fit to arc radius*
2. *select largest circular punch that satisfies the required roughness & min. allowable feed rule*

External and internal profiles that do not fall within the standard profile categories are classified as general profiles. The above-mentioned selection rules are applicable to both external and internal general profiles. Since tools are selected independently for each side of the general profile, the selected tools may overcut into the adjacent material that is wanted. Therefore, it is necessary to conduct overcutting checks during tool selections. However, conducting such tests required lengthy computations. To minimise computing time, general profiles are checked for critical features to determine if overcutting tests are required before tool selections commence. The general check detects for short geometry, any straight edge that is shorter than the smallest available punch and concave vertices that are in consecutive sequence. Only profiles with such features will undergo overcutting tests during selection for each edge. In the overcutting test, two lines, L1 and L2 (see Figure 2.60), with length equal to the largest punch dimension and normal to the edge are projected out from the end-points of the edge, E1. A third line, L3, which is parallel to E1 is constructed from the end-points of L1 and L2. Arcs have to be approximated to straight lines for the check. If any edges of the same profile intersect with either L1 or L2, then overcutting is detected. The perpendicular distance, D1, from the intersection point to the edge E1, from which the tool path is projected, is calculated. Similarly, if any edges of the same profile intersect with L3, then overcutting is detected. The overcut portion must contain at least one end-point of the edge that is cut off. The perpendicular distance, D2, from this end-point to the edge E1 is calculated. Therefore, any punch to be selected for this edge has to have a similar or smaller dimension than D1 and D2.

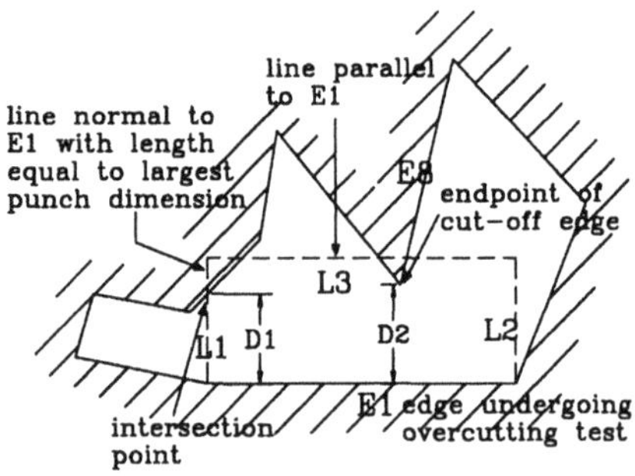

Figure 2.60: The overcut test.

Since overcutting checks are time-consuming, the user is given an option to turn off the check if he or she thinks that all general profiles do not have any critical features. The default is to conduct overcutting checks.

2.6.5.2 Secondary tool selection

The secondary tool selection criteria are based on the analysis of the characteristics of three different types of nibbling machines, viz. the AMADA Pega 375 [15], [16], TRUMPF Trumatic 240 Rotation [17] and TRUMPF Trumatic Laserpress 240 [17].

AMADA Pega 375 has a numerical control turret head that houses 58 tool stations. It has two auto-index stations which can rotate tools to different orientations during the nibbling process. Other stations can accommodate tools that can be prefixed at certain slant angles but cannot be changed thereafter during the nibbling process.

Trumatic Rotation 240 has one stamping head and ten automatic tool change stations. This stamping head can rotate tools to any orientation angles during the nibbling process.

Trumatic Laserpress 240, a hybrid CNC nibbling and laser cutting machine, has one stamping head, nine automatic tool change stations and one laser cut head. The stamping head cannot be rotated due to the existence of a laser cutting head. However, punches can be prefixed at any desirable slant angles. Both the stamping and laser cutting heads are stationary during cutting while the table moves.

Two steps are involved in secondary tool selection. Step one reduces the total number of primary selections to the number of tool stations that can be accommodated. This is to achieve a single tool set-up to produce the complete workpiece. Tools selected in the primary selection are re-selected based on the following criteria, listed in order of priority:

1. Tools that are manually selected for certain parts of a profile.
2. For lines with small vertex angles
   2a) Triangular tools if the machine has no laser-cuts.
   2b) Lasercuts if the machine has laser-cuts.
3. Tools that have been selected most frequently and with the highest priority (i.e., the first choice for a profile).

Based on the tools selected in step one, step two selects only one solution from the few primary selections obtained for every profile. At this step, some profiles may end up without any suitable punches. These profiles will be highlighted in the CAD environment, in this case in the AutoCAD drawing editor. By picking any of these profiles, the user will be able to find out the reason for the highlight and the recommended changes. The following examples illustrate some of the reasons:

1. A circular hole that cannot find a punch that matches the diameter and cannot find a smaller circular punch to profile around it.
   Reason: *no circular punch that can achieve the required surface roughness without violating the minimum feed rule*
   Recommendation: *change hole size or obtain a punch that matches the diameter.*
2. A line with sharp vertex angles that cannot find a suitable triangular punch
   Reason: *vertex angles too sharp, no suitable triangular punch*
   Recommendation: *increase the vertex angles or obtain triangular punch with angles <= vertex angles*

2.6.5.3 Manual tool selection

Manual tool selection is provided for the user to control the type and size of punches to use for certain profiles. Expert advice is provided by the software to assist the user to select suitable punches for the selected profile, and check that they do not violate any technological rules like allowable feed length and required surface roughness.

In the process of manual tool selection, the user first selects a profile. If this is a general profile, the user is required to indicate the exact edge for tool selection. Subsequently, the selected tool will be used to nibble this edge only. However, for square, rectangular, triangular, circular and arc-slot profiles, the selected tool will be used to punch or nibble the entire hole. For an oblong hole, two punches have to be selected because this type of hole is constructed by two types of standard profiles, i.e., two circular holes and one square or rectangular hole.

After the required profile is selected, the manual tool selection user interface will be loaded for the user's input. The user starts by selecting either a standard punch library or the special punch library to list all the available tools.

After the tool is selected, the program will check if any technological rules have been violated. The user will be prompted an error message to replace the selected tool. If the user ignores the error message and quits from the manual tool selection program, the program ignores the incorrectly selected tool and selects the right tool for the selected profile instead.

2.6.6 A case study

To illustrate the automatic tool selection algorithm and to highlight the difference between primary and secondary tool selection amongst different nibbling machines, a sample part with geometry as shown in Figure 2.61 is used. The results of the secondary tool selection were generated based on the configurations of three types of nibbling machines, i.e. Trumatic Rotation 240, Trumatic Laserpress 240 and AMADA NCT Pega 375. Three sets of final solutions were generated for comparison and discussion. Only the AMADA NCT Pega 375 solution is shown in this section. Details of the other solutions can be found in Choong et al [18].

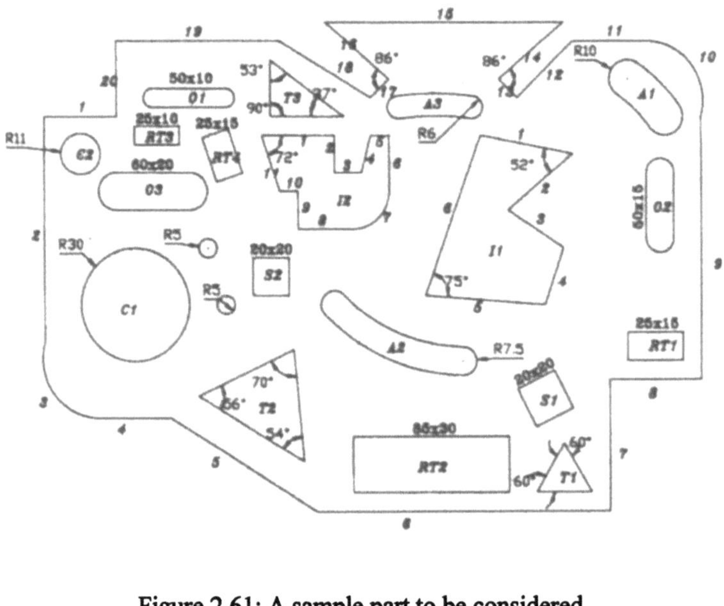

Figure 2.61: A sample part to be considered.

Based on the results generated, profiles for which punches are not selected in the primary selection eventually will not have a solution. For example, the triangular hole *T3*, for which no suitable punch was selected in the primary selection, eventually has no solution. Similarly, the two smallest circular holes and the two oblong holes *O1* (radius 5 mm) cannot be nibbled. Based on the circular punches available in the circular punch library, no single punch can be used to create the hole in one stroke. While smaller punches can be used to nibble the profile, the two smaller punches (diameters 6 and 8 mm) available cannot generate the profile within the required surface roughness without violating the minimum allowable feed constraint. Circular punches of diameter 10 mm or less than 6 mm will have to be added to the punch library if this profile is to be manufactured to the tolerance required.

The AMADA NCT Pega 375 machine has 56 non-indexable tool stations and two auto-indexing stations. Triangular punches which are selected more than once in the primary selection are assigned to the auto-indexing stations. This is to ensure that all triangular profiles or straight edges with acute vertices can be manufactured at any orientations. The large number of tool stations available allows all the punches selected in the primary selection to be accommodated. However, some of the primary selections may not be used eventually as higher priority punches will be selected first. Therefore, only 23 punches are selected finally. All profiles that have punches selected in the primary tool selection have final solutions. The punches selected for standard internal profiles can either generate the profiles in one stroke or through profiling.

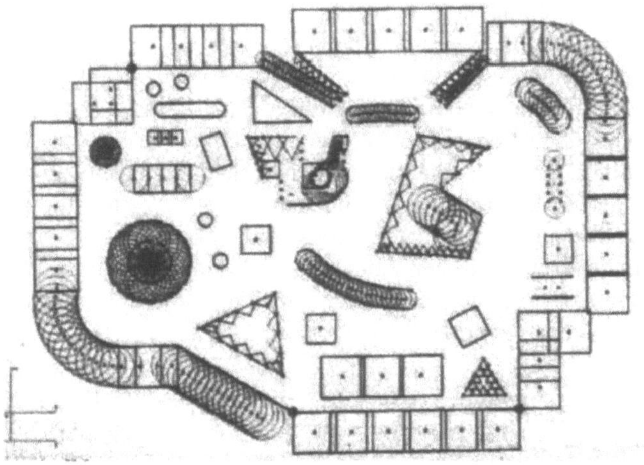

Figure 2.62: Generated tool paths for AMADA NCT simulated in AutoCAD.

It is observed that the final results of the tool selection process are different for different types of nibbling machines. The results are very much dependent on the configurations of the machine itself. Therefore, in the two stages of automatic punch selection, the secondary punch selection is necessary to generate the feasible solution based on the machine configurations from the generic solution generated in the primary punch selection. This approach allows the implementation of a generic system for different types of machine as only the secondary punch selection program needs to be modified. Figure 2.62 shows the generated tool paths for AMADA NCT in AutoCAD. There are several profiles such as the rotated square shapes do not have punch selection solutions due to unavailability of punches and hence these profiles such cannot be produced. However, on another machine where the turret head can be indexed, the shape can be produced. A user can view the reasons and recommendations to eliminate the problems at the end of the punch selection process.

2.7 Die construction design

Standardisation of dies is very crucial for establishing a CAD/CAM system for die design. The CAD/CAM system for blanking dies developed by Li et al [19] is facilitated by using stamping die standards GB2851 - GB2875-81, which include 14 typical die assemblies, 12 die sets, and standard parts such as holders, die plates, guide pillars and guide bushings.

Since stamping die standards are adopted in the system, the main tasks of die construction design are the selection of the die assembly, die set and other standard devices, and the design of the punches, dies, etc. Die parts and units are classified into two groups, standard parts and units, such as the backing plate, guide unit, and non-standard parts and units such as the punch, die and ejecting plate. Drawing data of standard parts are stored in the database. They can be retrieved during die construction design.

The die construction design software consists of eight modules, such as the basic construction design module, assembly design module, etc. The basic construction design module is used for selecting the type of die construction. Using the working part design module, the die and punch can be interactively designed to obtain a list of parts. The design and editing module for plates allows users to design, insert and remove die plates. Rods and plates can be automatically assembled using the assembly design module. Other modules perform the design of the die set, stripping device, and fastening and auxiliary devices. Part and assembly drawings are generated based on the design results.

2.7.1 Design of die and punch

The design of the die and punch includes the calculation of edge dimensions and the design of the shapes of the die and punch. In edge dimension calculation, dimensions of the die edge (A) are used as references in blanking, while dimensions of the punch edge (B) are used as references in hole punching. Wear of edges (Δ) should be considered in edge dimension calculation as the dimensions of the blanked part change with the wear of the edges. Based on the wear conditions, edge dimensions fall into three groups, *viz.*, increasing, decreasing and constant. The following formulae are used to determine edge dimensions.

$$A = (A_{max} - \Delta) + \delta/4 \qquad (2.30)$$

$$B = (B_{min} + \Delta) - \delta/4 \qquad (2.31)$$

$$C = (C_{min} + 0.5\Delta) \pm \delta/4 \quad \text{(when part dimension is } C + \delta \text{ or } C - \delta) \qquad (2.32)$$

$$C = C_{min} \pm \delta/8 \qquad \text{(when part dimension is } C \pm \delta) \qquad (2.33)$$

In these formulae, A, B and C are the three types of edge dimensions mentioned above. A_{max}, B_{min} and C_{min} are their maximum and minimum dimensions respectively. δ is the tolerance of the part dimension.

In die construction design, the dimensions of the dies are the key dimensions. For a typical die assembly, the other dimensions of the die parts and units, such as the closed height of the die set and the length of the punch, can be determined accordingly when the dimensions of the dies have been determined.

The overall dimensions of a die should ensure that the die has sufficient strength to resist the stresses during stamping. The design method adopted in the system is such that the height and wall thickness of the die are determined based on the maximum dimensions of the required part and its thickness. Consequently, the overall dimensions of the die can be determined. The overall dimensions of the die are determined by the shape and thickness of the workpiece, the rotation angle of the layout, the strip width and other factors. The procedure for die design is shown in Figure 2.63, where k, l, g and t represent the type of die assembly, layout parameters, geometry and thickness of the workpiece [19]. The shape, either circular or rectangular, and the material of the die can be selected from a menu or provided by a user.

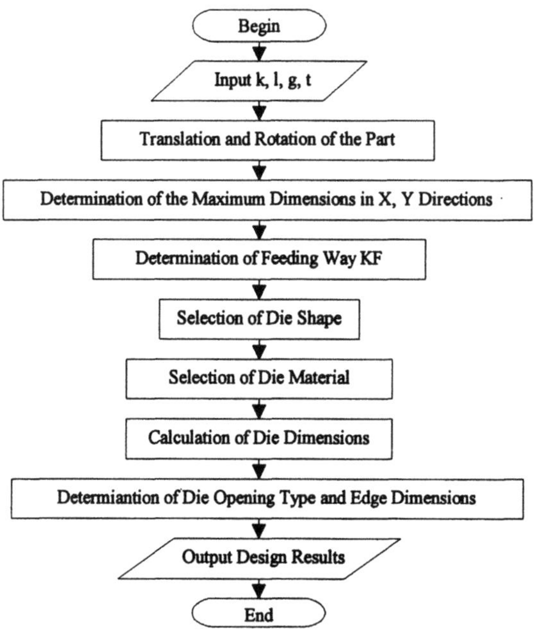

Figure 2.63: Flowchart for die design.

Four types of die openings are shown in Figure 2.64. A graphic menu is displayed on the screen during design. The user can select the type of die opening from the menu. Parameters of the die opening, such as the step height and the cone angle, are automatically determined by the program based on the selected type of die opening.

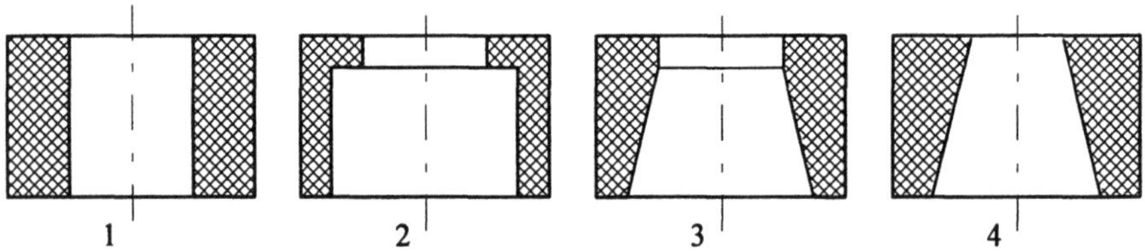

Figure 2.64: Types of die opening.

Figure 2.65 shows four types of punches. The length of the punch is determined based on the type of die assembly and the die dimensions. Punch material is interactively selected. The program can check for interference of the punches in the fixing plate, and determine the dimensions of the larger end of the punch to be removed.

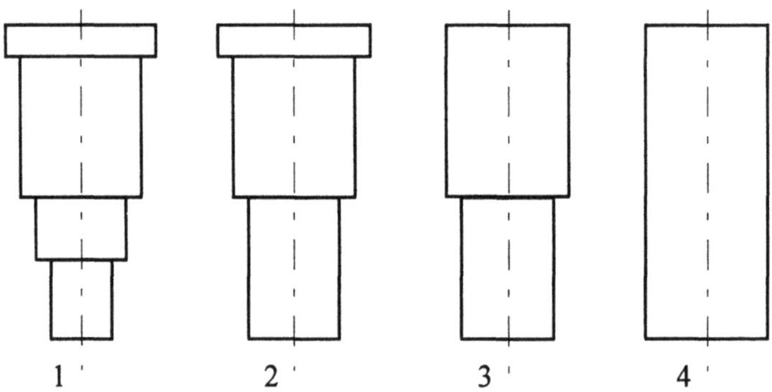

Figure 2.65: Types of punches.

When the type of die assembly has been selected and the die dimensions have been determined, the dimensions of other die parts, such as the stationary plate, backing plate and the stripper plate, can be determined based on standards. Hence, the calculation of the die edge dimensions and the determination of the overall die dimensions are important in die design. For certain parts, such as the stripper plate and the lower backing plate, the shape of the blanked part should be considered to determine their inner contours, i.e., hole shapes.

2.7.2 Layout of ejectors

Ejecting devices such as the ejecting plate, ejector pin, knockout plate and the knockout pin, should be designed to push the workpiece or scrap out of the die opening. Determination of the optimum positions of the ejectors is often difficult. In the system developed, the layout of ejectors can be optimised.

Figure 2.66 illustrates two types of ejecting devices, the upper and the lower ejecting devices. Rational layout of ejectors should satisfy the following requirements.

1. The resultant force of the ejectors should be as close to the pressure centre of the blanked part as possible.
2. The ejectors should be uniformly arranged along the periphery of the part.
3. A few ejectors should be arranged at special places, such as the long narrow area of a part.
4. The diameter and number of ejectors should be properly determined.

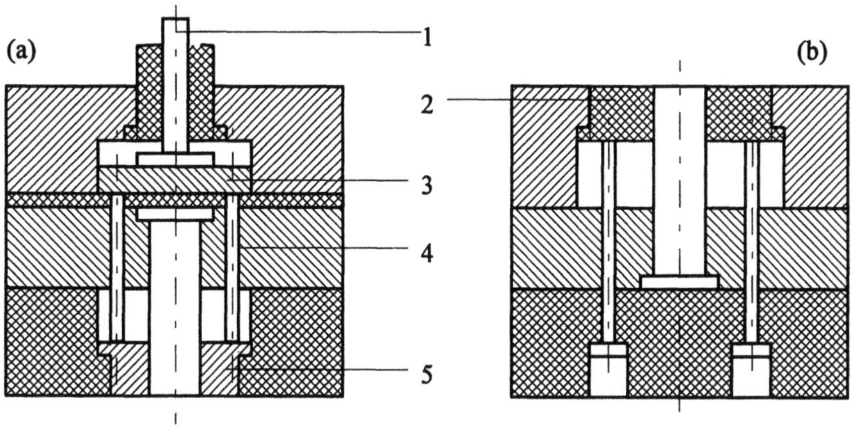

Figure 2.66: Ejecting devices: (a) upper ejecting device, (b) lower ejecting device, (1) knockout pin, (2) ejecting plate, (3) knockout plate, (4) ejecting pin, (5) ejecting plate.

The diameter and number of ejectors required are affected by the dimensions of the part, ejecting force, material property, etc., according to the strength or instability conditions. A user can interactively decide the diameter and number of ejectors. Alternatively, the user can select the diameter of the ejectors, and the number of ejectors (n) is then determined by the program based on the following formulae.

$$n = \begin{cases} \dfrac{64K_sP_eH^2}{\pi^2ED^4} & \lambda \geq 100 \\ \dfrac{4K_sP_e}{\pi D^2(A - B\lambda)} & 60 \leq \lambda \langle 100 \\ \dfrac{4P_e}{\sigma_b \pi D^2} & \lambda \langle 60 \end{cases} \tag{2.34}$$

D and H are the diameter and length of the ejectors respectively. K_s is a safety coefficient. P_e is the ejecting force, and A and B are the coefficients related to the material property. E is the Young's modulus. When $\sigma_b > 480$ N/mm^2, A = 46.9 N/mm^2 and B = 26 N/mm^2. In the formulae, $\lambda = H/(J/F)^{1/2}$, where J and F are the minimum moment of inertia and cross-sectional area of the ejectors respectively.

In the following section, the determination of the optimal positions of the ejectors, which is a key issue in the design of the ejecting device, will be discussed. The layout of ejectors is a multi-objective optimisation problem subject to constraints. Two objective functions have been established based on the requirements of ejectors layout [20].

A. Distance between the resultant force of ejectors and the pressure centre of the part

$$f_1(x, y) = \left[\left[\frac{\sum_{i=1}^{n} x_i}{n} - x_0\right]^2 + \left[\frac{\sum_{i=1}^{n} y_i}{n} - y_0\right]^2\right]^{1/2} \tag{2.35}$$

B. Reciprocal of the perimeter of the polygon enclosed by the ejectors

$$f_2(x, y) = \frac{1}{\sum_{i=1}^{n-1}\left[\left((x_i - x_{i+1})^2 + (y_i - y_{i+1})^2\right)^{1/2} + \left((x_1 - x_n)^2 + (y_1 - y_n)^2\right)^{1/2}\right]} \tag{2.36}$$

x_0 and y_0 are the co-ordinates of the pressure centre of the part. x_i and y_i are the co-ordinates of the centre point of ejector i, and n is the number of ejectors.

An evaluation function is defined to convert the multi-objective optimisation problem into a single objective problem.

$$U(X)=\sum_{j=1}^{2}\lambda_j f_j(X) \tag{2.37}$$

U(X) is the evaluation function. $f_j(X)$ is a single objective function. λ_j is the weight of the objective function. The evaluation function reflects the deviation of the objective functions from their optimal values.

$$\lambda_j=\left(\min_{X\in D} f_j(X)\right)^{-1} \tag{2.38}$$

The layout of the ejectors is subject to constraints that cannot be represented analytically due to the shape complexity of the required parts. A search method is adopted so that the constrained optimisation problem is converted into a problem without constraints to cope with this difficulty.

The procedures to remove the constraints are as follows:

1. The outer and inner contours of a part are discretised, i.e., the contours are approximated by a series of equi-distant points (Figures 2.67a and 2.67b).
2. The enveloping rectangle of the part is meshed, and the length of the mesh side is equal to the diameter of the ejectors (Figure 2.67c).
3. The meshes are grouped into boundary, inner and outer meshes based on their positions relative to the contour of the part. The inner meshes, which are called feasible meshes, are the possible positions of the ejectors.

The key to the removal of constraints is the determination of the feasible meshes (Figure 2.67d). In the case of a step ejecting plate, the contour of the ejecting plate is taken as a reference for the classification of the meshes.

Different layouts of the ejectors correspond to different combinations of the feasible meshes. Hence, the recognition of feasible meshes can speed up the arrangement of the ejectors. A three dimensional array WG(n,m,3) is used to record the attributes of the meshes. The first and second dimensions of the array record the row and column numbers of the meshes respectively. The third dimension is used to label the meshes, i.e., to record the signs of the inner, outer and boundary meshes. For boundary meshes, their contour numbers are recorded. 1 represents outer contour, and 2, 3, etc., represent inner contours.

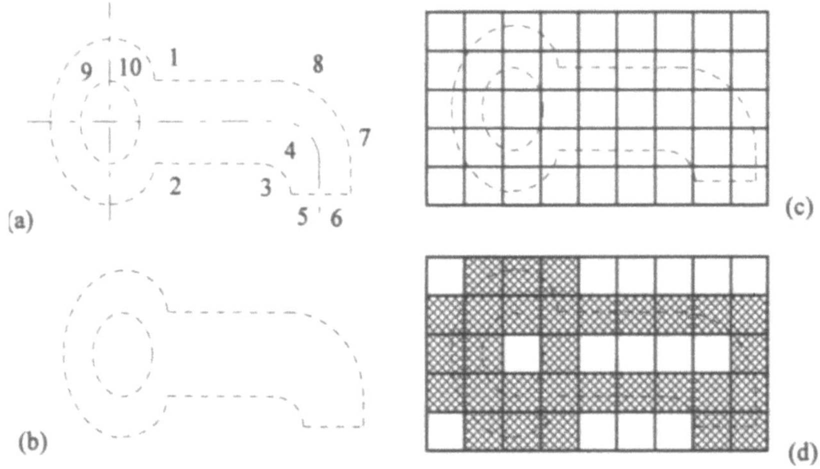

Figure 2.67: Meshing of the enveloping rectangle.

The feasible meshes are the domain for the optimisation of the ejector layout. This feasible domain is represented by the occupancy state of the meshes in order to avoid the difficulty of representing the domain analytically. After the determination of the feasible meshes, the program searches for an optimal layout of the ejectors.

The following method is adopted in the program to arrange the ejectors uniformly.

A. The vertical and horizontal lines that pass through the centre point of the part dividing the feasible meshes into four regions. The numbers of meshes in the four regions are assumed to be k_1, k_2, k_3 and k_4, and the numbers of ejectors arranged in the regions are n_1, n_2, n_3 and n_4. In the arrangement of the ejectors, the following condition should be satisfied.

$$\frac{n_1}{k_1} = \frac{n_2}{k_2} = \frac{n_3}{k_3} = \frac{n_4}{k_4}$$

B. If there are more feasible meshes and an ejector has already been allocated to a feasible mesh, other ejectors will not be assigned to the adjacent meshes any more.

C. Feasible meshes that are obviously not suitable for the layout of ejectors are discarded before optimisation to ensure the rationality of the ejector, and increase the design efficiency.

Figure 2.68 is the flowchart of the program [20]. As shown in Figure 2.68, after the input of the shape information of a part and other parameters such as the ejecting force, the program determines the diameter and number of the ejectors required. If the user does not agree with the results, he can input his own values. The program can design a step ejecting plate according to the choice of the user. Before the automatic optimisation of the ejector layout, the user is allowed to place ejectors at certain places. The optimisation process consists of three phases, *viz.*, finding the optimal values for each objective function, establishing the evaluation function and final optimisation. The ejectors layout that can be obtained by the program can be modified if the user is not satisfied with it.

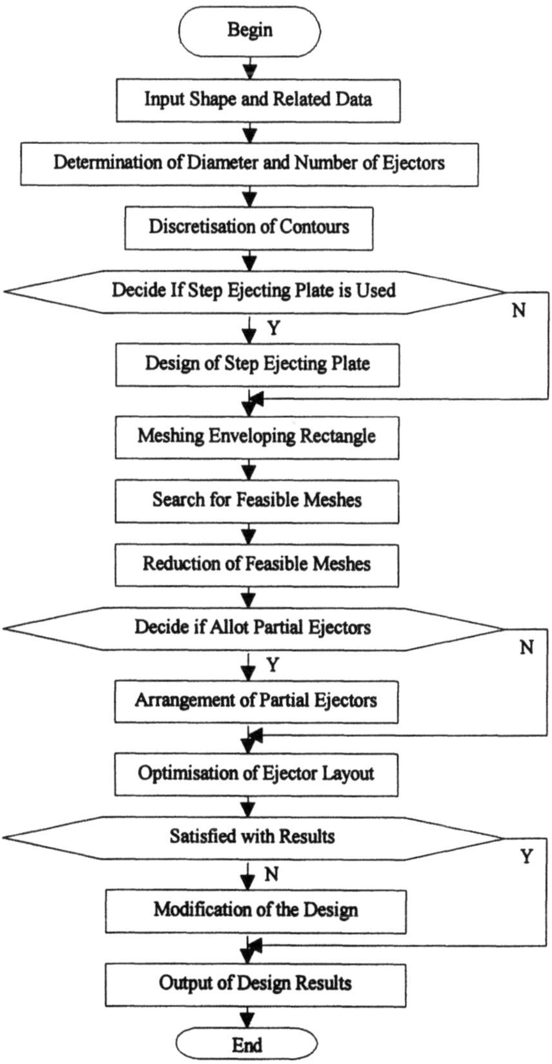

Figure 2.68: Flowchart for ejector layout.

2.8 NC programming of wire EDM

2.8.1 Process consideration for NC programming of wire EDM

Wire EDM is widely used to manufacture stamping dies. Conventional manual NC programming for wire EDM is tedious and error prone. In the CAD/CAM system for stamping dies developed by Yu et al [21], die construction design and NC programming are integrated. The design data can be used in NC programming. Geometric computation and NC programming can be automatically carried out in the system such that the productivity and quality of die manufacture are greatly improved.

In NC programming for wire EDM, the wire offset is equal to the wire radius plus the electric discharge gap, which is influenced by many factors, such as material, cutting speed, tension condition of the wire, coolant and power. The determination of the value of the electric discharge gap depends on the experience of the programmer. For precision workpieces, a simple shape is often cut beforehand to determine the exact gap needed. If there are successive operations after the wire EDM, a certain allowance should be added to the offset, i.e.,

$$f = {d_w}/{2} + z + \Delta \qquad (2.39)$$

f, d_w, z and Δ represent the wire offset, wire diameter, electric discharge gap and allowance for successive operations respectively.

The accuracy of the required part is influenced by its material structure and internal stresses. Therefore, these factors should be considered when determining the wire path.

For example, the part shown in Figure 2.69 can be cut off from the plate through the following two paths, namely, (a) A-a-b-c-d-e-f-a-A, and (b) A-a-f-e-d-c-b-a-A.

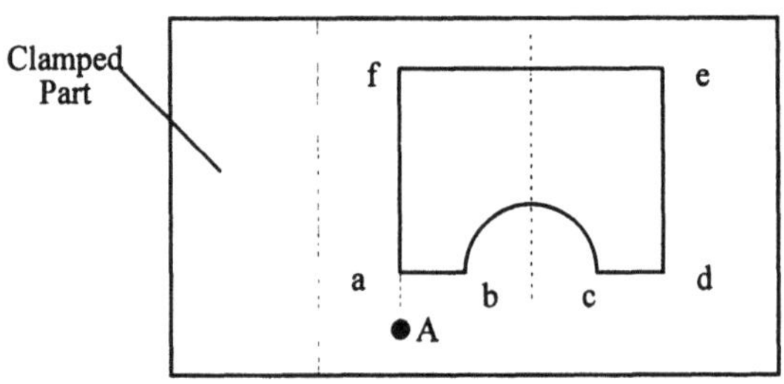

Figure 2.69: Influence of the cutting path on the accuracy of workpiece.

The accuracy of the workpiece produced using path (a) is better than using path (b) as the side area of the plate deforms easily.

In NC programming for wire EDM, the cut-in-point should be properly selected to avoid any scar on the surface. In general, the cut-in-point is selected from corner points on the periphery of the part, e.g., point a in Figure 2.69.

Usually, the starting point (e.g., point A in Figure 2.69) and the cut-in-point are not the same point. The starting point is selected outside the periphery of a part for punches, or inside for dies. A wire-hole is made at the starting point for the wire to pass through before cutting. An NC program is needed to move the wire to the cut-in-point at the start of cutting, and return it to the starting point after cutting.

2.8.2 Geometric computation

Extensive geometric computation is needed to prepare data for NC programming, such as finding intersection points, co-ordinate transformation and enlargement or reduction of the contours. For part drawings that are composed of arcs and lines, the following methods are used to simplify the computation and improve programming efficiency [21].

A. Orientation of the contour elements

To simplify the geometric computation, geometric elements of the contours are oriented such that the outer contour is counter clockwise, and the inner contours are clockwise, as shown in Figure 2.70. Thus, the enlargement and reduction of the contours are converted to the right and left translations of their geometric elements.

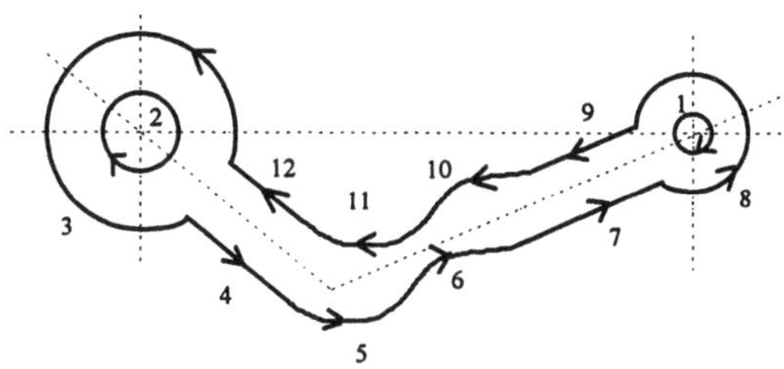

Figure 2.70: Orientation of the contour elements.

B. Parameter formalisation of geometric elements

Four parameters, *viz.*, A, B, C and T are used to represent the position, direction and type of geometric elements. For lines, their equations usually take the following form:

$$Ax + By + C = 0 \tag{2.40}$$

The normal equation is adopted to facilitate the movement of the lines.

$$x\cos\alpha + y\sin\alpha + p = 0 \tag{2.41}$$

α is the included angle of the normal of a line with the x-axis (Figure 2.71), and p is the distance from the origin to the line. $\cos\alpha$, $\sin\alpha$ and p in the equation describe the position and direction of the line. For line 12 shown in Figure 2.71, its direction is from point I to point II. The parameters used to represent a line can be determined as follows:

$$A = \cos\alpha = -\sin\theta = \frac{(y_1 - y_2)}{\left[(x_2 - x_1)^2 + (y_1 - y_2)^2\right]^{1/2}} \tag{2.42}$$

$$B = \sin\alpha = \cos\theta = \frac{(x_2 - x_1)}{\left[(x_2 - x_1)^2 + (y_1 - y_2)^2\right]^{1/2}} \tag{2.43}$$

$$C = -p = -Ax_1 - By_1 \tag{2.44}$$

$$T = 1$$

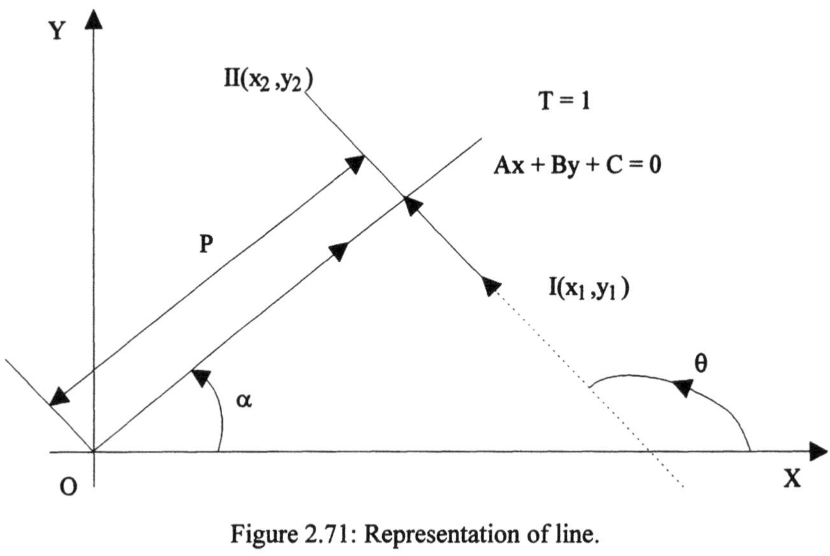

Figure 2.71: Representation of line.

Parameters for arcs are specified as the co-ordinates of the centre (A, B), radius C and type T (Figure 2.72). T = 2 for counter clockwise arcs, and T = 3 for clockwise arcs.

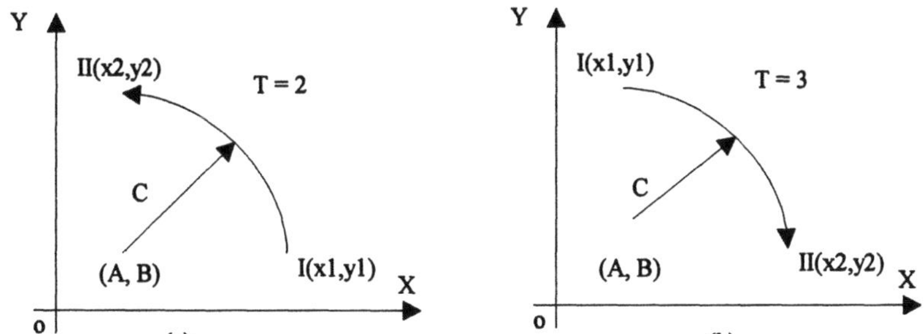

Figure 2.72: Representation of arcs, (a) counter clockwise arc, (b) clockwise arc.

After the parameter formalisation of the geometric elements, the movement of the elements can be simplified using the following calculation.

$$T' = T \tag{2.45}$$

$$A' = A \tag{2.46}$$

$$B' = B \tag{2.47}$$

For movement to right, i.e., enlargement,

$$C' = \begin{cases} C + W & (T = 1, 2) \\ C - W & (T = 3) \end{cases} \tag{2.48}$$

For movement to left, i.e., reduction,

$$C' = \begin{cases} C - W & (T = 1, 2) \\ C + W & (T = 3) \end{cases} \tag{2.49}$$

A', B', C' and T' are the parameters of the elements after the movement. W is the movement distance, i.e., the offset in the calculation of the wire path.

After the input part drawing has been processed, a list of geometric elements is obtained, as shown in Table 2.2. Each record in the list contains the information of a geometric element, which includes the co-ordinates of the nodes and the element parameters.

Table 2.2: Geometric element list.

node co-ordinates		element parameters			
X_1	Y_1	A_1	B_1	C_1	T_1
X_2	Y_2	A_2	B_2	C_2	T_2
X_3	Y_3	A_3	B_3	C_3	T_3
.	.	.	.	.	.
.	.	.	.	.	.
.	.	.	.	.	.

2.8.3 NC programming procedure for wire EDM

NC programming procedure for wire EDM consists of the following main steps:

1. Processing of graphic information of the required part,
2. Selection of wire diameter,
3. Determination of starting point and cut-in-point,
4. Calculation of cutting path,
5. Programming according to the assigned format, and
6. Output of the NC program.

The graphic information of the required part comes from the results of the die construction design. After this information has been input, the geometric elements of the part contour are oriented, and their parameters formalised. A list of elements as shown in Table 2.2 is prepared for NC programming.

The wire diameter, electric discharge gap, starting point and cut-in-point are interactively selected. Next, the offset needed for calculating the wire cutting path can be determined. With the formalisation of the geometric elements, calculations for the equi-distant enlargement or reduction of the part contours for determining the cutting path are simplified by moving the elements to the right or left respectively.

NC programming of wire EDM should be carried out according to an assigned format, which is dependent on the EDM machine to be used.

Table 2.3 shows a format used for NC programming of wire EDM. In this format, B is the discriminator, and X and Y are the co-ordinates. J and G are the counted length and counting direction respectively. Z is the instruction type. X, Y and J have micrometres as their unit.

Table 2.3: 3B programming format.

B	X	B	Y	B	J	G	Z

When an arc is cut, its centre is taken as the origin, and X and Y are the co-ordinates of its starting point. For a line, the starting point is taken as the origin, and X and Y are the co-ordinates of its end-point.

When the cutting length is counted in the X-direction, the cutting direction is represented by GX. When the cutting length is counted in the Y-direction, the counting direction is denoted by GY. The counted length is the total movement of the wire, which is equal to the sum of the projections of the arc (or line). For example, if the arc shown in Figure 2.73 is to be cut, and the counting direction is GY, the counted length J is equal to $J_{y1} + J_{y2} + J_{y3}$.

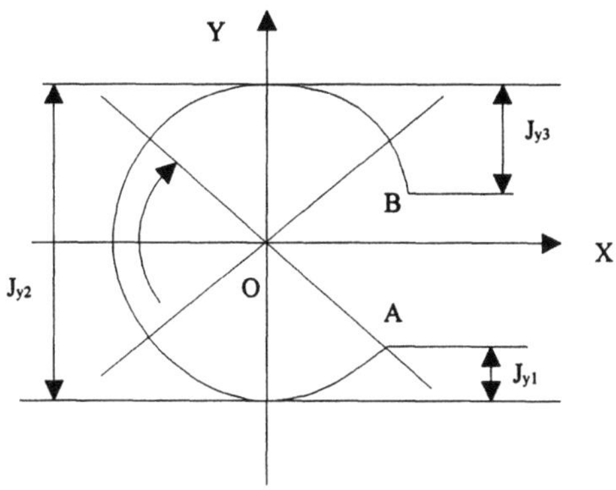

Figure 2.73: Counted length of an arc.

There are 12 instructions in the programming format. Instructions L1, L2, L3 and L4 (Figure 2.74a) are used to cut lines in the first, second, third and fourth quadrant respectively. Instructions SR1, SR2, SR3 and SR4 (Figure 2.74b) are assigned to cut clockwise arcs, while instructions NR1, NR2, NR3 and NR4 (Figure 2.74c) are specified for counter clockwise arcs.

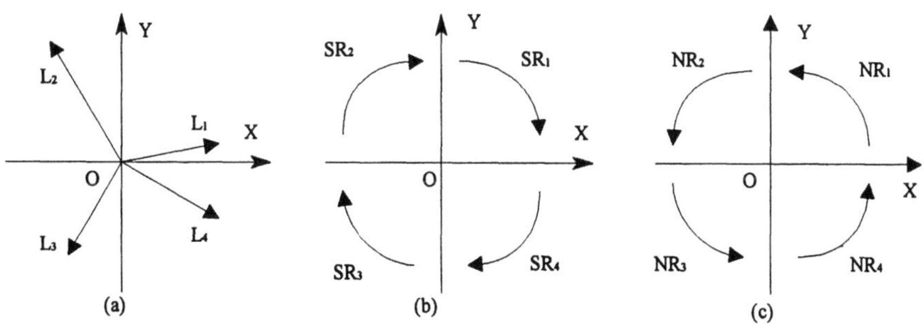

Figure 2.74: Instructions used to cut lines and arcs.

To cut the part shown in Figure 2.75, a wire of diameter 0.12 mm is used. The electric discharge gap is selected as 0.01 mm. Thus, the offset of the wire is found to be 0.07 mm. The cutting path is determined to be A-a-b-c-d-e-f-g-h-i-a-A. The wire EDM machine used has an offset compensation function. The NC program obtained for the part is shown in Table 2.4. The letter D in the last line of the program is a stop code to stop the cutting process.

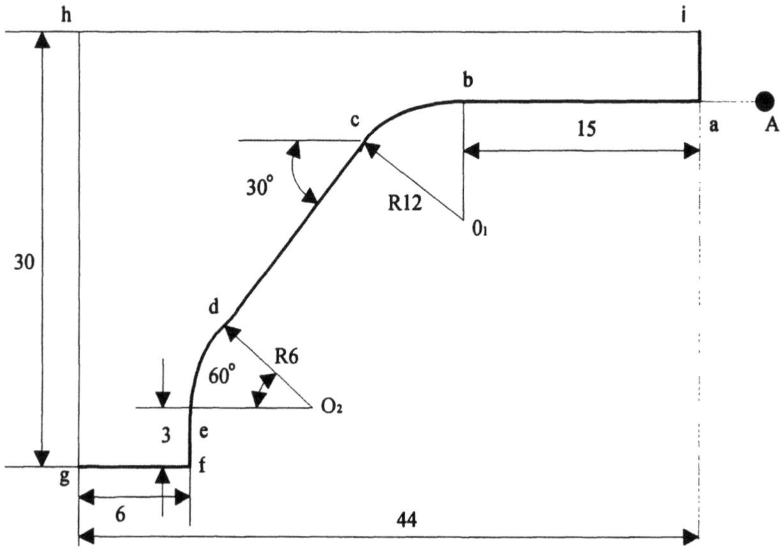

Figure 2.75: An example part.

Table 2.4: A NC program list.

B	0	B	0	B	10000	GX	L3
B	0	B	0	B	15000	GX	L3
B	0	B	12000	B	6000	GX	NR2
B	14000	B	8080	B	14000	GX	L3
B	3000	B	5190	B	5190	GY	NR2
B	0	B	0	B	3000	GY	L4
B	0	B	0	B	6000	GX	L3
B	0	B	0	B	30000	GY	L2
B	0	B	0	B	44000	GX	L1
B	0	B	0	B	10000	GY	L4
B	0	B	0	B	10000	GX	L1
D							

References

1. Shifman O F 1973 The PDDC System: A CAD-CAM Application to the Design and Fabrication of Press Tools. *SME TE73-107*, 1-15
2. Shirai K, Murakami H 1989 A Compact and Practical CAD/CAM System for Progressive Dies. *Bull Japan Soc. of Prec. Engg.* 23(1):25-30
3. Li Z G et al 1991 Geometric Feature Modelling of Stampings and Its Application. In: *Proceedings of the 11th ICPR*, 1251-1254
4. Eshel G et al 1986 Rule Based Modelling for Planning Axisymmetrical Deep-Drawing. *Journal of Mechanical Working Technology* 14:1-115
5. Reha U et al 1991 An Experimental Expert System for Process Planning of Sheet-Metal Parts. *Computer Industrial Engineering* 20(1):59-69
6. Xiao X Z et al 1993 An Intelligent Approach to the Strip Layout of Progressive Die Design. In: *Advanced Technology of Plasticity*. International Academic Publishers, China, 1817-1821
7. Li Z G et al 1984 Optimisation of Blank Layout. *Journal of Huazhong University of Science and Technology* 12(6):21-24 (in Chinese)
8. Yu H G et al 1987 Optimal Design of Blank Layout. *Journal of Huazhong University of Science and Technology*, 15(1):5-10 (in Chinese)
9. Xiao X Z et al 1984 On Fine Blanking Technique. *Journal of Huazhong Institute of Technology*, 12(6):7-12 (in Chinese)
10. Xiao J R et al 1984 Computer Aided Design and Manufacture of Fine Blanking Dies. *Journal of Huazhong Institute of Technology* 12(6):1-6 (in Chinese)
11. Li D Q et al 1984 Numerical Model and Program for the Shape and Dimension of Blanked Parts. *Journal of Huazhong Institute of Technology* 12(6):13-16 (in Chinese)
12. Li Z G et al 1992 CAD/CAM of Progressive Dies for Household Electronic Appliances. In: *Proceedings of International Conference on Advanced Technology and Machinery in Metal Forming*, Huazhong University of Science and Technology Press, Wuhan, 71-77
13. Xiao X Z et al 1990 Computer Aided Design of Progressive Blanking Dies. Advanced Technology of Plasticity. In: *Proceedings of the Third International Conference on Technology of Plasticity*, Kyoto, 1423-1427
14. Li Z G et al 1990 Computer Aided Design of Progressive Die Construction. In: *Proceedings of International Conference on Die and Mould Technology*, Shanghai, 356-363
15. Operating Manual for Pega 1986. AMADA
16. AMADA Sheet Metal Working Research Association 1987. *The NCT Handbook*. Machinist Publishing Co Ltd
17. TRUMPF Programming Manual for TC240 CC200, TC240 Laser CC200 and TC240 Rotation CC300, 1990, TRUMPF, Chapter 2, 9-10
18. Choong N F, Nee A Y C and Loh H T 1993 The Implementation of an automatic tool selection system for CNC nibbling. *Computers in Industry*, 2:205-222

19. Li Z G, Li D Q and Xiao J R 1990 *Computer Aided Design of Dies and Moulds*. Huazhong University of Science and Technology Press, Wuhan, 114-116 (in Chinese)
20. Li D Q et al 1986 Automatic Design of Ejecting Device of Compound Dies. *Forging and Stamping Technology* 11(5):48-52 (in Chinese)
21. Yu H G et al 1984 CAM of Fine Blanking Dies and Automatic Programming of NC Wire EDM. *Journal of Huazhong Institute of Technology*, 12(6):51-56 (in Chinese)

Bibliography

Anon 1987 CAD/CAM systems for sheet metal design and manufacture. *Sheet Metal Industries* 598

Daniels H R 1988 *Press Brake and Shear Handbook*. Cahners Books, Boston, Massachusetts

de Vin L J, de Vries J, Streppel A H, Kals H J J 1992 PART-S, a CAPP system for small batch manufacturing of sheet metal components. *Proc 24th CIRP Int Seminar on Manufacturing Systems*, Copenhagen, 171-182

de Vin L J, de Vries J, Streppel A H, Klaassen E J W, Kals H J J 1994 The generation of bending sequences in a CAPP system for sheet-metal components. *Journal of Material Processing Technology* 41:331-339

de Vin L J, Streppel A H, Kals H J J 1994 Tolerancing and sheet bending in small batch part manufacturing. *Annals of the CIRP* 43(1):421-424

Dobbins D B 1990 Putting control into your bending operation. *Tooling and Production* 55 (11):59-62, 65-66

Downie D 1991 Open systems CAE for sheet metal. *Sheet Metal Industries* 169-170

Ehrismann R, Reissner J 1988 Intelligent manufacture of laser cutting, punching and bending parts. *Robotics & Computer-Integrated Manufacturing* 4 (3/4): 511-515

Hoffmann M, Geißler U, Geiger M 1992 Computer-aided generation of bending sequences for die-bending machines. *Journal of Material Processing Technology* 30:1-12

Inui M, Kinosada A, Suzuki H, Kimura F, Sata S 1987 Automatic process planning for sheet metal parts with bending simulation. *Intelligent & Integrated Manufacturing Analysis and Synthesis*, PED 25:245-258

Janjua M S, Rathmill K, Allen D M 1983 A technical and manual appraisal of NC laser machine in a profile cutting application. In: *Proceeding of the First International Conference on Lasers in Manufacturing*, 41-52, Brighton, UK

Ling Z K, Chase T R 1987 Simulation of pressbrake bending using constructive solid geometry. In: *Proceeding of the International Computers in Engineering Conference & Exhibition* 2:321-326

Loh H T, Nee A Y C, Choong N F 1990 A PC-based computer software package for automated punch selection and tool path generation for a CNC nibbling machine. *Journal of Material Processing Technology*, 23:6(2),107-119

Loh H T, Nee A Y C, Choong N F 1990 Automated punch selecion and tool path generation for a CNC nibbling machine. In: *Asia-Pacific Industrial Automation '90, Conf Proc* 360-371

Merry Mechanization Inc 1988 Affordable graphics for punch press programming. *Fabricators & Manufacturers Association's Journal of the Fabricator*

Naessens D 1981 Numerical control of press-brake operations. *Sheet Metal Industries* 188:175-182

Nnaji B O, Kang T S, Yeh S, Chen J P 1991 Feature reasoning for sheet metal components. *Int J Prod Res* 29: (9) 1867-1896

Noaker M P 1991 Software that Packs a Punch. *Manufacturing Engineering* 108:(4) 93-96

Origami Technology Corp 1990 Expert system aids sheet-metal fabricating. *Industrial Laser Review* 4 (10): 4-9

Owen J V 1992 Give yourself a brake. *Manufacturing Engineering* 109 (3):37-44

Pourboghrat F, Stelson K A 1988 Bend allowance calculation in a flexibly automated sheet metal fabrication system. *Proc of the USA-Japan Symposium on Flexible Automation*, NY, USA, 999-1007

Prasad Y K D V 1992 Some studies on problems associated with automated design of cutting dies for sheet metal. PhD thesis, Indian Institute of Technology, Bombay

Raggenbass A, Reissner J 1989 Stamping-laser combination in sheet processing. *Annals of the CIRP*, 38(1):291-294

Raggenbass A, Reissner J 1991 An expert system as a link between computer-aided design and combined stamping-laser manufacture. *Proceedings of the Institution of Mechanical Engineers,* 205:25-34

Raggenbass A, Reissner J 1991 Automatic generation of NC production plans in stamping and laser cutting. *Annals of the CIRP*, 40(1):247-250

Sachs G 1976 *Principles and Methods of Sheet Metal Fabrication.* Reinhold Publishers, New York

See Toh K H, Loh H T, Nee A Y C , Lee K S 1995 A feature-based flat pattern development system for sheet metal parts. *Journal of Materials Processing Technology* 48:89-95

Shimozono M, Okada M, Kobayashi H, Tanabe K 1978 Computer aided manufacturing system for sheet metal parts. *NEC Research and Development* 49:76-85

Shpitalni M 1993 A new concept for design of sheet metal products. *Annals of the CIRP,* 42(1):123-126

Shpitalni M, Saddan D 1994 Automatic determination of bending sequence in sheet metal products. *Annals of the CIRP,* 43(1):23-26

Stamp R J, Earl C F 1992 Production of sheet metal components by an automatically planned robot assisted press brake. *Proc. 24th CIRP Int. Seminar on Manufacturing Systems*, Copenhagen, 211-220

Streppel A H, de Vin L J, Brinkman J, Kals H J J 1993 Suitability of sheet bending modelling techniques in CAPP applications. *J of Material Processing Technology* 36:339-356

Uzsoy R, Ramcharan D J, Martin-Vega L A 1991 An experimental expert system for process planning of sheet-metal parts. *Computers Ind Engng* 20 (1):59-69

Wilson F W, Harvey P D, Gump C B 1965 *Die Design Handbook*, Second Edition, McGraw Hill

Yamada K, Makino S, Jinnai N 1986 A flexible automated system in sheet metal production by CAD/CAM and shop floor control. *Japan-U.S.A. Symposium on Flexible Automation*, 639-642

Yares E 1990 Autofold, spindle or mutilate - from formed to flat in moments. *CADalyst*, 59-60

Zopf R 1998 Computer aided design for sheet metal parts. *Sheet Metal Industries* 584

Chapter 3

CAD/CAM for massive (bulk) metal forming

J.C. Xia

3.1 Introduction

Research in the CAD and CAM of forging dies started in early 70's. Battelle-Columbus laboratory has done much work in this field. Many CAD/CAM systems have been developed for blade-shaped forming dies, aluminium forging dies, precision forming dies and pre-forming dies, and CAE systems for simulating metal forming processes. The system for blade-shaped forging dies consists of modules for geometric description, finish die design, pre-form die design and automatic NC program generation. This system is capable of calculating forging stresses, loads and energy, optimising the die construction and generating NC program. The system for aluminium forging dies includes two modules: ALEXTR [1] and EXTCAM. The ALEXTR module is used for die design. It is capable of determining the optimal parameters of die cavities, the arrangement of the cavities in the die, stresses in the die and the compensation for bending during forging. The EXTCAM module is used for NC machining. It is capable of generating NC machining tapes, and machining the die cavity.

When the system for precision forging dies is used for designing bevel gear dies, process parameters such as the geometric shape of the gear, dimensions of the insert, pressure rings, forging temperature, die temperature, coefficient of friction, material properties and forging velocity, are input to calculate the elastic bending caused by thermal inflation and the forming stress to modify the dies. Based on the modified dimensions, the graphite electrodes and die impressions of the gears are machined. The CAE system for simulating forging processes, which uses an established ALPID [2] finite-element analysis program, is capable of predicting geometric shapes of forging parts, local strain, strain rate and stresses. This 2-D simulation system is capable of displaying the cross-sections of parts, automatic selection and generation of the original mesh, mesh rezoning based on deformation, and overall organisation of all the simulated sections.

In the mid 70's, Birmingham University in UK was successful in the research and development of a CAD system for axi-symmetric forging dies. This system, which uses the geometry of half of the axial section of a part as input, is capable of designing finish and pre-form dies. However, it is incapable of generating NC programs automatically. This system cannot be used for mechanical parts with holes. Recently, the new system CTRCON, reported by the Birmingham University has the following features:

1. The new system is interactive and has greater flexibility.
2. It is capable of providing NC machining tapes that are used to machine electrodes and die impressions.
3. The system can be used with both solid and hollow parts.

These improvements have made CTRCON a more practical system. The processing of CAD programs for forging dies, and the automatic generation of NC programs are based on the geometric information of a part and other non-geometric information. Therefore, a 3-D geometric configuration is crucial in the CAD/CAM for forging dies. In 1978, Birmingham University set up a geometric configuration system, MODCON, for the design and manufacture of forging dies. This system uses common geometric shapes in forging parts as volume elements. The system is capable of describing 75% of common forging parts. Birmingham University also developed an axial forging die CAD/CAM system.

Russian researchers have also done much research work in the CAD of forging dies. For example, Teterin [3] proposed advanced methods for the calculation of the geometric complexity of forging parts and die design. These calculation methods have been proven in practice, and adopted in many CAD/CAM systems for forging dies.

Some companies in Japan and Germany have widely used NC and CNC machine tools for machining forging dies to improve the precision of the dies and enhance productivity. In recent years, some companies have begun to use CAD/CAM to improve die design and manufacture. In 1981, a steel-iron plant in Japan developed a CAD/CAM system for hot forging dies. This system can be used for upsetting and press die-forging, and for designing blockers, pre-form and finish cavities, trimming die and precision shaping die. It has a 3-D NC machining software that can be used to machine cavities.

Research on the CAD for forging die has been mainly confined to geometric design. Benefits from the application of CAD for forging die stem mainly from computer-aided draughting and computerisation of design calculations. However, a significant advantage of CAD for forging die can be derived from the computer simulation of metal flow in forging. This is because forging process tryouts can be simulated on computers, reducing and/or removing the time of real tryouts.

Research in the simulation of the forging process is limited by the complexity of plastic deformation of the metal during hot forging. At present, simulation programs for the forging process are only applicable to two simple states of deformation, namely plane strain and axi-symmetric deformation. For complex forging parts, metal flow can only be simulated in a few selected regions (plane strain or axi-symmetric deformation). Research in the simulation of 3-D flow is still at the development stage.

Research on CAD/CAM for forging in PR China started in the late 70's, when many universities, institutes and factories started research work in this field. Some theoretical research, methods and system development have been reported. It is necessary to develop the CAD/CAM technology for forging dies in two main directions:

1. Geometric modelling of complex forging parts, e.g., 3-D description, automatic draughting and sectioning, and NC machining.
2. Design analysis methods, e.g., calculation of forging stresses and stresses in the dies, prediction of elastic deflections in the dies, metal flow analysis, blockers and pre-form cavities design.

At present, CAD/CAM applications in forging are focused on draughting and NC machining of forging dies. The application of CAD/CAM in forging is increasing rapidly with the modernisation of the forging industry, and the decrease of the price of draughting and machining software.

3.2 Cold upsetting

3.2.1 Determination of operations and sequences

Cold upsetting is used mainly for producing standard parts, such as bolts and nuts, where the quantity is large, the shapes are variations of each other and the manufacturing processes are fundamentally the same. The main operations for bolts are bar cutting, pre-form upsetting, finished upsetting and hexagon shearing. For nuts, the main operations are bar cutting, upsetting bar into the bowl, hexagon pressing and hole punching. Upsetting bar into the bowl is a typical upsetting operation. Pressing hexagon is a complex deformation operation of upsetting and backward extrusion, in which upsetting is the main deforming operation. As can be readily seen, bar cutting, hexagon shearing and punching are fixed operations. The material is not deformed in these operations. The main deformation operations are pre-form upsetting and finished upsetting. The number of pre-form upsetting operations needed, and the shapes and sizes of the pre-formed workpieces and finished parts are different for different types of bolts and nuts. In the cold upsetting operation, the main objectives are the determination of the number of operations and their sequence, and the distribution of the deformation to each operation and sequence. The number of upsetting operations required is mainly dependent on the upsetting rule. The amount of deformation for each operation depends on the allowable deformation degree.

3.2.1.1 Expression of upsetting rule and deformation degree

Figure 3.1 shows the upsetting operation in a cone die. Based on the constant volume principle, the volume of the cone is equal to the volume of the deformed part in the bar, as illustrated in equation (3.1).

$$\frac{\pi}{12}\left(D_G^2 + D_G D_S + D_S^2\right)L = \frac{\pi}{4} D_O^2 L_O \tag{3.1}$$

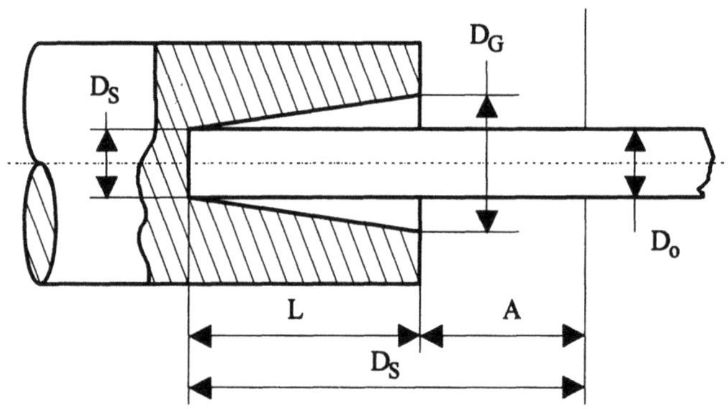

Figure. 3.1: Upsetting in a cone die.

Assuming $D_G/D_O = \varepsilon$, $D_S/D_O = \eta$, $L_O/D_O = \varphi$, $L/D_O = \lambda$ and $A/D_O = \beta$, equation (3.1) becomes:

$$\varepsilon = \sqrt{\frac{3\varphi}{\varphi - \beta} - \frac{3}{4}\eta^2} - \frac{\eta}{2} \tag{3.2}$$

Based on experience, ε is 1 ~ 1.05. However, it is necessary to establish the relation between ε and β in order to solve equation (3.2). The design nomogram of the storing dimensions in the cone cavity as shown in Figure 3.2 can be obtained experimentally. If the critical point, at which there is loss of steady state or when a crack is produced, lies in the curve, the relation for the common materials for producing bolts and nuts can be determined as follows.

$$\varepsilon = f(\beta) \tag{3.3}$$

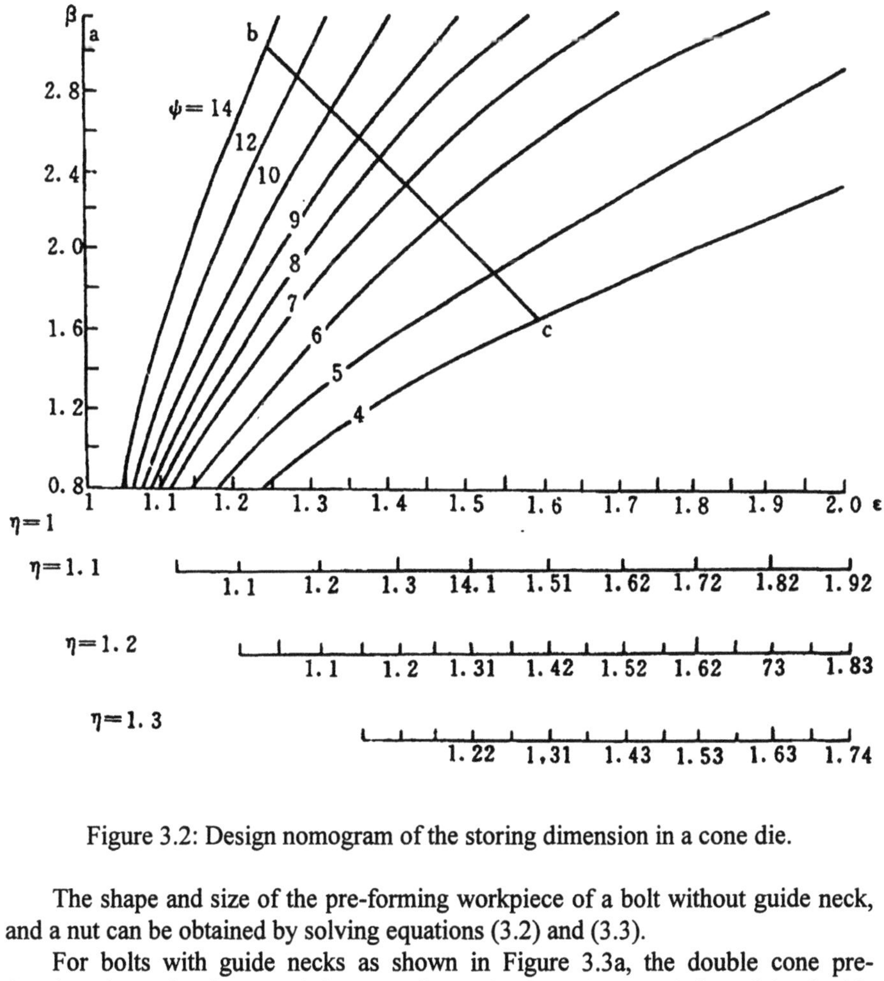

Figure 3.2: Design nomogram of the storing dimension in a cone die.

The shape and size of the pre-forming workpiece of a bolt without guide neck, and a nut can be obtained by solving equations (3.2) and (3.3).

For bolts with guide necks as shown in Figure 3.3a, the double cone pre-forming shown in Figure 3.3b is generally used. The volume and size of the double cone pre-forming workpiece are obtained from the constant volume principle.

From Figure 3.3b, the volume of the double cone pre-form is given by equation (3.4).

$$V_D = \frac{\pi}{12}\left(D_S^2 + D_S D_M + D_M^2\right)L_C + \frac{\pi}{12}\left(D_N^2 + D_N D_m + D_m^2\right)h + \frac{\pi}{4}\left(\frac{D_M + D_m}{2}\right)^2 m \tag{3.4}$$

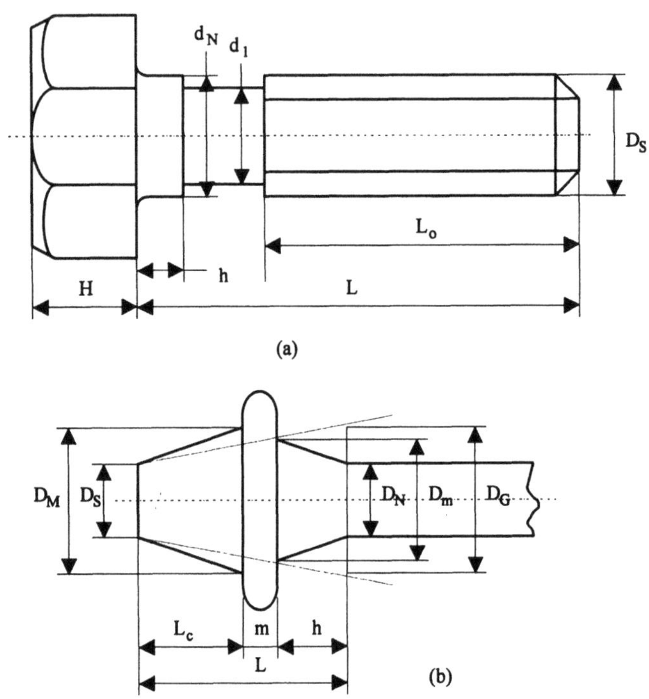

Figure 3.3: (a) Bolt with guide neck, and (b) double cone pre-forming.

For a single cone pre-form corresponding with the double cone pre-form, which is denoted in broken line in Figure 3.3b, its volume is given by equation (3.5).

$$V_S = \frac{\pi}{12}\left(D_S^2 + D_S D_G + D_G^2\right)L \tag{3.5}$$

The cone with the rod shown in Figure 3.3b is eventually formed into the guide neck of the bolt shown in Figure 3.3a. Therefore,

$$\frac{\pi}{12}\left(D_N^2 + D_N D_m + D_m^2\right)h = \frac{\pi}{12} d_N^2 h \tag{3.6}$$

d_N represents the diameter of the bolt. D_N can be found using equation (3.6). The pre-forming shape and size of the double cone can be determined from the condition that equation (3.4) is equal to equation (3.5), i.e., $V_D = V_S$.

The upsetting deformation degree, i.e., the upsetting ratio δ is denoted by equation (3.7).

$$\delta = \frac{H_0 - H_1}{H_0} \times 100\% \tag{3.7}$$

H_0 is the original height of the upset part in the bar and H_1 is the upsetting height in the workpiece.

3.2.1.2 Technical constraint and determination of number of steps

Each step is constrained by a technical constraint, which is the allowable deformation degree that leads to no defect. For example, the maximal length must be determined from its length to diameter ratio for a single upsetting. If $\varphi \le \varphi_o \le 2 \sim 2.5$, there is no loss of steady state. If $\varphi \ge \varphi_o = 2 \sim 2.5$, inter-upsetting steps, i.e., pre-forming steps are necessary. The pre-forming workpiece corresponding to the pre-forming step can be made into single cone or double cone.

The number of pre-forming upset operations needed is determined by the following method. Let the length of the deformed part in the bar be L_0 and its diameter D_0. If $\varphi(L_0/D_0) \le \varphi_0$, the original stock can be made into any shape in one upsetting operation. If $\varphi \rangle \varphi_0$, pre-forming operations are needed, and the shape and size of the pre-forming workpiece can be determined from the upsetting rule. For the second condition, i.e., if the cone head-part is as shown in Figure 3.1, its average diameter is:

$$D_A = \sqrt{\frac{1}{3}\left(D_S^2 + D_S D_G + D_G^2\right)}$$

Subsequently, the length to diameter ratio can be found, i.e., $\varphi_1 = L/D_A$, where L is the deformation length of the bolt head-part in the first upsetting operation.

If $\varphi_1 \le \varphi_0$, the second pre-forming upset operation is not needed. If $\varphi_1 \rangle \varphi_0$, further performing upsets are required until the upsetting rule is satisfied. Using this method, the total numbers of the pre-forming upset steps can be obtained.

3.2.1.3 Application of equal deformation principle

The equal deformation principle states that the amount of deformation in each step is equal or approximately equal in the cold upsetting processes of bolts or nuts. When this principle is used, the deformation amount is distributed uniformly such that the stress concentration in the deformed body and the die bearing pressure of each step are reduced. Therefore, die life and production quality are improved.

The deformation processes from the original stock to the last pre-formed workpiece are shown in Figure 3.4. Considering the effect of work hardening in cold upsetting deformation, a hardening coefficient M is introduced. M is determined experimentally. For the ease of computation, let the hardening coefficient of each step be equal. Then, the formula for this deformation principle can be obtained as equation (3.8).

$$\frac{L_{(1)}}{L_{(0)}} = M\frac{L_{(2)}}{L_{(1)}} = \ldots = M^{l-1}\frac{L_{(l)}}{L_{(l-1)}} = M^{n}\frac{L_{(H)}}{L_{(n)}} \quad (0 \leq M \leq 1) \qquad (3.8)$$

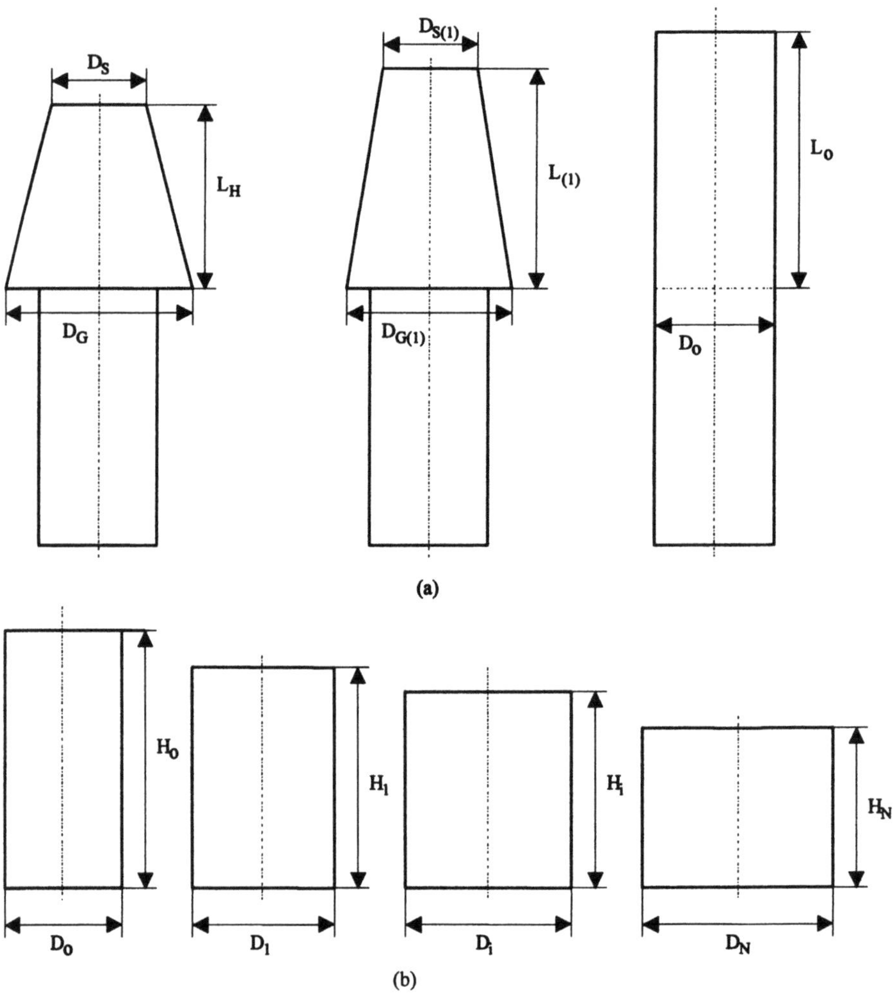

Figure 3.4: Principle of equal deformation amount for (a) bolt and (b) nut.

3.2.2 Calculation of process parameters

3.2.2.1 Calculation of process parameters for bolts

A. Determination of dimensions

Based on the finished product, the above rules and production experience, the size of the pre-forming workpiece for each step can be obtained.

The diameter d_l of a bolt is calculated using equation (3.9).

$$\left.\begin{aligned} d_{1max} &= d_{2max} - 0.03S^{0.25} \\ d_{1min} &= d_{2min} + 0.03S^{0.25} \end{aligned}\right\} \tag{3.9}$$

d_{2max} and d_{2min} represent the maximum and minimum mid-diameters of the screw thread respectively, and S represents the screw distance.

The diameter d_0 of a bar is obtained from equation (3.10).

$$\left.\begin{aligned} d_{0max} &= d_{2min} - 0.03 \\ d_{0min} &= d_{2max} - 0.06 \end{aligned}\right\} \tag{3.10}$$

The length L_H required for the upset forming of the head-part can be determined using equation (3.11).

$$L_H = 1.01\frac{D^2H}{d_{0min}^2} \tag{3.11}$$

D and H denote the diameter of the inside tangent circle and the height of the hexagon respectively.

The length L_h required for the upset forming of the guide neck is calculated using equation (3.12), where D_N is the diameter of the guide neck.

$$L_h = \frac{D_N^2 h}{d_{0min}^2} \tag{3.12}$$

The total length of the bar can be determined using equation (3.13).

$$L_t = 1.01\frac{DH}{d_{0min}^2} + \frac{D_N^2 h}{d_{0min}^2} + L - h \tag{3.13}$$

The dimensions of the pre-forming upset workpiece shown in Figure 3.3b are as follows.

$$\left.\begin{array}{ll} d_s = d_{0max} - 0.1, & D_M = 1.35 d_{0max} \\ L_c + m = 0.747 \dfrac{D^2 H}{d_{0max}}, & \alpha = 2\tan^{-1} \dfrac{D_M - d_s}{2(L_c + m)} \\ m = \dfrac{1}{3} H, & d = d_{0max} + 0.04 \\ d_m = 1.10\,(d_{0min} + 0.04) & \end{array}\right\} \quad (3.14)$$

The shape and dimensions of the finished upset workpiece in Figure 3.5 are as follows.

$$\left.\begin{array}{ll} d = d_{0min} \sim d_{0max}, & d_c = 0.85d \\ D_0 = 0.92S, & D_1 = D + 0.08 \\ C = \dfrac{1}{3} H, & m = \dfrac{1}{3} H \end{array}\right\} \quad (3.15)$$

D, H, D_N, h and L are illustrated in Figure 3.3.

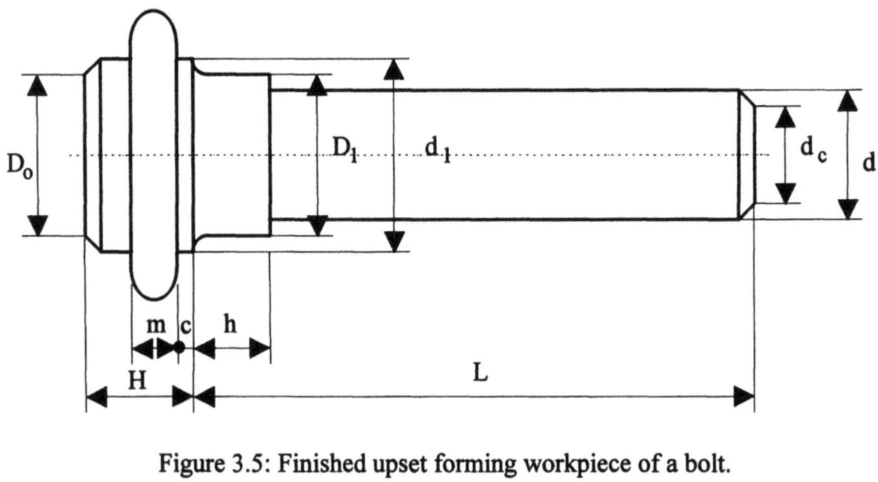

Figure 3.5: Finished upset forming workpiece of a bolt.

B. Calculation of cold upset forming force

Both pre-formation and final deformation of bolts are considered as half-closed upsetting as shown in Figure 3.6. At the point when the upsetting process is completed, the plastic region is considered as a circle plate which height h_0 is equal to 2h. h is the unclosed distance between the upper and lower dies. Since the die cavity surface is smooth and the deformation is in cold state, friction between the deformed metal and the die is in accordance with the Coulomb friction conditions. Since both the pre-formation and final deformation belong to axi-symmetric deformation, the slab method can be used to calculate the pre-formation and final deformation forces. The formula can be expressed as equation (3.16).

$$\left.\begin{aligned} P &= p_f F_f + p_b F_b \\ p_f &= \frac{2\pi}{F_b\left(\frac{2\mu}{h}\right)^2}\sigma_s\left[\left(\frac{\mu D}{h}+1\right)e^{\frac{2\mu b}{h}} - \frac{2\mu\left(\frac{D}{2}+b\right)}{h} - 1\right] \\ p_b &= \sigma_s\left[e^{\frac{2\mu b}{h}} + \left(\frac{1}{4h}-\frac{1}{b}\right)D\right] \end{aligned}\right\} \qquad (3.16)$$

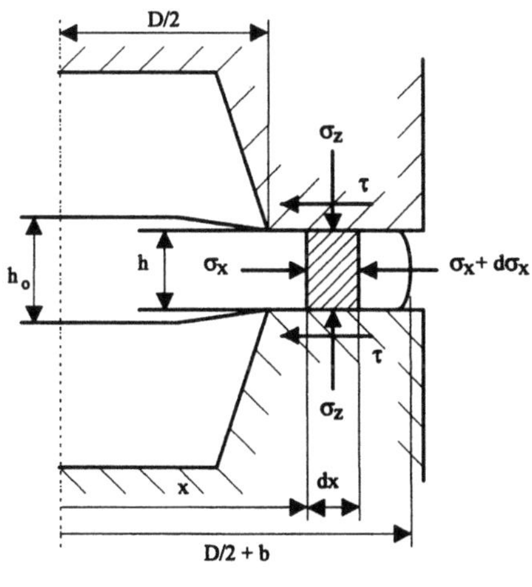

Figure 3.6: The stress state during upsetting.

p_f and p_b are the pressures applied to the flash and the body respectively. F_f and F_b are the projection areas of the flash and body respectively. e = 2.7183, and μ is the friction coefficient, and σ_s is the flow stress of the deformed material.

3.2.2.2 Calculation of process parameters for nuts

A. Determination of dimensions

Figure 3.7 shows the shapes and dimensions of workpieces corresponding to steps in the upsetting process of a nut.

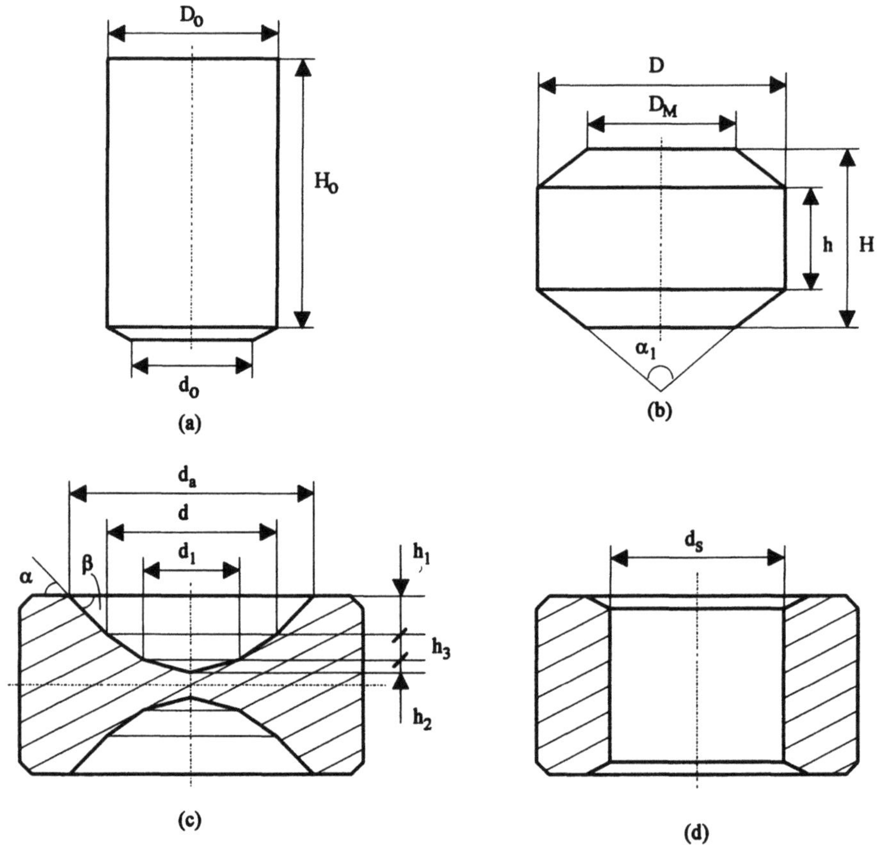

Figure 3.7: The working process of a nut.

The hole diameter of a nut blank (shown in Figure 3.7d) is given by equation (3.17). d_s denotes the diameter before threading.

$$d_s = (d_{1min} + 0.9TD_1) \tag{3.17}$$

d_{1min} and TD_1 represent the minimum diameter and minimum tolerance of the screw thread respectively.

$$d = d_{smax} + C_1 \tag{3.18}$$

d_{smax} represents the maximum inside diameter of the nut. C_1 is the corrected coefficient and C_1 is $0 \sim 0.03$.

$$d_a = C_2 M \tag{3.19}$$

M denotes the nominal inside diameter of the nut. C_2 is the corrected coefficient and C_2 is $1.05 \sim 1.10$.

$$\alpha = 55^\circ \sim 60^\circ, \qquad \beta = 5^\circ \sim 10^\circ \tag{3.20}$$

$$h_1 = \frac{d_a - d}{2}\tan\alpha, \qquad h_3 = \frac{d - d_1}{2}\tan\beta \tag{3.21}$$

The volume of the nut can be determined from equation (3.22).

$$V = V_1 - 2V_a - 2V_d \tag{3.22}$$

V_1 is the volume of the nut workpiece (entire cylindrical volume) before punching (Figure 3.7a). V is the volume of the nut calculated from S and H. V_a is the volume of the hollow. V_d is the volume of the outside chamfer of the nut.

The size of the upsetting bowl (Figure 3.7b) is calculated from equation (3.23).

$$D = S - \Delta \tag{3.23}$$

S denotes the diameter of the inside tangent circle. Δ is the corrected coefficient and Δ is $0.2 \sim 0.6$.

$$D_M = C_3 D \tag{3.24}$$

C_3 is the corrected coefficient and C_3 is $0.50 \sim 0.65$.

The size of the pre-form (Figure 3.7d) is determined by equation (3.25).

$$\left.\begin{aligned} d &= d_w + \Delta \\ \frac{d}{D} &= C_4 \end{aligned}\right\} \qquad (3.25)$$

d_w denotes the diameter of the bar according to the upsetting rule, and $0.70M \le d_w \le 0.90M$. C_4 is a coefficient and $C_4 = 0.85$.

B. Calculation of cold upset forming force

Figure 3.8 shows the deformation and stress conditions of a blank in an upsetting step. The deformed body can be divided into two deformation regions. The first region is a cylinder and the second is a ring. The cylinder is upset under the side pressure caused by the ring, and the ring is upset between two taper plates parallel to each other.

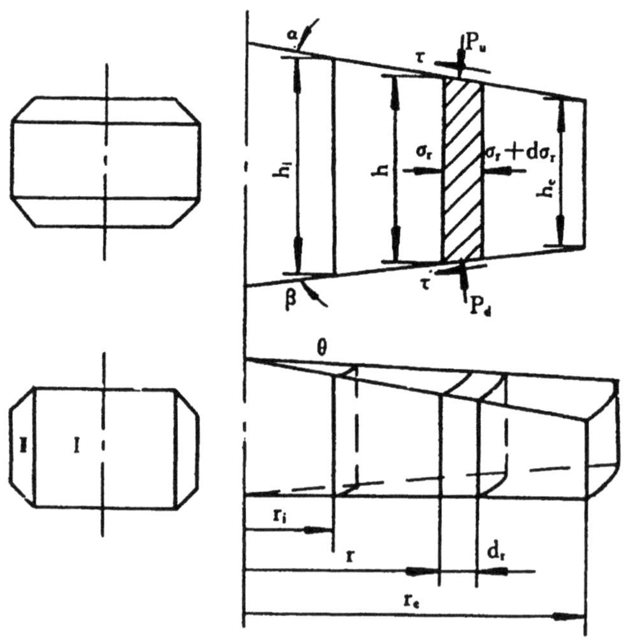

Figure 3.8: The deformation and stress state in the workpiece during the upsetting bowl operation.

The required forces P_1 and P_2 for upsetting the cylinder and ring are derived respectively using the slab method. The total upsetting forming force P can be obtained by adding P_1 and P_2, according to equation (3.26), where p_1 and p_2, and F_1 and F_2 are the stresses and areas of the cylinder and the ring respectively.

$$P = P_1 + P_2 = p_1 F_1 + p_2 F_2 \tag{3.26}$$

where $P_1 = p_1 F_1 = F_1\left(\frac{2}{3}\frac{\tau r_1}{h_1} + \sigma_{ze}\right)$,

$$P_2 = p_2 F_2 = F_2\left(\sigma_s + \frac{K_2}{K_1}\ln\frac{K_3 - K_1 r}{h_e}\right),$$

$$K_1 = \tan\alpha + \tan\beta,$$

$$K_2 = K_1\sigma_s + \left(2 + \tan^2\alpha + \tan^2\beta\right)\tau, \text{ and}$$

$$K_3 = h_i - K_1 r_i .$$

$\sigma_z = \sigma_{ze}$ at $r = r_e$. σ_s is the flow stress of the material.

The force of the final upsetting step should be calculated using the general evaluation formula for closed upset forming of any forging.

3.2.3 Design of dies cavities, general parts and combined dies

3.2.3.1 Design of cavities

Design of cavities consists of designing the pre-forming and final forming cavities. In general, the design of pre-forming cavities is more important than the design of the final forming cavity. This is because the shape and size of the pre-forming cavity affect the metal flow. In addition, the shape and sizes of the pre-formed workpiece directly affect the metal flow during the final formation, the forming quality and the life-span of the die. Die cavity design is based on the technique of analysis and calculation, i.e., optimum modelling from a combination of the upsetting rule and the equal deformation principle, and from experience. For bolts, the design tasks are the determination of the large and small diameters, and the depth of the taper cavity.

For nuts, the dimensions of the reshaping and upsetting bowl and the final upsetting dies need to be determined. The design of the cavity is explained in detail using the cold upsetting die for a nut as an example.

A. Design of reshaping die

A reshaping die is shown in Figure 3.9. The diameter d and length L_1 of the cavity are given by equation (3.27).

$$\left.\begin{aligned} d &= d_w + \Delta d \\ L_1 &= l_o + \Delta l \end{aligned}\right\} \tag{3.27}$$

d_w and l_0 represent the diameter and length of the bar respectively. Δd and Δl represent the corrected coefficients of the diameter and length of the cavity respectively. Δd is 0.10 ~ 0.15 and $\Delta l \geq 0.15$.

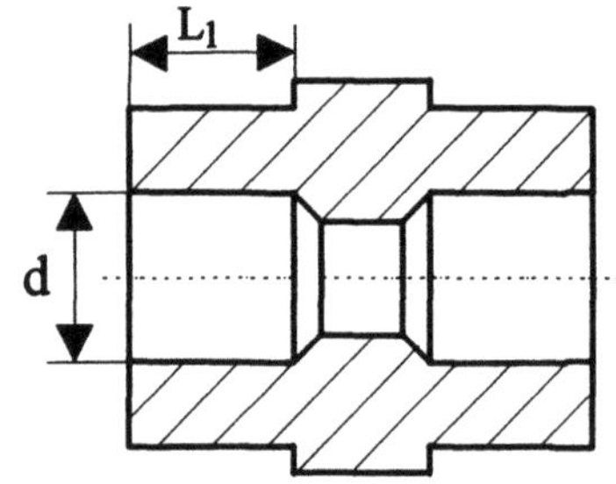

Figure 3.9: A reshaping die.

B. Design of upsetting bowl die

An upsetting bowl upper die is shown in Figure 3.10a. It is designed according to the workpiece shown in Figure 3.6b.

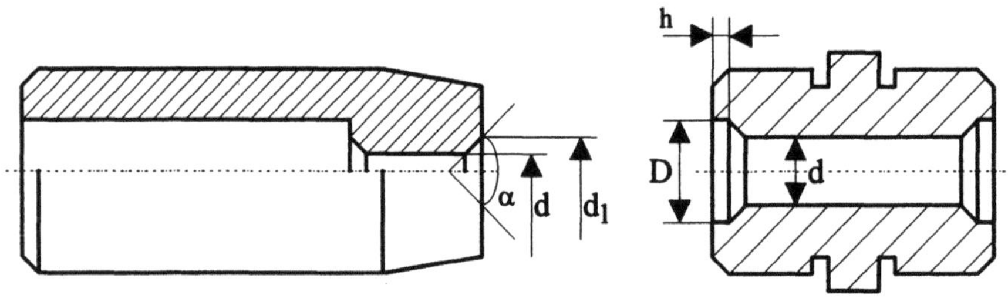

Figure 3.10: The upper and lower dies for upset bowl.

$$\left.\begin{aligned} d &= d_M \\ d_1 &= D \\ \alpha &= 100^\circ \end{aligned}\right\} \tag{3.28}$$

Dimensions of the upsetting bowl lower die shown in Figure 3.10b are calculated using equation (3.29). S is shown in Figure 3.3a. h_c is a calculated dimension, and Δd and Δh are the corrected values.

$$\left.\begin{aligned} D &= S - \Delta d \\ h &= h_c - \Delta h \end{aligned}\right\} \tag{3.29}$$

C. Design of final upset die for forming the six-angle

A six-angle die is shown in Figure 3.11, and its dimensions are determined using equation (3.30). H_n is the normal height of a nut. S is the same as in equation (3.29) and C_1 is 0.97 ~ 0.99.

$$\left.\begin{aligned} H &= (2 \sim 2.5)H_n \\ S_1 &= C_1 S_{min} \\ S &= S_1 + 2H\tan\alpha \\ \alpha &= 20' \sim 1^\circ \end{aligned}\right\} \tag{3.30}$$

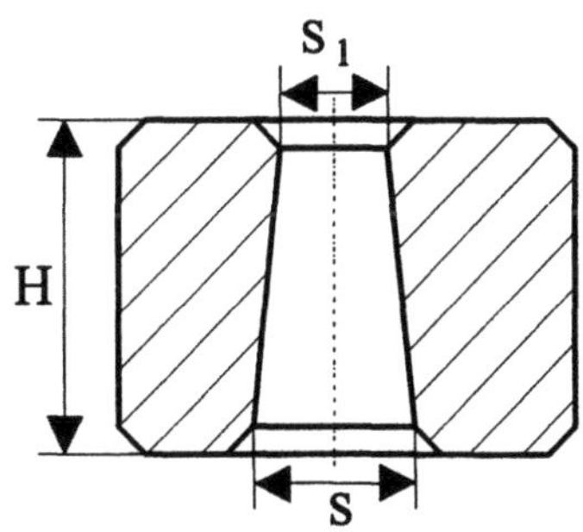

Figure 3.11: The six-angle die of a nut.

D. Design of punching hole dies

Figure 3.12 shows the punching hole die. The dies shown in Figure 3.12a and 3.12b are used for nuts that are smaller and larger than M10 respectively.

For nuts smaller than M10 (Figure 3.12a),

$$\left.\begin{aligned} d &= d_{K\max} + \Delta \\ D &= d + \Delta_1 \end{aligned}\right\} \tag{3.31}$$

For nuts larger than M10 (Figure 3.12b),

$$\left.\begin{aligned} d_1 &= d_{K\max} + \Delta \\ d &= d_1 + \Delta_1 \\ D &= d + \Delta_2 \end{aligned}\right\} \tag{3.32}$$

$d_{K\max}$ denotes the maximum inside diameter of the screw thread, and Δ, Δ_1 and Δ_2 are the corrected values.

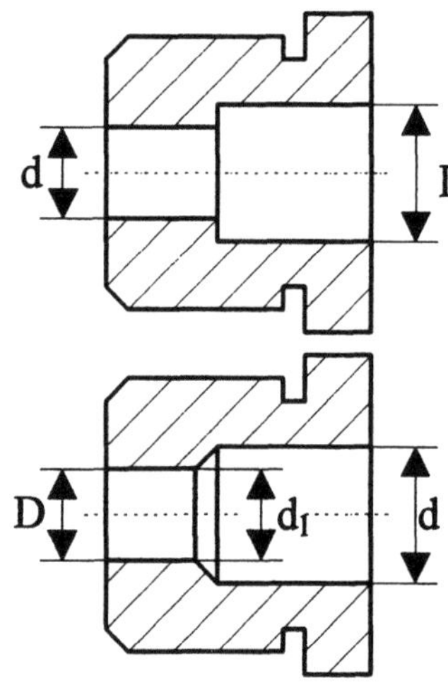

Figure 3.12: The punch hole lower die.

3.2.3.2 Design of general parts

General parts in dies such as the die holder, holder floor plate, punch, punch floor plate, centre-pike, centre-pike holder, etc., are designed based on experience and standardised.

3.2.3.3 Design of combined die

In general, cold upsetting dies are designed as combined dies with series connection shown in Figure 3.13. This construction is suitable for bolts of equal diameter and unequal lengths. There are relationships between the heights of the combined dies with the bolt productions, and the dimensions of the related equipment. The relationships are expressed as equation (3.33).

$$\left.\begin{aligned} H_0 + H_A + H_B + H_C + H_D = A \\ H_0 + h_0 + H_A + H_B + c + h = L \end{aligned}\right\} \tag{3.33}$$

A is determined by the cold upset equipment. L is the rod length of the bolt. H_0, h_0 and C are determined from production techniques.

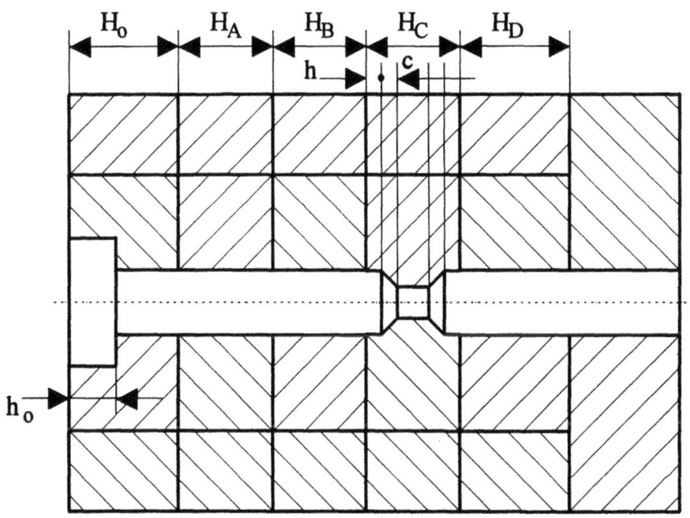

Figure 3.13: The cold upsetting die construction.

3.2.4 The BNC CAD system

The CAD system for bolts and nuts cold upsetting process and die design is named as BNC CAD. The program flowchart is shown in Figure 3.14. The system consists of four modules, *viz.*, process analysis and evaluation, die design, die strength calculation and drawing. The flowchart for process analysis and evaluation is shown in Figure 3.15. Figure 3.16 shows the flowchart for the calculation of the deformation forces and the determination of the equipment capacity.

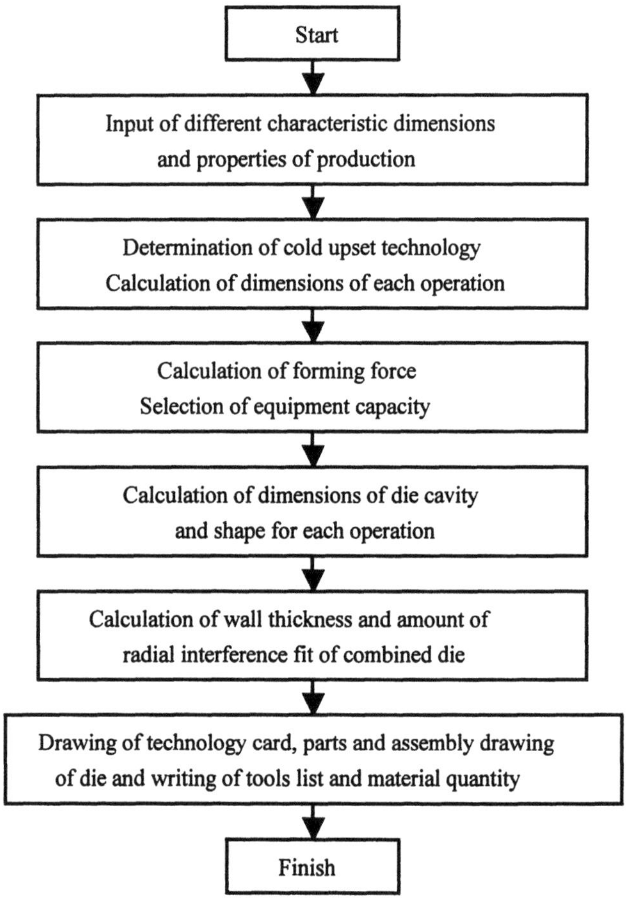

Figure 3.14: The program flowchart of the BNC CAD system.

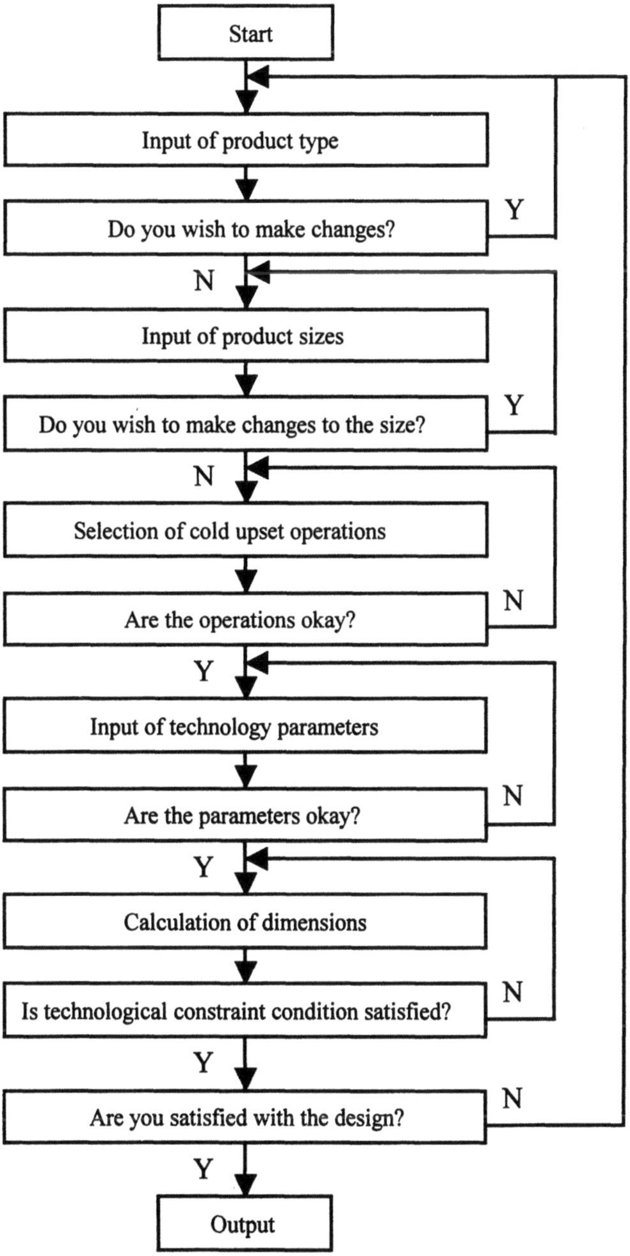

Figure 3.15: The module for process design.

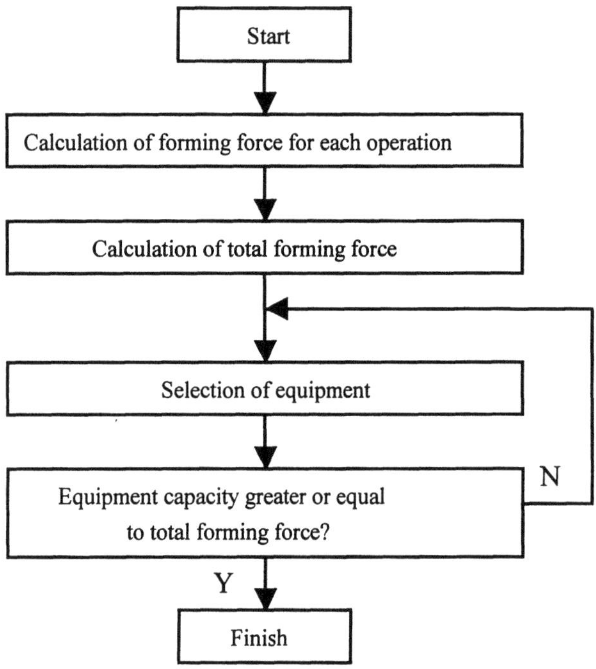

Figure 3.16: The module for calculation of deformation forces and determination of equipment capacity.

The BNC CAD system has the following functions: selection of optimum cold process parameters for bolts and nuts, determination of die dimensions, plotting of die constructions and parts drawings, and calculation and verification of the combined die strength. In addition, the system is suitable for both cold and hot upsetting processes and die design of other parts.

3.3 Closed-die forging

3.3.1 Selection and calculation of bars

Selection and calculation of bars are based on the geometry and dimensions of the forged part and the die forging process. The volume, cross-sectional dimensions and cutting length required are calculated. Generally, round steel bars are used as blanks.

3.3.1.1 Disc-shaped forgings

Disc-shaped forging parts usually require upsetting pre-forming operations. Therefore, the dimensions of the bar are calculated based on upsetting deformation.

Volume of blank:

$$V_b = (1+k)V_f \tag{3.34}$$

Diameter:

$$d_b = 1.13\sqrt[3]{\frac{(1+k)V_f}{m}} \tag{3.35}$$

k is an allowance factor, based on the influence of the complexity of the forging, the flash and fire waste. For round-shaped forging parts, k is 0.12 ~ 0.25. For non-round shaped forging parts, k is 0.2 ~ 0.35. V_f is the volume of the forged part excluding the flash. m is the height to diameter ratio of the bar. Generally, m is 1.8 ~ 2.2.

Cutting length:

$$L_b = \frac{V_b}{F_b} = 4\frac{V_b}{\pi d_b^2} \tag{3.36}$$

F_b is the cross-sectional area of the bar and d_b is diameter of the bar.

3.3.1.2 Long axis forgings

For a forging part with a long axis, the dimensions of the bar are determined from the average cross-sectional area on the mass distribution diagram of the bar, along the axis of the forging. Based on the requirements of different blockers and pre-forming operations, the cross-sectional areas of the required bar for all types of die forging processes can be calculated. Detailed methods are described as follows:

When no pre-forming operation is used,

$$F_b = (1.02 \sim 1.05)F_a \tag{3.37}$$

When pre-forming operations are used,

$$F_b = (1.05 \sim 1.3) F_a \tag{3.38}$$

F_b is the cross-sectional area of the bar, and F_a is the average cross-sectional area of the bar.

3.3.2 Determination of operations and sequences

All forging parts need a final forging operation. Therefore, the determination of the required operations and their sequence is mainly to assess the need for pre-forming and blocker operations.

3.3.2.1 Axi-symmetric forgings

For an axi-symmetric forging, the main objective is to determine the number of blocker operations required. For the design of die forging operations and sequences, the need for blocker operations depends on the following factors:

1. When the capacity of the forging equipment is insufficient to make the forged part directly out of the bar, blocker operations are needed.
2. Forging parts with complex geometry. Even when the capacity of the equipment is sufficient, a lack of blocker operations can also lead to inadequate filling of the corners and ribs of the cavities. In this case, a blocker operation is used to enhance the metal flow in the cavity during forging.
3. When there is more than one pre-forming operation, the economic benefits derived from the prolonged life span of the finished die and the reduction of the material used, is sufficient to cover the extra cost of die machining and forging parts translation. Therefore, the production and life span of the finished die should be considered simultaneously.

In the case of manual design, blocker die design is greatly dependent on the experience and skills of the designers. The designed dies are usually modified through tryouts under the production conditions. Empirical rules that are applied in blocker die design are summarised as follows:

1. The cross-sectional area of the blocker cavity should be equal to that of the finished cavity.
2. The horizontal size of the cross-section of the pre-formed cavity must be slightly smaller than that of the finished cavity so that the pre-formed forging fits into the finished cavity easily.

3. The pre-formed cavity should have larger fillets and corner radii to enhance metal flow.
4. It is useful to provide an opening taper from the centre of the webs towards the ribs to facilitate the flow of the metal towards the ribs.
5. For a forging with higher hulls or ribs, the corresponding part in the pre-formed cavity is often reduced in width and height, while the web thickness in the pre-formed cavity is expanded.
6. For steel forging, whenever possible, sections of ribs in the blocker should be narrower but slightly larger than those in the finished sections. This would ensure that the ribs are deformed by upsetting and the wear of the die is reduced.

Obviously, it is very difficult to embed the above empirical rules in CAD programs. Therefore, several CAD systems have adopted exponential curves to describe the pre-formed forging outline, and design the pre-formed cavity using the analysis method.

This design method is explained with reference to Figure 3.17. The fillet radius R_p between the pre-formed cavity and the flash gap is influenced by the depth of the cavity nearby. It is determined by the following formula, where C_l is selected according to Table 3.1.

$$R_p = R_f + C_l \tag{3.39}$$

Table 3.1: Relationship of C_l and impression depth.

Impression Depth (mm)	C_l(mm)
≤ 10	2
$>10 \sim 25$	3
$> 25 \sim 50$	4
> 50	5

For hubs or ribs, the outline of the pre-form is designed using the exponential curve. The curve AB in Figure 3.17 is:

$$y = h \exp(\frac{-4x}{md}) \tag{3.40}$$

m is a factor and its value should satisfy the following boundary conditions.

$$\begin{cases} \frac{d}{2} > 0 \\ r \geq C_2 \end{cases} \tag{3.41}$$

C_2 is a limit of the forgeable corner radius for the forging material used in the given forging conditions.

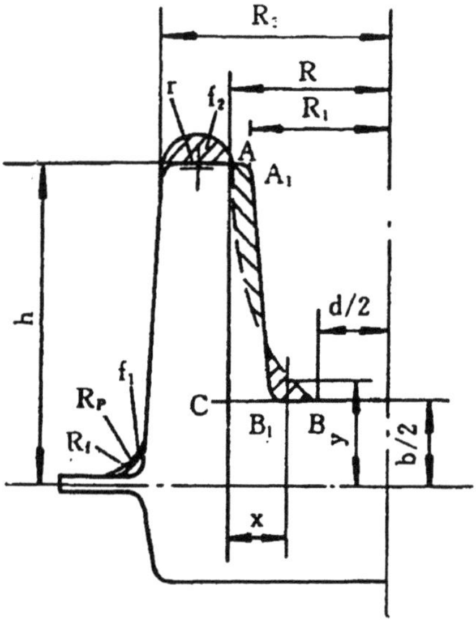

Figure 3.17: Using the exponential curve as the outline of pre-formed forging.

The intersection point B of the curve with the upper plane of the forging is calculated using the following formula.

$$d = \frac{2R}{1 + \left| \frac{m}{2} \ln \frac{b}{2x} \right|} \tag{3.42}$$

The value of R must satisfy the condition that the cross-sectional areas are equal, i.e.,

$$\text{Area}(f_1 + f_2) + \text{Area}(ABC) = \text{Area}(AA_1, B_1, C) \tag{3.43}$$

In the above equation,

$$\mathrm{Area}(ABC) = \frac{-mdh}{4}\left[\exp\left(-\frac{4\left(R-\frac{d}{2}\right)}{md}\right)-1\right]-\frac{b}{2}\left(r-\frac{d}{2}\right) \tag{3.44}$$

The value of r should satisfy the contact of the upper arc f_2 with the curve AB. The value of point A in the X-direction is as follows.

$$X_A = \frac{(R_2 - R)}{2} \tag{3.45}$$

In this procedure, W of the pre-forming forging for the cross-section shown in Figure 3.17 must be calculated before the shape of the cross-section of the pre-formed cavity is designed. The value of W is calculated using equation (3.46).

$$W = \frac{R_2 - R_1}{h - \frac{b}{2}} \tag{3.46}$$

If there is more than one pre-formed operation, the cavities of all the pre-forming operations are designed based on the principle of area equality, and the different factors W of the pre-forming operations. The required pre-forming operation time depends on the limiting value of the pre-forming operation factor. Forging is more difficult with increasing W as shown in Figure 3.18. Based on different values of W, the program designs several cross-sectional shapes of the pre-formings, and displays them on the computer screen. The designer can select a design scheme.

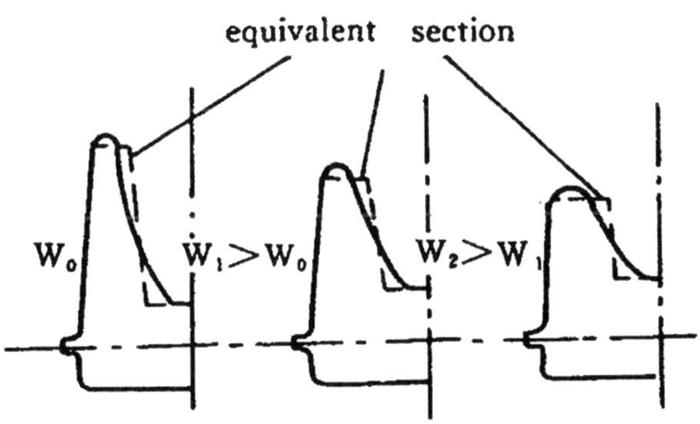

Figure 3.18: Degree of difficulty of forging increases with W.

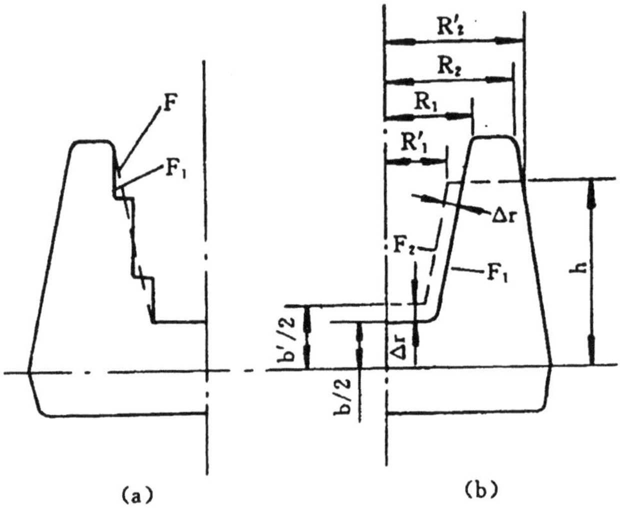

Figure 3.19: Simplification of the section shape.

For a given W, the cross-sectional shape is designed according to the following steps.

1. If the final cross-sectional shape is more complex, it is necessary to simplify the shape, as shown in Figure 3.19a. During simplification, the area equality of the cross-sections must be ensured, i.e., $F_1 = F_2$. The design of the pre-form shape of the cross-section is based on the simplified cross-section.
2. If W of the pre-formed forging factor exceeds the limiting value, it is modified to an acceptable range. As shown in Figure 3.19b, the modified shape satisfies the following relationship.

$$W = \frac{R_2' - R_1}{h - \frac{b'}{2}} \tag{3.47}$$

3. The value of m is determined based on boundary conditions.
4. The shapes of the pre-formed cavities are designed based on the above described formulae.

For a forging with complex cross-sectional shape, the cross-section is first divided into several parts similar to the sections shown in Figure 3.17. The shapes of all the parts that have been designed are then integrated to obtain the overall cross-sectional shape of the pre-forming impression.

3.3.2.2 Long axis forgings

For forged parts with long axis, the main task is to determine the need for blockers and pre-forming operations.

A. Determination of blocker operations and sequence

The determination of blockers and pre-forming operations for long-axis forgings is based on the change of the cross-sectional area along the axis. This will enhance the metal mass distribution along its axis according to the need of the forging. Based on the efficiency of the metal flow, the optimising order of blocker operations is as follows: drawing and close rolling, drawing out, closed rolling, and open rolling. Blocker operations are always determined using the curves in Figure 3.20. X in Figure 3.20 is the ratio of the forging length to the average edge width of the equivalent bar drawing. Y is square root of the ratio of the maximum cross-sectional area to average cross-sectional area in the mass distribution drawing.

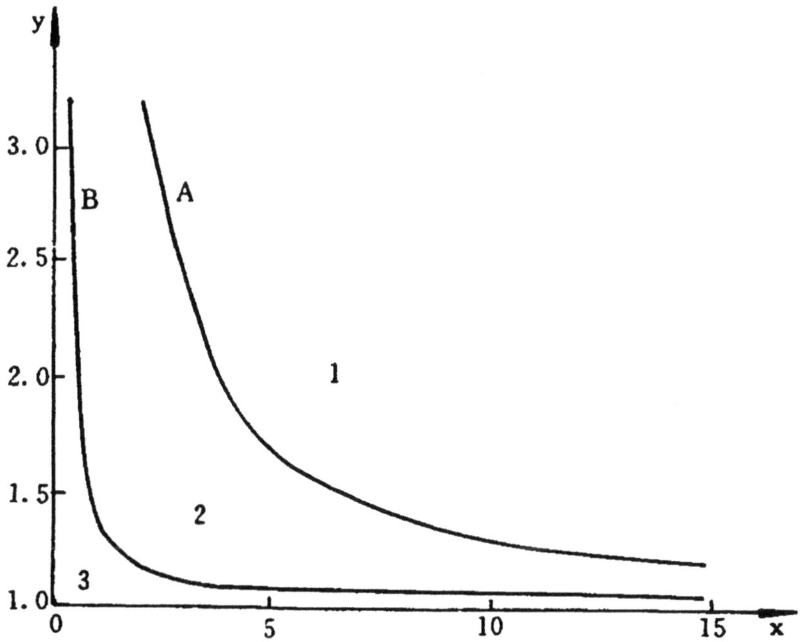

Figure 3.20: The curve for determining blocker operations.

Region 1 in Figure 3.20 indicates that drawing out and rolling are needed in pre-forming. Region 2 requires rolling and region 3 does not need any blocker operations. Here, the design rules are given by means of diagrams. Equations of A and B can be obtained using the regression analysis method. The curves in the diagram are discretized initially. The curves of A and B are then fitted using a fourth order polynomial. They are as follows:

$$A: Y = 0.677 \times 10^{-3} x^4 - 0.0243x^3 + 0.323x^2 - 1.948x + 5.907 \quad (3.48)$$

$$B: Y = 0.131x^4 - 1.965x^3 + 7.465x^2 - 9.570x + 5.056 \quad (3.49)$$

B. Basis for selection of pre-forming operation

Complex forging parts such as the connecting rod, shifting fork, blade, and forging with high ribs or thin webs, are difficult to deform. Pre-forming operation is used to avoid folding cracks or other defects. With mass production, pre-forming operations can reduce the load and wear of the finished cavity, and prolong the life-span of the die. For mechanical press die forging, pre-forming operations are mainly used to ensure that the shape of the blank approximates the shape of the forged part. This is to ensure the metal material fills the finished cavity better during finish forging, and defects such as folding can be avoided.

C. Rolling as the blocker operation

$$F_b = F_r = (1.05 \sim 1.2) F_a \quad (3.50)$$

F_b denotes the blank cross-sectional area and F_a is the average cross-sectional area in calculating the blank diagram. When the forging has only one head and one rod, the coefficient should be smaller. When the forging has two heads and one rod, a larger value is selected.

D. Drawing out as the blocker operation

$$F_b = F_d = \frac{V_h}{L_h} \quad (3.51)$$

V_h is the volume of the forging head including the fire waste, and L_h is the length of the forging head.

E. Drawing out together with rolling as the blocker operation

$$F_b = F_d - k(F_d - F_r) \quad (3.52)$$

It should be noted that for blocker operations, drawing out precedes rolling. The metal flows along the axis, resulting in an increased length during drawing out, while rolling causes the material to concentrate at the head of the forging. Therefore, the reduction of the cross-sectional area of the billet stock resulting from rolling must be considered before determining the cross-sectional area of the billet. Based on an analysis of the above equation, the reduced part is $k(F_d - F_r)$. F_b can be found after F_d and F_r have been calculated. k is the draft inclination ratio of the rod part in calculating the billet stock diagram. It should be noted that the value of the coefficient is 1.2 when calculating F_r.

Based on the above conditions, a standard bar that is close to the calculated bar is selected after the cross-sectional area of the billet stock has been calculated. The following equation is used to calculate the cutting length.

$$L_b = \frac{V_b}{F_b} + L_f \tag{3.53}$$

V_b represents the volume of the billet stock including the flash.

$$V_b = (V_p + V_f)(1 + \delta) \tag{3.54}$$

δ is the fire waste percentage, as shown in Table 3.2. F_b is the cross-sectional area of the standard bar and L_f is the length of the grip holder. V_f is the volumn of the flash and V_p is the volumn of the part.

Table 3.2: Fire waste percentage δ%.

Heating method	δ%
Room type furnace	3 ~ 2.5
Continuous type furnace	2.5 ~ 3.0
Room type gas furnace	2.5 ~ 2.0
Continuous type gas furnace	2.5 ~ 1.5
Electric resistance furnace	1.5 ~ 1.0
High frequency heated furnace	1.0 ~ 0.5
Contact electric heated furnace	1.0 ~ 0.5
Room type gas furnace	2.5 ~ 4.0

3.3.3 Calculation of forging load and stress

Forging load is the basis for forging equipment selection and forging die strength design. In computer-aided calculation of the forging load, a forged part is first divided into a number of sections with some definite deformation types. The forging load in every deforming section is calculated. Next, the total forging load is obtained by superposition as in the "slab" analysis method, based on the main stresses method.

Based on the characteristics of forging deformation, the following four types of deformation are defined: (1) Plane strain, lateral flow: upsetting, (2) Plane strain, longitudinal flow: extrusion, (3) Axi-symmetric, lateral flow: upsetting, and (4) Axi-symmetric, longitudinal flow: extrusion. All forged parts can be divided into deformation regions with these deformation types.

3.3.3.1 Axi-symmetric, lateral flow

Figure 3.21 shows the metal flow between the upper and lower draft planes. Stress and load applied on the fan-shaped deformation element with an angle θ (in radian) are given as follows, where $\bar{\sigma}$ is effective stress and τ is shearing stress [4].

$$\frac{d\sigma_z}{dr} = -\frac{1}{(k_3 + k_1 r)} k_2$$

$$\sigma_z = \frac{k_2}{k_1} \ln\left(\frac{h_e}{(k_3 + k_1 r)}\right) + \sigma_{ze} \tag{3.55}$$

$$P = \int_{r_b}^{r_e} \sigma_z \theta \, r dr \tag{3.56}$$

where $k_1 = \tan\alpha + \tan\beta$,

$k_2 = -\bar{\sigma} k_1 + \tau(2 + \tan^2\alpha + \tan^2\beta)$, and

$k_3 = h_b - r_b k_1$.

For parallel planes, it can be shown that

$$\sigma_z = \frac{2\tau}{h}(r_e - r) + \sigma_{ze} \tag{3.57}$$

For a circular element, the forging load P is given by

$$P = \theta\left[-\frac{2\tau}{h}\left(\frac{r_e^3 - r_b^3}{3}\right) + \left(\frac{2\tau\, r_e}{h} + \sigma_{ze}\right)\left(\frac{r_e^2 - r_b^2}{2}\right)\right] \tag{3.58}$$

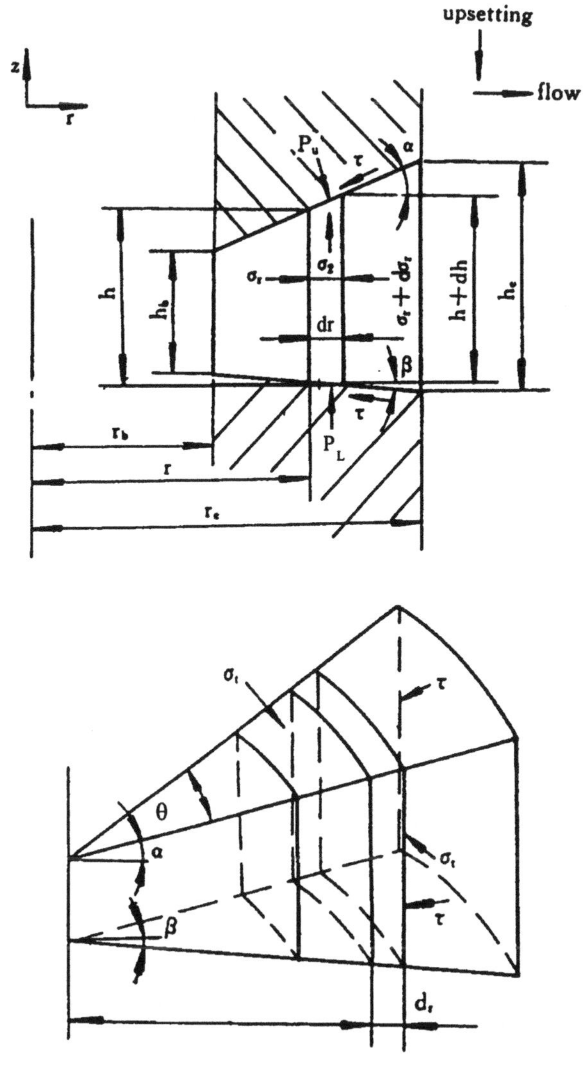

Figure 3.21: Metal flow between upper and lower inclined planes.

3.3.3.2 Axi-symmetric, axial flow

Axi-symmetric axial metal flow occurs when the direction of the metal flow is parallel to that of the die movement during extruding, as shown in Figure 3.22.

Axial stress is:

$$\sigma_z = k\ln\left(\frac{r_b - Z\tan\gamma}{r_b - Z_e\tan\gamma}\right) \tag{3.59}$$

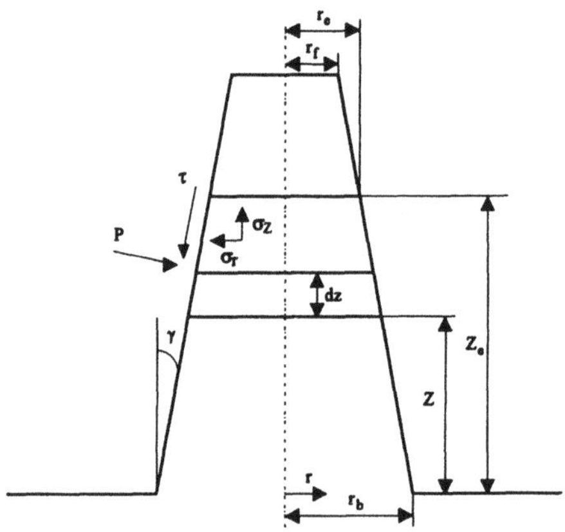

Figure 3.22: Metal flow between upper and lower inclined planes.

Load applied on a fan-shaped element with an angle θ is:

$$P = \frac{\theta}{2} r_b^2 k\ln\left(\frac{r_b}{r_b - Z_e\tan\gamma}\right) \tag{3.60}$$

where $k = \dfrac{2\left[\tau\left(1+\tan^2\gamma\right)+\bar{\sigma}\tan\gamma\right]}{\tan\gamma}$.

3.3.3.3 Plane strain, lateral flow (Figure 3.23)

Vertical stress distribution is:

$$\sigma_y = \frac{K_2}{K_1}\ln\left(\frac{h_e}{h_b + K_1 x}\right) + \sigma_{ye} \tag{3.61}$$

where $K_1 = \tan\alpha + \tan\beta$, and

$$K_2 = \frac{-2}{\sqrt{3}}\bar{\sigma} K_1 + \tau\left(2 + \tan^2\alpha + \tan^2\beta\right).$$

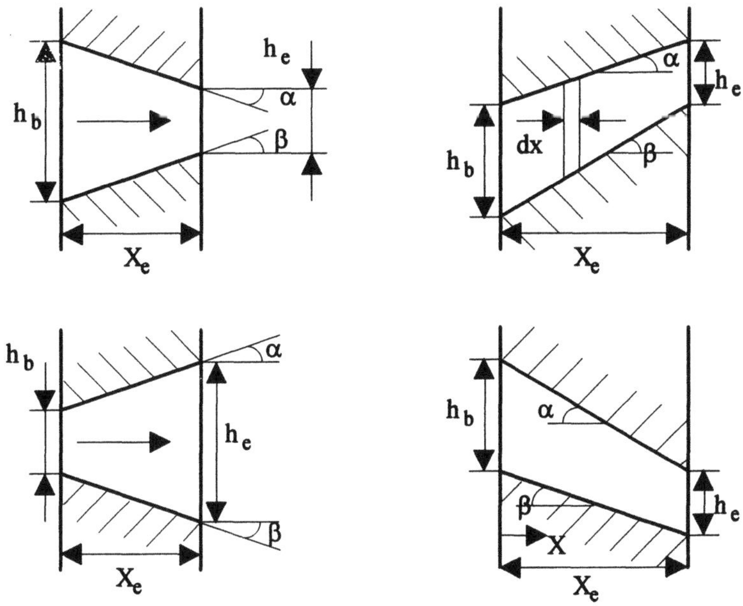

Figure 3.23: Plain strain lateral flow.

Load applied on a deformation element with unit thickness is:

$$P = \frac{-K_2}{K_1^{\,2}}\left[h_e\left(\ln(h_e)-1\right) - h_b\left(\ln(h_b)-1\right)\right] + \left(\sigma_{ye} + \frac{K_2}{K_1}\ln(h_e)\right)x_e \tag{3.62}$$

For plane strain between two horizontal planes, the equations are as follows.

$$\sigma_y = \frac{2\tau}{h}\left(x_e - x\right) + \sigma_{ye} \tag{3.63}$$

$$P = x_e\left(\sigma_{ye} + \frac{\tau\, x_e}{h}\right) \tag{3.64}$$

3.3.3.4 Plane strain, axial flow forming ribs (Figure 3.24)

Horizontal stress distribution is:

$$\sigma_x = \frac{K_2}{K_1} \ln\left(\frac{W_e}{W_b + K_1 Y}\right) + \sigma_{xe} \tag{3.65}$$

where $W_e = W_b + K_1 Y_e$,

$K_1 = -(\tan\gamma + \tan\delta)$, and

$K_2 = \frac{-2}{\sqrt{3}} \bar{\sigma} K_1 + \tau\left(2 + \tan^2\gamma + \tan^2\delta\right)$.

Using plastic condition, it is obtained as:

$$\sigma_y = \sigma_x - \frac{2}{\sqrt{3}} \bar{\sigma} \tag{3.66}$$

Load applied on a deformation element at W_b with unit thickness is:

$$P = W_b\left[\frac{K_2}{K_1} \ln\left(\frac{W_e}{W_b}\right) + \sigma_{xe} - \frac{2}{\sqrt{3}} \bar{\sigma}\right] \tag{3.67}$$

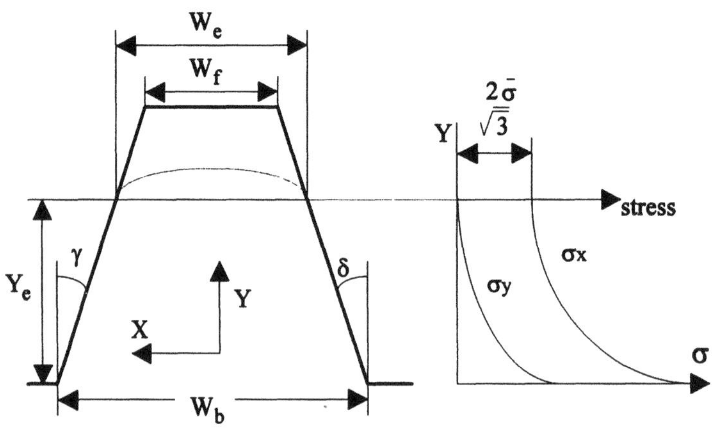

Figure 3.24: Stress distribution in rib forming.

It is necessary to determine the shearing surface in a forging when the metal flow does not take place as in the case when the forging is in contact with the die. Figure 3.25 shows an example of shearing deformation in a workpiece. In this example, the material flows into the flash cavity on the shearing surface. In order to establish the metal flow model, a representation of the stress σ_{yn} should first be found. Next, the angle and height h are found based the condition that σ_{yn} is minimum.

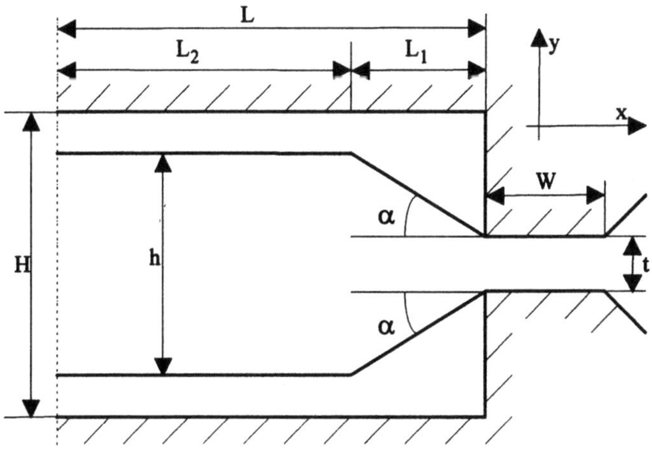

Figure 3.25: Shearing deformation produced in metal.

3.3.4 Design of dies

The main tasks of forging die design are, *viz.*, forging drawing design, flash cavity dimensions calculation, finish impression design, block impression design, pre-forming impression design and impression arrangement.

3.3.4.1 Design of forging drawing

The design of the forging drawing is based on the product part drawing. The design of the forging drawing involves the following steps:

1. Choose the location of the parting line,
2. Determine the wadding and matching allowance,
3. Determine the draft angles,
4. Add fillet and corner radii,
5. Determine the hollow gusset layer, and
6. Determine the tolerance.

3.3.4.2 Calculation of flash cavity dimensions

Figure 3.26 shows a common flash cavity comprising a land (1) and a gutter (2). The main function of a flash cavity includes holding surplus metal, increasing resistance on metal crosswise flow, ensuring adequate die filling, and buffering the effect of the die hammer. The key to flash cavity design is to choose a reasonable land size. At present, the common practice in the CAD of forging die is to use the following formula to calculate the height and width of the land.

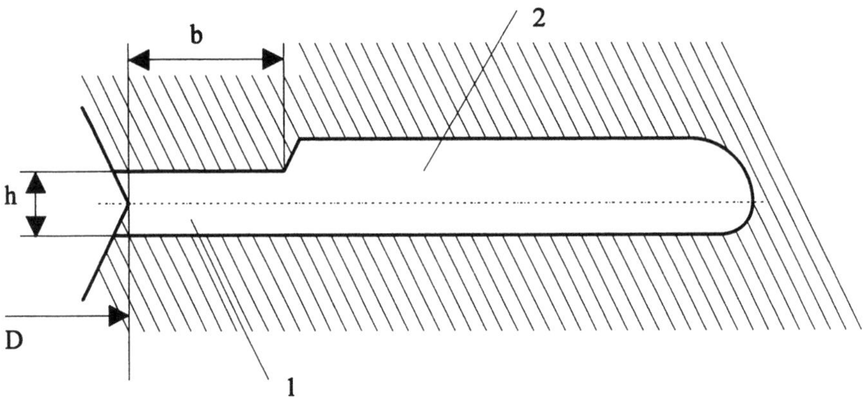

Figure 3.26: Shape of flash cavity.

For the die hammer:

$$h = -0.09 + 2\sqrt[3]{Q_f} - 0.01Q_f \tag{3.68}$$

$$\frac{b}{h} = -0.02 + 0.0038S\frac{D_0}{h} + \frac{4.93}{Q_f^{0.2}} \tag{3.69}$$

For the hot die forging press:

$$h = 2.17 + 1.39Q_f^{0.2} \tag{3.70}$$

$$\frac{b}{h} = -1.985 + 5.258Q_f^{-0.1} + 2.56x10^{-2}\frac{D_0}{h} \tag{3.71}$$

S is the complex shape coefficient of the forging, Q_f is the weight of the forging, and D_0 is the maximum block diameter.

The forging shape is first transformed into the cross-sectional shape of the finished forging cavity. The flash cavity is then located at the side of the finished forging cavity so that a complete finished die can be constructed.

3.3.4.3 Design of blocker impression

The designs of the drawing out and the rolling impressions are introduced in this section.

A. Design of drawing out impression

The drawing out impression comprises the ridge and the gutter. The stock is deformed in the ridge and the formed metal is held in the gutter. The vertical outline of a ridge can be classified into the flat and flange arc types. Its cross-sectional outline falls into two types, namely the closed and open types, as shown in Figure 3.27. The design of the drawing out impression includes determining the drawing out procedures, and designing the impression.

A.1 Determination of drawing out procedures

In the determination of the drawing out procedures, it is necessary to first obtain the curve of the material distribution and the calculating block drawing (Figure 3.28). Based on the variation of the cross-section, the forging is divided into the head and rod parts. The volume and average cross-sectional area of every part are next calculated to obtain the simplified block diagram (Figure 3.28b).

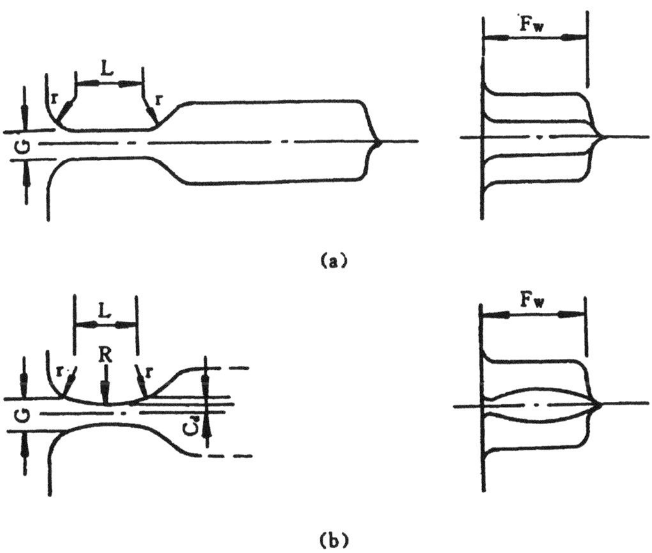

Figure 3.27: Two types of drawing out impressions.

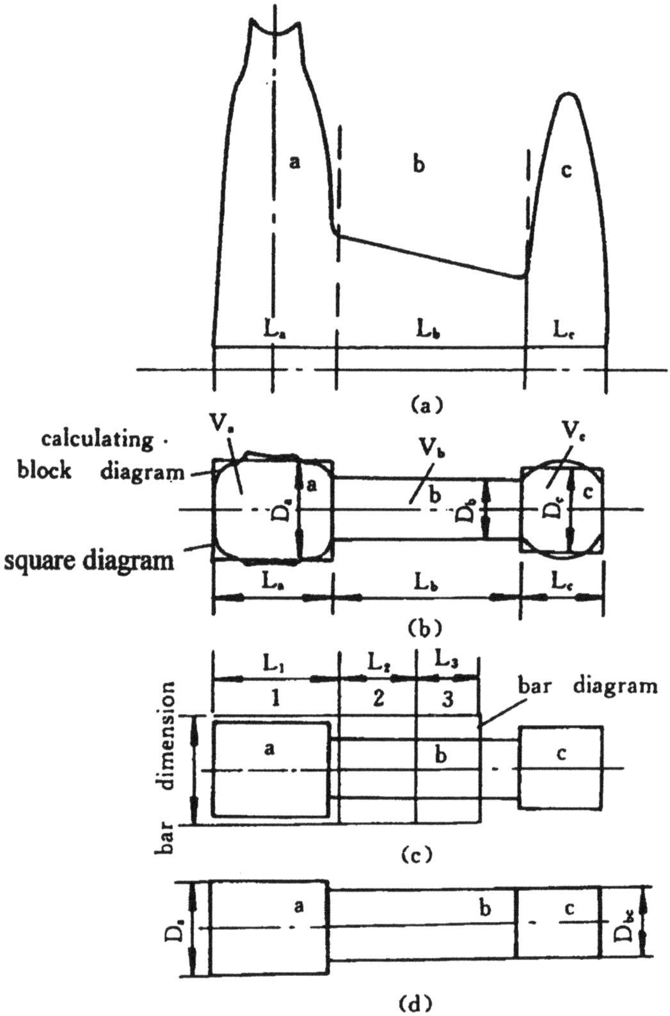

Figure 3.28: Design of drawing out impression.

Another factor that should be considered is the minimisation of the number of forging sequences required to obtain the forged shape, which is similar to that described in the calculating block diagram. The choice of the drawing out procedure is mainly determined by stock dimensions and the variation of the cross-sectional area shown in the simplified block diagram. Figure 3.28 illustrates the determination of the drawing out procedure. A piece b is first drawn out. The cross-sectional dimensions of piece b have to be made similar to piece c. If the difference between the two neighbouring cross-sectional areas is not large, i.e., D_c - D_b < d_m, and D_m is a given value, pieces b and c may be simplified into one piece. The size of the cross-section of this combined piece is calculated as follows.

$$D_{bc} = \left(\frac{V_b + V_c}{L_b + L_c}\right)^{\frac{1}{2}} \tag{3.72}$$

Under this condition, only one drawing out procedure is required to satisfy the requirement.

A.2 Design of drawing out impression

The design of the drawing out impression includes determining the ridge height and designing the longitudinal and cross-sectional outlines. In the CAD of forging dies, several methods have been used for designing the drawing out impression. Among others, the following formula is used to calculate the height G.

For flat longitudinal outline:

$$G = \sqrt{A} \tag{3.73}$$

For flange arc cross outline:

$$G = \sqrt{A} - B \tag{3.74}$$

A denotes the cross-sectional area that does not include the flash, and $1.6 \leq B \leq 3.2$. To prevent instability, G must not be smaller than one-third of the rod dimension. Otherwise, two drawing out operations are needed.

Formula for calculating the length of the ridge:

$$L = L_f - 13 \tag{3.75}$$

L_f expresses the drawing out length, i.e., the length of the stock that coincides with the part.

For flange arc type longitudinal outline, the formulae used to calculate the arc height G_d and the radius R of the main arc are as follows.

$$G_d = 3.2 - 1.6\left(\frac{25}{D_0}\right) \quad (D_0 \rangle 25) \tag{3.76}$$

$$G_d = 1.6 \quad (D_0 \leq 25) \tag{3.77}$$

$$R = \frac{G_d}{2} + \frac{L_f}{32} \tag{3.78}$$

Width of the drawing out ridge F_w:

$$F_w = \frac{0.75A}{G} \tag{3.79}$$

$$F_w \geq (1.4 \sim 1.5) D_0 \tag{3.80}$$

B. Design of fuller impression

The main objective of fuller impression design is to decrease the cross-sectional areas of some parts and increase the cross-sectional areas of other parts. This is to ensure that the metal mass distribution along the length approximates the shape of the calculating block configuration. A fuller impression consists of the gate, basic cavity and tail (Figure 3.29). A fuller impression can be classified into the open, close and mix types based on the shape of its cross-section.

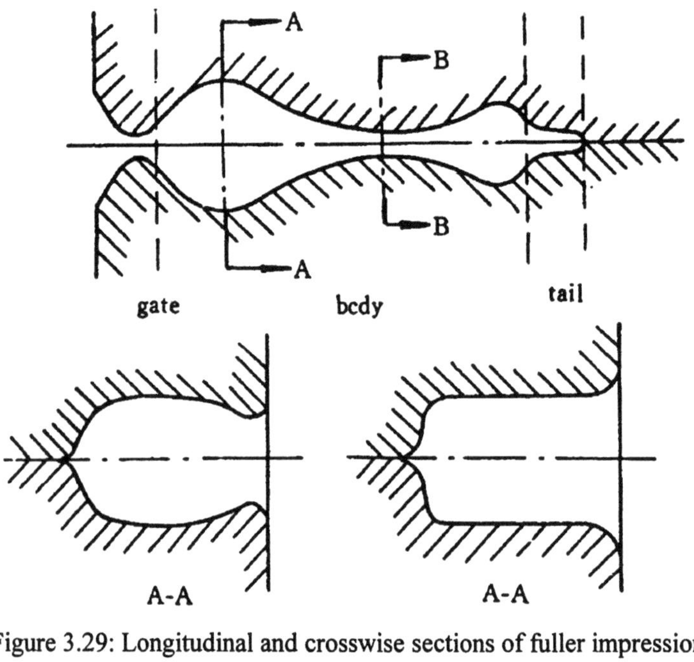

Figure 3.29: Longitudinal and crosswise sections of fuller impression.

The basic cavity design is the main task in the fuller impression design. The calculating block diagram is the basis for this design. The calculating block diagram is usually divided into several parts using the same dividing method as the design of the drawing out impression. The new configuration is then reconstructed using arcs and lines. The fillet radii of the linking neighbouring longitudinal configuration sections may not be smaller than 25 mm and should be as large as possible.

The process that uses the fitting curve method to design the basic configuration of the fuller impression is next explained. Figure 3.30a is a forging drawing of a link rod. Figure 3.30b shows the calculating block diagram and the simplified block diagram of the link rod based on the divided pieces method. The calculating block diagram is divided into three pieces, i.e., two heads and one rod, as shown in Figure 3.30b. In each part of the simplified block diagram, three points are defined and fitted with an arc. Figure 3.30c shows the blocking impression to obtain the block of the link rod.

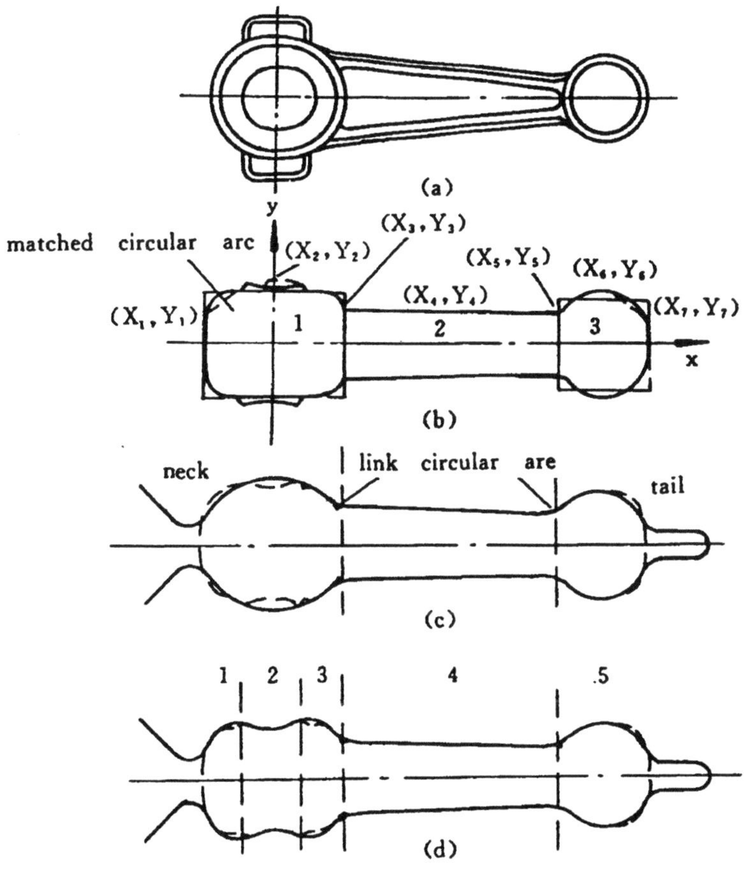

Figure 3.30: Design process of the basic configuration of fuller impression.

C. Design of pre-forming impression

The design of the pre-forming impression for axi-symmetric forging has been discussed earlier. As distinguished from axi-symmetric forging die, it is necessary to choose a series of typical cross-sections for the design of the pre-forming impression for long axial type forgings. Based on these cross-sections, the cross-sectional shapes of the pre-forming impressions are designed and put together to form the shape of the corresponding pre-forming impression.

For a H-shaped type pre-forming impression, this method is applied based on two conditions of the ratio of H_f and W_f, where H_f and W_f denote the height and width of the web respectively.

If $H_f/W_f \leq 2$ (Figure 3.31a), and the top and bottom of pre-forming impression are flat, its cross-sectional width is:

$$X_m = X_f - C \tag{3.81}$$

X_f represents the cross-sectional width of the impression. C is 2 ~ 10, which is determined by the designer.

Fillet radii of the pre-forming impression are:

$$R_{mc} = 1.25R_{fc} + 3.2 \tag{3.82}$$

$$R_{mf} = 1.25R_{ff} + 3.2 \tag{3.83}$$

The cross-sectional height is determined based on the principle that the areas must be equal.

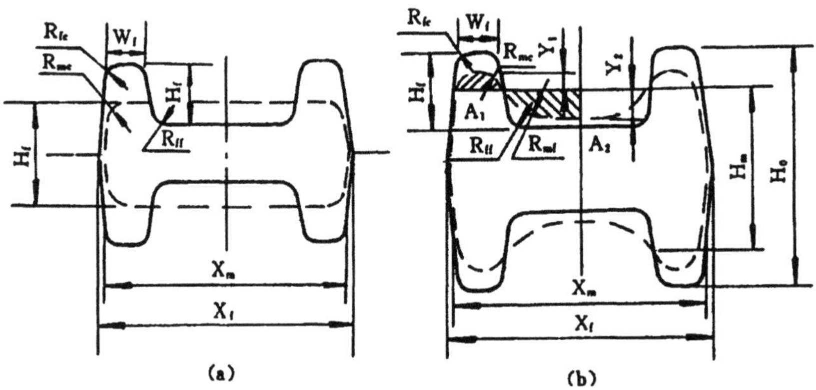

Figure 3.31: Calculation method for design of H-type section of pre-form impression.

If $H_f/W_f > 2$ (Figure 3.31b), R_{mc}, R_{mf} and the average height H_m of the cross-section are calculated using the above method. Y_1 is calculated as follows.

$$Y_1 = 0.25(H_0 - H_m) \tag{3.84}$$

Y_2 is determined based on the relation that A_1 must be equal to A_2.

In CAD, the method that uses the exponential curve to define the cross-sectional shape of pre-forming impression is sufficient. When this method is used to design pre-forming impressions, it is necessary to divide the cross-section into several L-shape units.

The formulae for the L-shape unit are discussed next. First, the cross-section is divided into several L-shape units as shown in Figure 3.32a. Next, the outline of the L-shape unit is determined based on the deformation type in the cross-section.

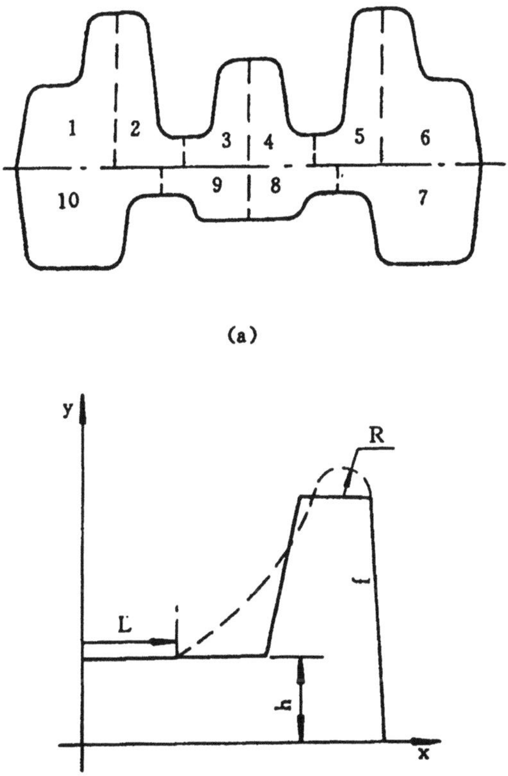

(b)

Figure 3.32: (a) Section is cut into several L-type elements, (b) method to handle L-type elements.

When the cross-section is a plane strain area (Figure 3.32b):

$$X = L\left[\ln\left(\frac{y}{h}\right) + 1\right] \tag{3.85}$$

When the cross-section is an axi-symmetric deformation area:

$$X = L\left[0.5\ln\left(\frac{y}{h}\right) + 1\right] \tag{3.86}$$

After the outlines of all the L-shape units have been determined, they are combined to construct the overall pre-forming impression.

D. Impression arrangement

The rule for impressions arrangement is described next. The centre of the finished impression must coincide with that of the side when a pre-forming impression is not required. The distance between the finished impression and the centre of the die swallow should be half the distance between the pre-forming impression and the centre-line of the die swallow, when they are simultaneously finished and pre-formed. Blocking impressions are arranged at the two sides of the pre-forming and finishing impressions. The first blocking procedure should be assigned near the heated equipment. During the arrangement of the impressions, it is essential to ensure that the operating path is the shortest during forging.

3.3.5 Program flowchart and description

3.3.5.1 Description of program flowchart

In this section, the CAD system for the long-shaft type forging die design will be introduced. The JDM system developed is suitable as a CAD/CAM application for mechanical press die forging. Its program flowchart is shown in Figure 3.33.

The JDM system includes modules for files operating, volume element editing, cross-section dividing, process analysis, die design, NC programming, etc.

1. The file operating module has the following major functions: file load, file save, DOS shell, exit JDM, etc.
2. The volume element editing module has the following major functions: define every volume element, smooth linking of elements, etc. A volume element in JDM is a geometry with inside and outside radii, and draft angle. Every volume element is a basic shape of the forging die or the forged part.

3. The cross-section dividing module has these functions: obtaining the geometric parameters of the volume element and the forging cross-section, calculating the cross- sectional area, etc.
4. The functions of the process analysis module include designing the calculating block diagram, designing the forging drawing, pre-forming workpiece and finished forging, roll forging, etc.
5. The functions of the die design module are forging die impression design and die construction design.
6. The functions of the NC programming module are to generate the NC tool path of the electrode for machining the die cavity and the NC codes.

The functions of the other modules are mainly to change the background and the colour of the screen, display the date and time, and exit the system.

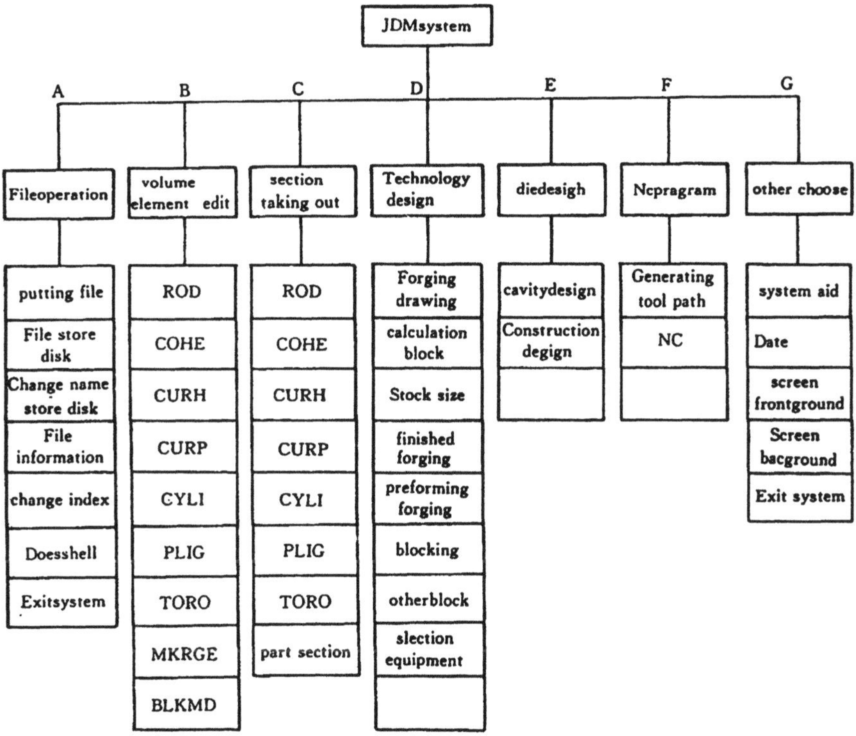

Figure 3.33: Flowchart of JDM system.

3.3.5.2 Example

The key parts of JDM are the volume element editing and the cross-section dividing modules. The automobile push-rod forging part is selected as an example to illustrate the description of the forging shape and the calculation of the cross-sectional area of the forging. The push-rod is shown in Figure 3.34. The die forging draft angle that is not indicated in the figure is 7. Non-specified radii are R5.

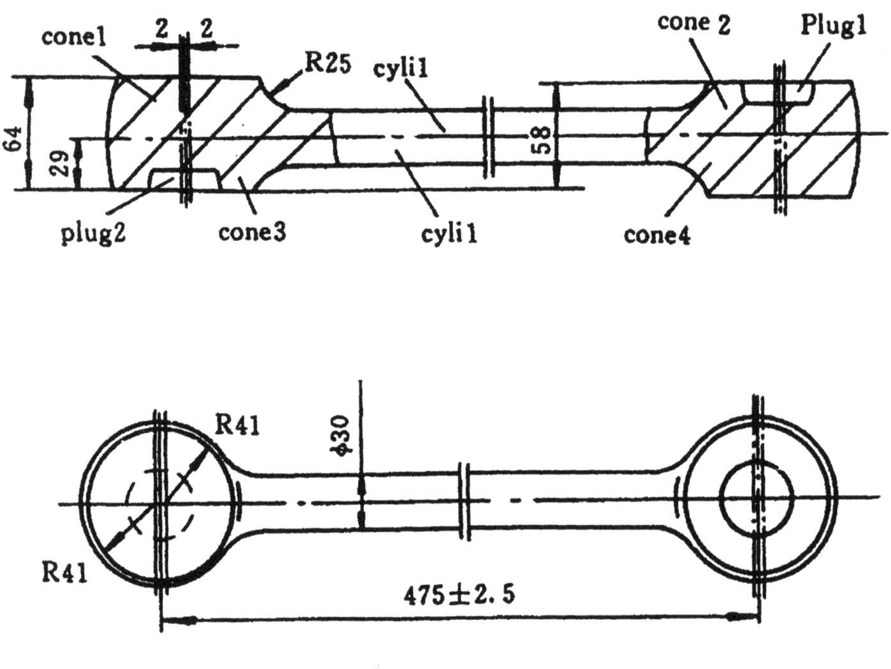

Figure 3.34: Push-rod forging.

A. Description of the forging shape

The push-rod can be divided into four cones, two cylinders and two plugs, and they are input individually. They are pieced together using the MERGE and BLEND commands.

Under the subsidiary directory of JDM, "JDM" is keyed in to enter the "JDM" system environment. The file to be edited is input by selecting the "inputting file" menu from the "file operating" module.

A.1 Input the parameters of the volume element

The "cone," "cylinder" and "plug" menu can be selected from the "volume element edit" module. Parameters of the volume elements shown in Table 3.3 are input according to the system prompts.

Table 3.3a: Parameters of CONE.

	H_v	R_u	R_l	α	R_{eu}	R_{el}	X_{sep}	Y_{lsep}	Y_{rsep}	Z_K	X_0	Y_0	Z_0	ϕ
1	35.0	41.0	/	7	5	0	4.0	0	0	1	0	0	0	0
2	29.0	41.0	49.6	/	5	0	4.0	0	0	1	475.0	0	0	0
3	29.0	41.0	49.6	/	5	0	4.0	0	0	-1	0	0	0	0
4	35.0	41.0	/	7	5	0	4.0	0	0	-1	475.0	0	0	0

Table 3.3b: Parameters of PLUG.

	H_d	R_u	R_l	α	R_{eu}	R_{el}	X_{sep}	Y_{lsep}	Y_{rsep}	Z_K	X_0	Y_0	Z_0	ϕ
1	12.0	21.0	/	7	5	10	0	0	0	1	475.0	0	29.0	0
2	12.0	21.0	/	7	5	10	0	0	0	-1	0	0	29.0	0

Table 3.3c: Parameters of CYLI.

	R	L	Sep	α_1	α_2	R_e	Z_K	X_0	Y_0	Z_0	ϕ
1	15.0	430.0	0	0	0	0	1	237.0	0	0	0
1	15.0	430.0	0	0	0	0	-1	237.0	0	0	0

A.2 MERGE and BLEND handling

The "MERGE" menu from the "volume element edit" module can be selected to piece the elements together. The type and order of each volume element that is required to be pieced together are input according to the system prompts. The "BLEND" menu is selected individually for each connection between two elements to smoothen the connection of these elements. The volume element types, order required for smoothening and the smoothening radii in the vertical and horizontal directions are input according to the system prompts.

In "MERGE" and "BLEND" handling, the forging should be divided into the upper part and the lower part with the parting lines as the boundary. The parameters of the push-rod forging are input as follows during the MERGE and BLEND handling.

Upper part on the parting line
MERGE: CONE1, CYLI1, CONE2, PLUG1, -1;
BLEND: CONE1, CYLI1, R25, R25;
BLEND: CYLI1, CONE2, R25, R25;
lower part under the parting line
MERGE: CONE 3, CYLI 2, CONE 4, PLUG 2, -1;
BLEND: CONE 3, CYLI 2, R25, R25;
BLEND: CYLI 2, CONE 4, R25, R25.

B. Calculation of cross-section area

The "cross-section of part" menu can be selected from the "cross-section taking out" module to input the direction axis of the cross-section. If the cross-section of the push-rod forging is vertical with the x-axis, "Y" is entered. Next, the location of the cross-section, i.e., the intersected distance of the cross-section on the x-axis, is input according to the system prompt. The shape and area of the cross-section at this location are shown in Figure 3.35.

After the cross-sectional area and shape of the forging have been obtained, the calculating block diagram and the process can be designed.

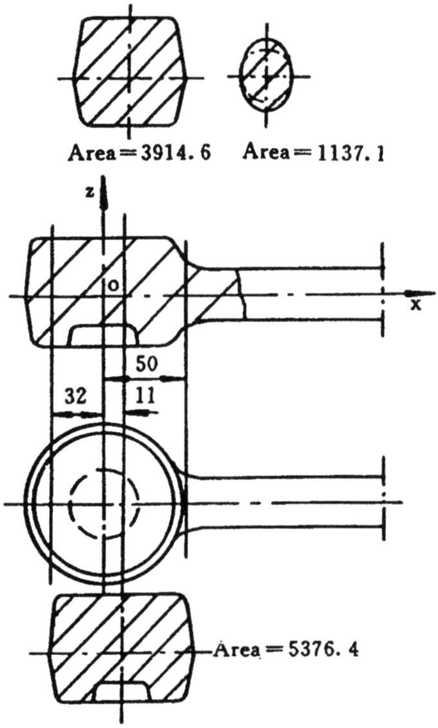

Figure 3.35: Calculation of cross-sectional area of push-rod.

References

1. Billhardt C F, Nagpal V, Altan T 1978 *A Computer Graphics System for CAD/CAM of Aluminum Extrusion Dies*. SME Paper MS78-957
2. Oh S I, Lahoti G D, Altan T 1981 ALPID - A General Purpose FEM Program for Metal Forming. In: *Proceedings of NAMRC-IX*, May 1981, State College, PA. pp 83
3. Teterin G P, Tarnovskij I J 1968 Calculation of Plastic Dimensions in Forging Axissymmetric Parts in Hammers. *Kuznechno-Stampovochnoe Proizvodstvo*, 5:6 (In Russian)
4. Boër C R, Rebelo N, Rydstad H, Schröder G 1986 *Process Modelling of Metal Forming and Thermomechanical Treatment*. Springer-Verlag, New York

Chapter 4

CAD/CAE/CAM for injection moulding

D.Q. Li and X.G. Ye

4.1 Introduction

4.1.1 Brief history of development

Common polymer materials such as PVC, PE, PS and ABS appeared only in the 1940s. The history of plastic injection moulding is less than 70 years. However, the development of plastic injection moulding is very rapid. The number of plastic injection moulding companies in USA and Japan is more than 20 000 presently. In the past, mould design and manufacturing are heavily dependent on trial-and-error approaches. This "rule-of-thumb" method that has been used in the past has now been taken over by computerised and sophisticated analysis tools, which are CAD, CAE and CAM.

The use of computers in manufacturing dates back to the 1950s when the objective was to control metal cutting machine tools using computers. During that period, only extremely expensive mainframe computers were available. Programming was accomplished via a punched card medium. It was tedious and time-consuming to develop and debug the programs.

Using a graphics display device to visually display the programmed cutter path was proposed during 1960s. Mini-computers were used during that period. This created a drastic increase in the acceptance of computers. In 1980s, the introduction of personal computers and workstations brought a totally new system approach to computer graphics. During the last twenty years, rapid progress in polymer rheology, geometric modelling, numerical control and computer technology have resulted in the development of CAD/CAE/CAM systems for plastic injection moulding. Beginning in the 1960s, many researchers such as Tool (USA, 1960), Pearson (UK, 1966), Kamal (Canada, 1972), Stevenson (USA, 1977) and Wang

(USA, 1977), have carried out a series of work to simulate the mould-filling process numerically. Based on reasonable simplifications and assumptions, general 1-D and 2-D flowing and cooling simulation programs were reported in 1970s. In 1980s, 3-D flowing, packing and cooling simulation programs were reported. Research has also been expanded to predict material orientation directions and warpage of injection moulded parts. In 1990s, integrated research on the complete injection moulding process, including flowing, packing, cooling, stress analysis and warpage, has been widely applied.

Remarkable progress on geometric modelling and NC machining has also been made in recent years. Surface modelling and solid modelling provided the enabling CAD/CAM technology for plastic injection mould making. Presently, there are many commercially available CAD/CAM systems for plastic injection moulding. The use of these CAD/CAM systems as tools for designing and manufacturing plastic injection moulds has become widespread.

In recent years, the development of CAD/CAE/CAM systems for plastic injection moulding has become one of the most popular research topics. Studies have become more focused, systematically researched and developed. The most representative research groups are the Cornell Injection Molding Project headed by Professor K.K. Wang of the Cornell University in U.S.A. with their C-Mold system, Chemical Development of the McGill University in Canada headed by Professor M.R. Kamal with their McKAM-II system, and the IKV Institute of Aachen University in Germany headed by Professor G. Menges with their CADMOULD system.

4.1.2 Technological characteristics

The emphasis on the application of CAD/CAM for plastic injection moulds lies in plastic products modelling, mould design, mould draughting and NC data generation. These CAE systems contain much more comprehensive engineering analysis, linking engineering design, test, analysis and manufacturing together throughout every stage of product development.

Manual design of plastic injection moulds depends heavily on the experience of mould designers. Drawings are used as the main communication medium for subsequent manufacturing operations. Moulds are interactively re-designed after they have been built to achieve the functionality of a product. In the CAD/CAE/CAM approach, the product model database serves as the communication medium. Analysis programs help mould designers improve their design. Iteration is mainly conducted in the design stage before machining the steel cavity. There are generally fewer moulding trials.

A computer-based model is usually a complete 3-D representation, as compared to a 2-D model, which appears as a series of views on a product drawing in the manual method. Information contained in the 3-D model is organised differently from that contained on a product drawing. This information may be extracted directly from the database without scaling errors or other mistakes.

In the manual method, selection of the gate location relies heavily on the experience of the mould designer. This can lead to major mistakes when the gate location is not correct. With the aid of the flow analysis program, the gate location can be determined optimally to minimise moulding stresses in local areas, provide the desired fill pattern and position weld lines. Flow analysis programs can be used to identify a range of process parameters such as the melt temperature, mould temperature and cavity pressure. When the gate location has been established, selections of the mould layout and mould base are carried out. Economic analysis program is available to assist designers in determining the number and sizes of the cavities required, and the size of the injection machine needed to achieve the lowest cost.

With the number of cavities and gate location optimised, the CAD software can be used to establish the minimum size of the mould base required to accomplish the cavity layout. Data such as the expected cavity insert sizes and runner layout may be input for the software to recommend the size of the mould base needed. Next, the user can easily retrieve the required mould base from a standard library. This standard mould base is available for editing the mould assembly.

The next step is to merge the part product model with the mould product model. The part geometry is usually defined by a product designer and stored in the database. This geometry may be copied from the database and automatically installed in the mould model at a location specified by the designer. At the same time, a material shrinkage allowance may be incorporated in the product model. If the plastic material exhibits a uniform and consistent shrinkage, a simple scaling command can be used to rectify the model based on its shrinkage factor.

Once the product geometry has been merged into the mould and corrected for material shrinkage, the cavity blocks can be defined by using the editing function provided by the CAD software. Subsequent extraction of the entities that describe the mould components is easy when the blocks have been defined. This facilitates the creation of detailed drawings for these components.

The runner system design is next performed. Flowing and packing analysis packages may be used at this stage to successfully design runner systems such as artificially balanced runner systems and constant pressure drop runner systems.

The cooling system is designed after the runner design has been completed. In manual design, relationships among the waterline geometry, water temperature and waterline location have rarely been studied or optimised. By using a cooling analysis software, the performance characteristics of waterlines and their economic impacts can be understood before trial moulding begins. Results from the cooling analysis software can be used to recommend appropriate water temperatures and flow rates for proper mould cooling, determine the projected cooling time, and evaluate the capability of the cooling circuit in preventing hot spots in the mould cavity. It can also determine if the same water temperature can be used in both halves of the mould, so that parts may be moulded with a minimum cycle time without thermally induced warpage. Once the cooling system has been designed, remaining components such as ejector pins and support pillars may be incorporated into the design.

When the basic design layout has been completed, some moving mould components can be checked for functionality. Several of the new CAE/CAD/CAM software systems available provide true kinematics analysis for designers to check the mechanical action of the design for a desired motion path, while simultaneously checking for interference with other mould components.

The last steps in mould making include creating the detailed drawings needed and generating NC tool paths to machine the actual mould. Using CAD/CAM systems, the drawing requirements can be significantly reduced, and NC tool paths generation from the existing geometry is relatively easy, especially when the shrinkage-corrected product model contains a complete detailed description of the product.

As mentioned above, CAD/CAE/CAM technology for moulding is based on scientific principles. These computer-aided software systems provide users with powerful tools to design, modify, optimise and manufacture products and moulds. This increases the probability of success and the optimality of mould making in terms of quality, performance and economy.

4.2 Graphic input and geometry construction of injection moulded products

Three predominant methods for geometric construction of injection moulded products are used today. They are wire-frame modelling, surface modelling and solid modelling. In the following sections, brief overviews of these methods and the benefits and costs associated with them are presented.

4.2.1 Wire-frame modelling

Wire-frame modelling is the simplest of the three modelling methods used for graphic inputs of injection moulded products. Using this method, the product is constructed as a collection of geometric entities such as points, lines, arcs and B-splines. In a wire-frame model, lines and other entities represent the edges of the physical surfaces of the product. The computing efficiency and storage requirements are minimal because of the simplicity of this method and its database. Manipulation of the geometry for displaying different views is relatively fast and easy, as only a small number of data elements would require mathematical transposition.

The major disadvantage of this method is the simultaneous display of all the entities that describe the geometry. For products with complicated geometry, this results in a complex and confusing image on the screen. Hidden lines may not be removed in this method. An example of a wire-frame is shown in Figure 4.1.

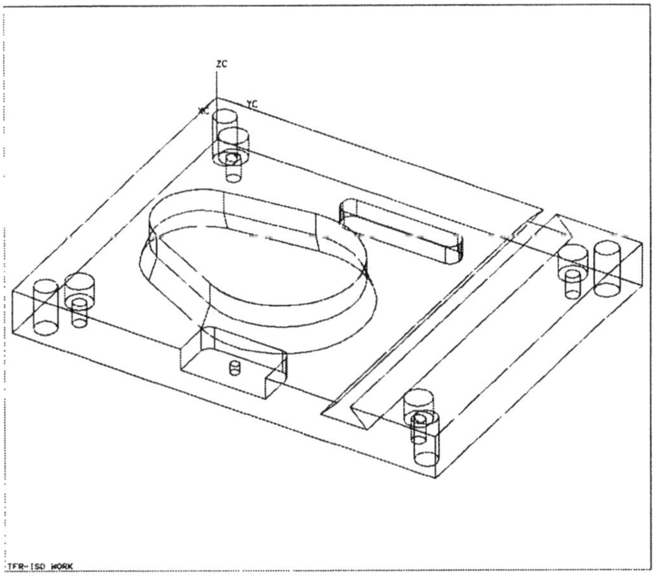

Figure 4.1: A product modelled in wire-frame geometry.

The wire-frame modelling method is more suitable for rectilinear objects without complex surface geometry, such as gears, cam brackets, etc. However, such plastic products are relatively few as most plastic products have complicated geometry. Therefore, wire-frame modelling is unsuitable for graphic inputs of injection moulded products. In mould design, this method is used for objects such as mould bases, cavity blocks and leader pins.

4.2.2 Surface modelling

Surface modelling differs from wire-frame modelling as all the faces of the geometry are described and represented. By using surface modelling, ambiguities present in the wire-frame method are eliminated. With a surface model, every point on the product surface is definable with explicit co-ordinates of a key point, or can be interpolated using an explicit set of parametric equations to specify the points between key points. Therefore, surface modelling is particularly appropriate when complex 3-D geometry needs to be described without any ambiguities.

Many plastic products require not only basic functions but also considerable aesthetic appeal to enhance its market attractiveness. Surface modelling provides the means to define the surfaces and evaluate the aesthetic appeal of a product.

Tool paths for a three- to five-axis milling machine can be created as every point on the surface may be explicitly defined. This is of great practical significance in mould design and manufacturing. In addition, surface modelling can create shaded images, calculate mass-related properties such as volume, surface area and moments of inertia, and generate sectional views automatically.

With these benefits, surface modelling has become a main method for describing geometry that is typical of injection moulded products. However, there is a cost associated with the benefits of surface modelling. It requires far more computing resources and database storage space than those required by wire-frame modelling.

The same example shown in Figure 4.1 is now modelled with surfaces and shown in Figure 4.2, with hidden lines removed.

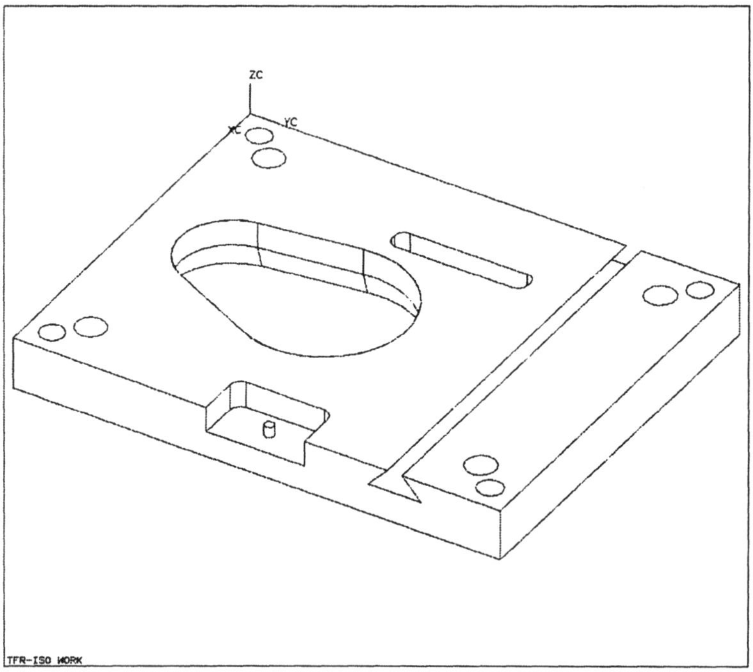

Figure 4.2: A product modelled in surfaces with hidden line removed.

4.2.3 Solid modelling

Solid modelling has many ways of creating a product and assuring that the product being modelled is valid and realisable compared to surface modelling. A solid model uses a variety of operations, such as Boolean additions and subtractions of primitive shapes. Other solid modelling operations include the sweeping of 2-D profiles through space and the sewing together of edges of surface models. Geometric primitives include objects such as cubes, spheres and cylinders. Using a solid model, the mass and boundaries of a product are represented in completely defined terms.

In surface modelling, the inside and outside of a product being modelled cannot be distinguished. In solid modelling, this difference can be recognised. During design, inconsistencies that may be present with the use of surface mathematics can be avoided. Every point in space can be determined to be either inside or outside the object being modelled in solid modelling, while surface modelling defines every point on a surface. Therefore, solid modelling is a more attractive method, allowing solutions to some very complex problems to be determined, such as automatic interference checking.

However, solid modelling requires greater computing power for practical designs. In many cases, solid models are based on the addition and subtraction of simple primitive geometric shapes that can be easily defined. While this speeds up the design process, it is inadequate to describe solid objects with sculptured surfaces. Since many injection moulded products are thin-walled with complex surfaces, some difficulties are encountered with the solid modelling approach.

In recent years, several advanced CAD/CAM systems have incorporated solid modelling with surface modelling to construct complex objects with sculptured surfaces. These systems are more suitable for graphic inputs and geometry construction of injection moulded products.

4.3 CAD for construction design of plastic injection moulds

4.3.1 Program flowchart

An integrated CAD/CAE/CAM system for plastic injection moulds has been developed and successfully implemented by the CAD group of the Huazhong University of Science and Technology (HUST), P.R. China. The program flowchart of the CAD modules in this system for the design of plastic injection moulds is shown in Figure 4.3.

The first step in the CAD for mould structure design is the creation of the model of the product to be moulded. In this system, this is provided via the product drawing. Surface modelling is used to develop the cavity and core, based on parting line selection and material shrinkage rate. Based on the cavity and core, fixed and moving mould assemblies can be created based on the experience of the mould designers. These assemblies have three functions, namely (1) standard mould base selection, (2) general mould assembly creation, and (3) cavity or core division, which is sometimes necessary to generate the mould components during mould design.

After the cooling line design, ejection pins design and mould component dimensioning have been completed, the general mould assembly can be designed. The mould drawings and data for injection moulding process simulation and NC modules are provided by the graphics editor. The program flowchart shown in Figure 4.3 gives the relationship and flow direction of the steps in the CAD stage.

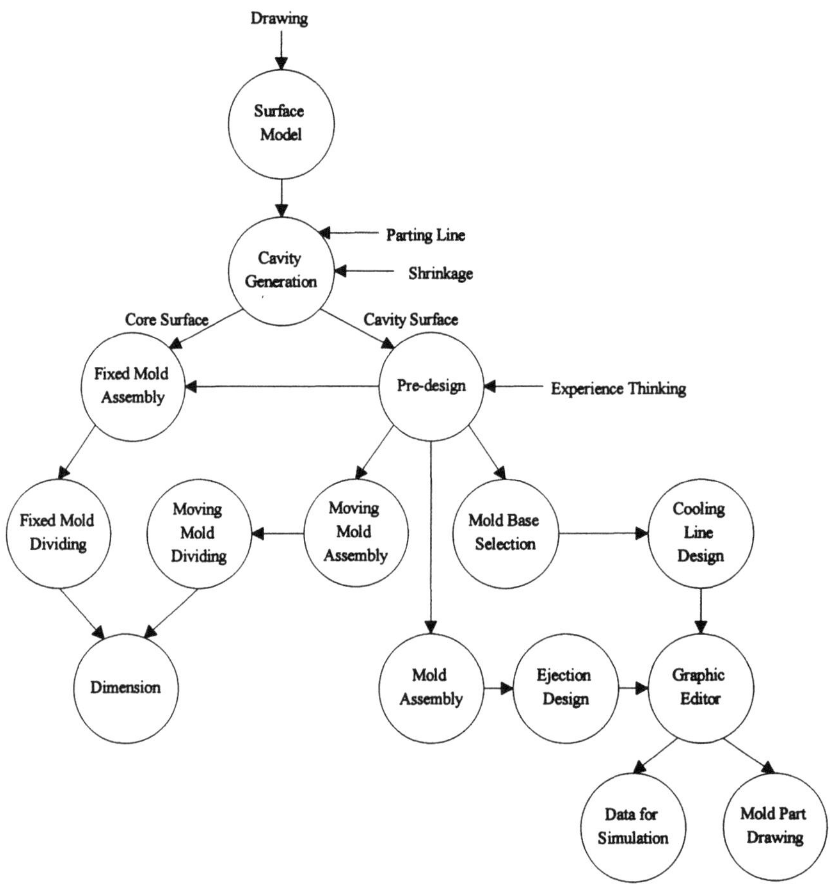

Figure 4.3: The program flowchart of CAD modules.

4.3.2 Standard mould base design

Selection of the standard mould base is very important for mould structure design. The components of a standard mould base are shown in Figure 4.4.

Many countries have their own standard mould bases, such as the D.M.E of U.S.A., D.M.S of U.K., Hasco of Germany and Futoba of Japan. In D.M.E, there are seven typical series of mould assemblies. The A-series, AR-series and B-series mould assemblies belong to the two-plate mould base, while the X5-series, X6-series and AX-series belong to the three-plate mould base.

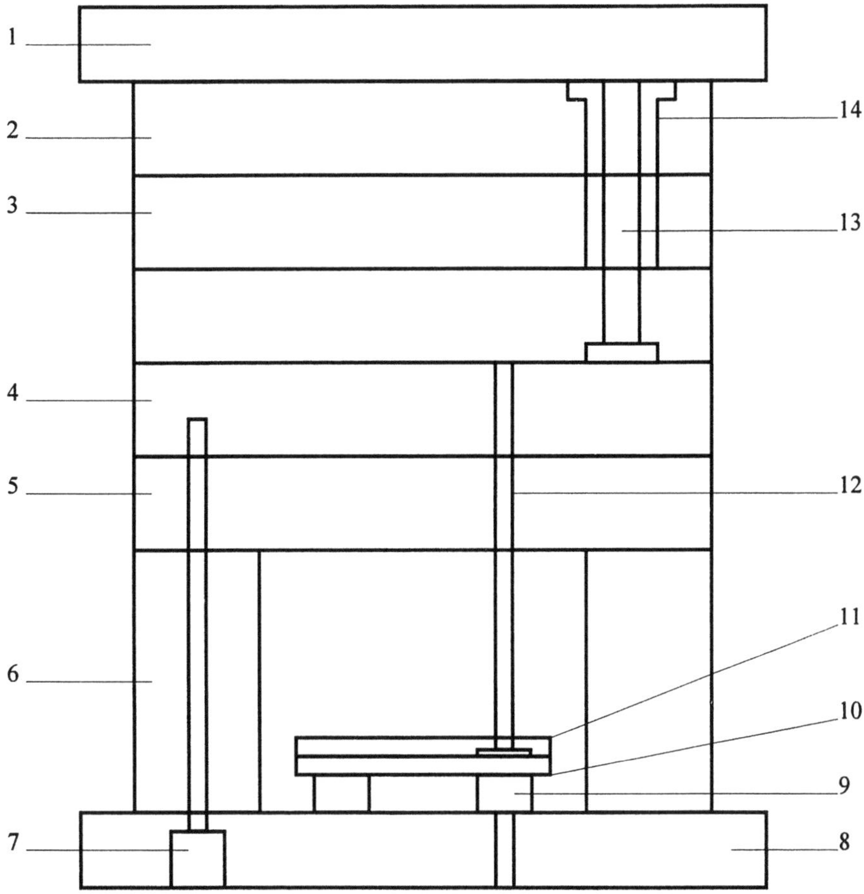

Figure 4.4: The components of standard mould base: (1) fixed clamp plate, (2) fixed plate, (3) stripper plate, (4) moving plate, (5) support plate, (6) spacer parallel, (7) screw, (8) moving clamp plate, (9) stop pin, (10) ejector plate, (11) ejector retainer plate, (12) return pin, (13) guide pillar straight, and (14) guide bush.

The A-series is the most frequently used mould assembly. The AR-series is identical to the A-series, except that the leader pin and bushings are reversed. For the A-series and AR-series, the cavities or cores can be firmly mounted in the mould plates as inserts. The B-series is similar to the A-series. However, it is used when the cavities or cores are machined directly into the mould plates. The X5- and X6-series have stripper plates. The difference between the X5- and X6-series is that the X5-series does not have a support plate while the latter has. The AX-series is used when the mould requires a floating plate for ejecting the sprue and runner separately from the injection moulded part. The T-series mould assembly is used for hot runner moulds that require two floating plates, which stay on the stationary half of the assembly.

Using the flowchart shown in Figure 4.5, the program can easily select an appropriate type of mould assembly.

A database is established to store all information, such as the type of mould assembly, sizes and prices of mould bases, and detailed dimensions, to automate the mould base selection process. Basic shapes of commercial standard mould components consist mainly of rectangular blocks and cylinders. Therefore, a general purpose coding scheme is inappropriate for this application because of its heavy requirement for computer memory. For the purpose of efficient storage and classification of the large number of mould components, a special coding scheme has been established and used in the mould selection program at the Cornell University and the HUST.

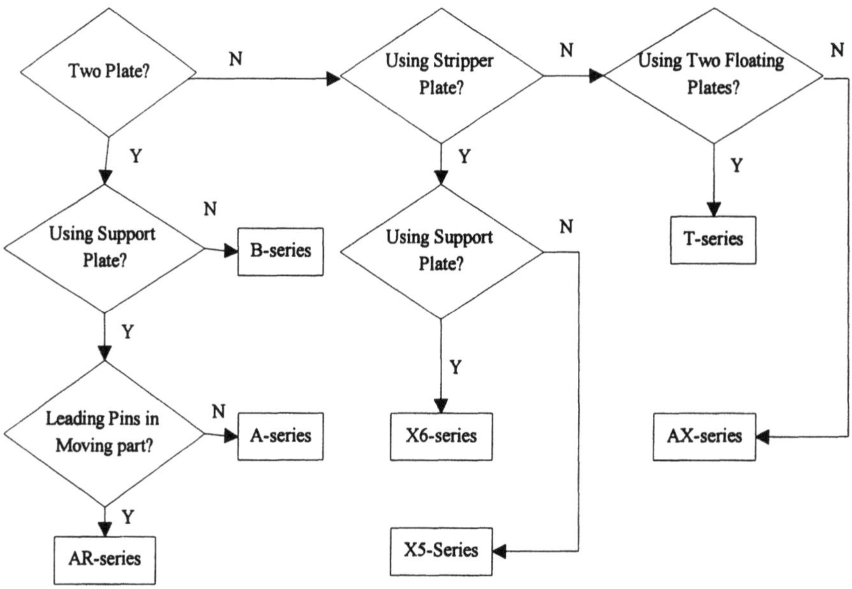

Figure 4.5: The flowchart for selecting mould assembly.

This coding scheme makes use of hierarchical and chain type structures. It stores the geometric data of every component in separate dimensioned variables or data tables. The data tables consist mainly of 16-bit integer words that contain several pieces of information to make full use of the available bit positions. Dimensions of components, which can be fractions of an inch, would normally require real variables (32-bit floating point number) to store them. However, they are coded in this scheme as an integer number by combining the integer and the fractional part to form a one integer number string. For example, dimensions $4\frac{5}{8}$ inch and $12\frac{7}{12}$ inch are coded as 45 and 1207 respectively. This highly-packed database occupies a total of 1.71 kilo-bytes of memory for representing 40 000 numbers in the D.M.E. catalogue, which would ordinarily require 160 kilo-bytes of memory. The coding scheme is also flexible. It is possible to add, delete or modify existing entries in the database.

4.3.3 Cavity and core design

In manual design, the cavity and core can be generated by multiplying every dimension in the part drawing by a shrinkage factor, and re-draw them for incorporation into the mould layout. For example, the formula for the radial dimension of a cavity is given by equation (4.1).

$$D_M = \left(D_S + D_S \times S_a - K\Delta\right) \tag{4.1}$$

D_M is a radial dimension for the cavity. D_S is a dimension of the part drawing corresponding to D_M. S_a is the average shrinkage factor of the polymer. K is a correcting coefficient (between 0.5 and 0.8), and Δ is the dimensional tolerance of the part drawing.

For complex moulds with many dimensions, the scaling of all the dimensions is a tedious, time-consuming and error-prone process. Using the CAD method, the geometry of a part has been previously defined by a product designer and stored in a computer. All the part geometry may be copied and displayed on the computer screen to generate the cavity and core. If the plastic material exhibits a uniform shrinkage, a simple scaling command can be used to multiply all the dimensions of the part by the shrinkage factor, and re-displaying the corrected graphics that define the cavity and core of the mould. If the polymer material is more complex in nature, such as materials with different shrinkage values in the flow and transverse directions, a slightly more complex approach is required. In general, a more complex shrinkage factor is applied by defining a local co-ordinate system within the part and axes that correspond to the flow and transverse shrinkage directions. A command may be issued to scale the part with independent scaling constants for the

X-, Y- and Z-directions. This manipulation does not precisely model the actual shrinkage. Many R&D activities are currently underway to provide calculations for creating the geometry of cavity and core efficiently.

Recently, an algorithm has been developed for creating the cavity and core by the CAD group of the HUST. Based on dimensions superposition, co-ordinates of the points on the cavity and core can be calculated using their dimensions, which can be obtained from the dimensions of the injection moulded part using transforming formulae, such as equation (4.1).

The basic formulae for dimensions superposition can be written as equation (4.2).

$$\left.\begin{aligned} X_B &= X_A + \sum_{i=1}^{n} K_i \times D_i \\ Y_B &= Y_A + \sum_{i=1}^{n} K_i \times D_i \\ Z_B &= Z_A + \sum_{i=1}^{n} K_i \times D_i \end{aligned}\right\} \qquad (4.2)$$

(X_B, Y_B, Z_B) are new co-ordinates of points on the cavity or core, while (X_A, Y_A, Z_A) are co-ordinates of points that have been input earlier. D_i is the dimension involved in the calculation of point B from point A. K_i is a constant or a triangular function based on the different dimension arrangement of the part drawing.

4.3.4 Runner bar design

The runner bar in a mould consists of runners and gates. Runners are the flow paths that deliver the polymer melt to the cavity, and gates are small openings that control the flow of the polymer into the cavity.

The design of the runner bar has very important influence on the quality and production cost of the moulded part. For multi-cavity moulds, balancing the filling time is often the key to a successful design. The traditional way of designing the runner bar is based on experience and trial-and-error. This is time-consuming, inefficient and costly. The CAD approach is much easier and more reliable for determining the dimensions of the runner bar. For a balanced runner bar design, the design criterion is that the flow front in each flow path should reach the gate at the same time.

In Figure 4.6, runners a_r and a_s have a common upstream runner. Based on the assumption that the pressure drop in an upstream runner is independent of the downstream runners, the relation between a_r and a_s has been derived by the CIMP group of the Cornell University as equations (4.3) and (4.4).

$$\frac{a_r}{a_s} = \frac{L_r + \sum_{k=1}^{m}\left[\prod_{j=1}^{k}\left(1+\sum_{i=1}^{d_j-1} r_{i,j}\right)^{1/3}\right]L_k}{L_s + \sum_{l=1}^{n}\left[\prod_{j=1}^{l}\left(1+\sum_{i=1}^{d_j-1} r_{i,j}\right)^{1/3}\right]L_l} \tag{4.3}$$

$$a_u = \sum_{i=1}^{d} a_i^3 \tag{4.4}$$

L_i is the length of runner i. d_j is the number of downstream flow branches of runner j. $r_{i,j} = a_i^3 / a_j^3$, i = 1, 2, ..., d-1. a_u is the upstream radius of the runner while a_d is the downstream radius of the runner. d is the number of downstream flow branches of runner u.

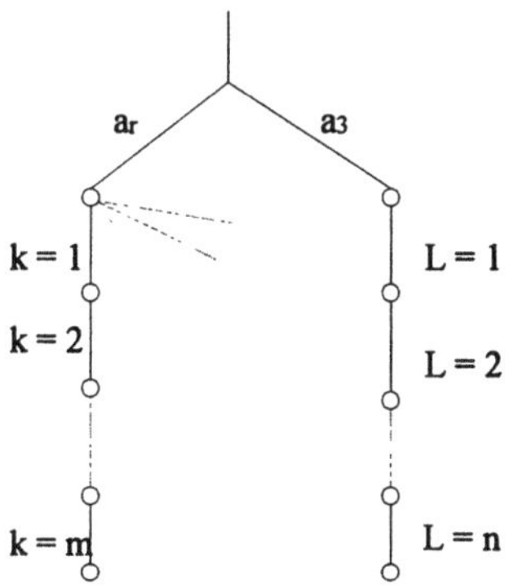

Figure 4.6: Diagram of relations between various flow paths.

A simple illustration of the above result is given in Figure 4.7, where the total volume of the runner system is given as 10 cm³. From equation (4.3),

$$\frac{a_8}{a_9} = \frac{a_5}{a_6} = \frac{a_2}{a_3} = 1$$

$$\frac{a_7}{a_5} = \frac{L_7 + \left(1+1^3\right)^{1/3} L_8}{L_5} = 3.26$$

$$\frac{a_4}{a_2} = \frac{L_4 + \left(1 + 1^3 + 3.26^3\right)^{1/3}}{L_2} = 5.32$$

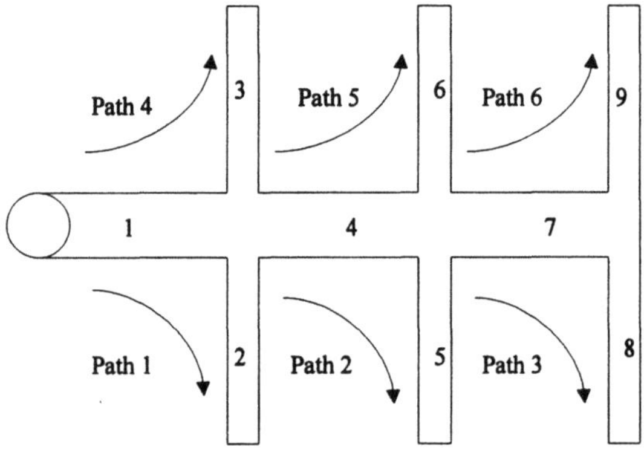

Figure 4.7: An example for runner bar design.

From equation (4.4),

$$a_7^3 = a_8^3 + a_9^3 = 2a_8^3$$

$$a_4^3 = a_5^3 + a_6^3 + a_7^3 = \left(1^3 + 1^3 + 3.26^3\right) a_5^3 = 36.65 a_5^3$$

$$a_1^3 = a_2^3 + a_3^3 + a_4^3 = \left(1^3 + 1^3 + 5.32^3\right) a_2^3 = 152.57 a_2^3$$

There are eight relations among the nine variables, a_1 to a_9. These nine variables can be solved directly when the relations are combined with a specified volume constraint. The results are given in Table 4.1.

$$V = 10\,\text{cm}^3 = 3.1416\begin{pmatrix} a_1^2 L_1 + a_2^2 L_2 + a_3^2 L_3 + a_4^2 L_4 + \\ a_5^2 L_5 + a_6^2 L_6 + a_7^2 L_7 + a_8^2 L_8 + a_9^2 L_9 \end{pmatrix}$$

Table 4.1 (Unit: cm).

No.	1	2	3	4	5	6	7	8	9
Length	2	1	1	2	1	1	2	1	1
Radius	0.77	0.14	0.14	0.76	0.23	0.23	0.75	0.60	0.60

It can be noted that the radii of runners 1, 4 and 7 (Figure 4.7) are unequal. This is because the calculations are based on equation (4.4), which leads to unequal radii of the runners along the flow path. In practice, equal radii of runners are often used when they are in a line, for the ease of manufacturing. Therefore, a set of equations has been derived to guarantee an equal radius for runners that are in a line. The forms of the equations are different depending on the number of cavities.

For a mould with eight cavities as shown in Figure 4.8,

$$a_1 = \frac{a_3^{\,3} L_1}{a_3^{\,2} L_3 + 2a_4^{\,2} L_4} \tag{4.5}$$

$$R_1^{\,3n+1} = \frac{(a_1 a_3 a_4 R_4)^{3n+1} L_g}{2^n (a_4 a_1 R_4)^{3n+1} + (a_3 a_1 R_4)^{3n+1} L_4 + (a_1 a_3 a_4)^{3n+1} L_g - (a_3 a_4 R_4)^{3n+1} L_1} \tag{4.6}$$

a denotes the radius of the runner. R denotes the radius of the gate. L_g is the length of the gate. n is the index of the power-law of the polymer melt. When a_4 or R_4 has been determined, a_1 or R_1 can be solved directly. The reverse is also true.

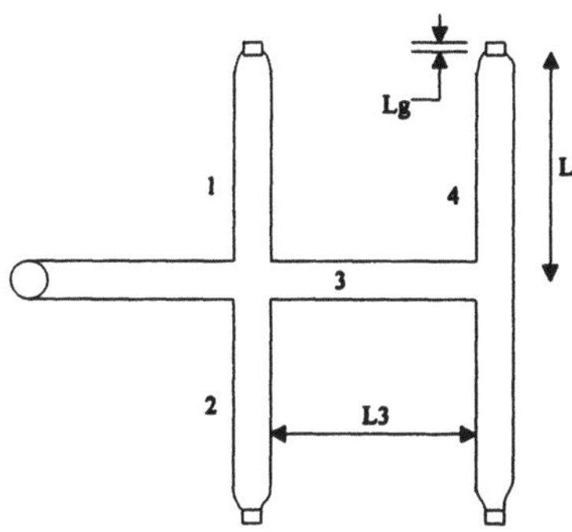

Figure 4.8: Example a mould with eight cavities (only half shown).

4.4 Flow simulation of plastic injection moulding

4.4.1 One-dimensional flow analysis

A 1-D flow means that the mould filling is characterised by a single directional flow of the polymer melt. There are three basic forms of 1-D flow, *viz.*, the tube flow, channel flow and radial flow as shown in Figure 4.9. The tube flow is used mainly for the simulation of mould filling in runners and gates, while the channel flow is used mainly for the simulation of the mould filling in cavities.

The governing equations for the momentum, energy and continuity of 1-D flow are given in equations (4.7), (4.8) and (4.9).

$$\frac{\partial}{\partial Z}\left(\eta\frac{\partial U}{\partial Z}\right)-\frac{\partial P}{\partial x}=0 \tag{4.7}$$

$$\rho C_P\left(\frac{\partial T}{\partial t}+U\frac{\partial T}{\partial X}\right)=K\frac{\partial^2 T}{\partial Z^2}+\eta\left(\frac{\partial U}{\partial Z}\right)^2 \tag{4.8}$$

$$\Gamma(x)\int_{-b}^{b}U dZ=Q \tag{4.9}$$

$\Gamma(x)$ is a shape function of the mould cavity for a circular disk and W for a rectangular strip.

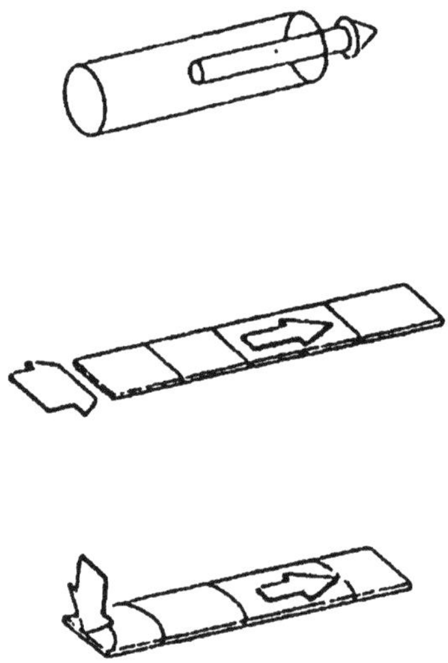

Figure 4.9: The three forms of 1-D flow.

As shown in Figure 4.10, the boundary conditions of the flow field are as follows.

$$U=0 \text{ at } Z=b;\ \frac{\partial U}{\partial Z}=0 \text{ at } Z=0 \tag{4.10}$$

The rheological property of the polymer is characterised by equation (4.11).

$$\eta = \frac{\eta_o(T)}{1 + \left(\eta_o \dot{\gamma} / \tau^* \right)^{1-n}} \qquad (4.11)$$

$\eta_o(T) = B \exp(T_b / T)$. The constants n, T_b, B and τ^* in equation (4.11) can be obtained through curve fitting of the viscosity data obtained using capillary rheometers.

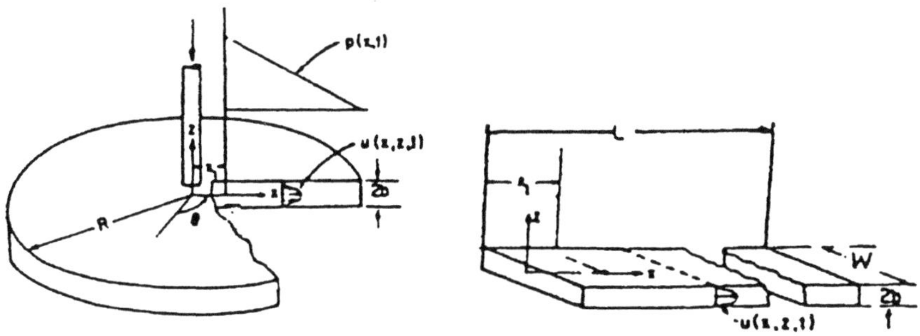

(a) Disk cavity (b) Strip cavity

Figure 4.10: Schematic diagrams and notations for 1-D flow.

For equations (4.7) and (4.8), it is assumed that the effects of inertia, stream-wise (x-direction) conduction, gap-wise (z-direction) convection, and elastic behaviour of the melt are negligible. Experiments have shown that the resulting pressure predictions agreed reasonably well with the measured data under these assumptions.

The boundary conditions of the temperature field are as follows.

$$T = T_e \text{ at } x = x_1,$$

$$T = T_c \text{ at } Z = \pm b,$$

$$\frac{\partial T}{\partial Z} = 0 \text{ at } Z = 0, \text{ and} \qquad (4.12)$$

$$\frac{\partial T}{\partial x} = 0 \text{ at } x = x_{int}(t).$$

$x_{int}(t)$ denotes the stream-wise location of the melt-front at time t. This is based on the following assumptions: (1) the melt enters the cavity at x_1, (2) the flow is far away from the junction and is fully developed, (3) the melt has a uniform temperature T_c during the filling process, (4) the temperature is symmetric about Z = 0, and (5) the heat lost to air at the advancing melt-front is negligible.

Equation (4.7) is integrated to solve the pressure field,

$$\int_0^z \frac{\partial}{\partial Z}\left(\eta \frac{\partial U}{\partial Z}\right) dZ = \int_0^z \frac{\partial P}{\partial x} dZ$$

Since $\frac{\partial P}{\partial x}$ is independent of the variable Z,

$$\therefore \qquad \frac{\partial U}{\partial Z} = \frac{\partial P}{\partial x}\frac{Z}{\eta}$$

The above equation is integrated as,

$$\int_z^b \frac{\partial U}{\partial Z} dZ = \int_z^b \frac{\partial P}{\partial x}\frac{Z}{\eta} dZ$$

Using the boundary condition U = 0 at Z = b,

$$U = -\frac{\partial P}{\partial x}\int_z^b \frac{Z}{\eta} dZ$$

The equation above is integrated again as,

$$\int_0^b U\, dZ = \int_0^b -\left[\frac{\partial P}{\partial x}\int_z^b \frac{Z}{\eta} dZ\right] dZ$$

Using integration by parts,

$$\int_0^b U\, dZ = -\frac{\partial P}{\partial x}\int_0^b \frac{Z^2}{\eta} dZ$$

The above equation is substituted into the equation (4.9) to obtain equation (4.13).

$$\Lambda = \frac{Q}{2\Gamma S} \tag{4.13}$$

where
$$\Lambda = -\frac{\partial P}{\partial x} \tag{4.14}$$

$$S = \int_0^b \frac{Z^2}{\eta} dZ \tag{4.15}$$

From the above,

$$\dot{\gamma} = \frac{\Lambda Z}{\eta} \tag{4.16}$$

$$U = \int_z^b \dot{\gamma}\, dZ \tag{4.17}$$

$$\Phi = \eta \dot{\gamma}^2 \tag{4.18}$$

Φ denotes the viscous-heating term in equation (4.8).

The finite difference form is used for the temperature field.

$$T_{i,j,k} \equiv T(X_i, Z_j, t_k)$$

(X_i, Z_j) denotes a fixed-grid system where

$$Z_j = (j-1)\Delta Z, \qquad \Delta Z = b/N$$

$$t_1 = 0, t_{k+1} = t_k + \Delta t_k \qquad (k \geq 1)$$

An appropriate finite difference representation of equation (4.8) is given as equation (4.19).

$$\rho C_p \left(\frac{T_{i,j,k} - T_{i,j,k-1}}{\Delta t_{k-1}} + U_{i,j,k-1} \frac{T_{i,j,k} - T_{i-1,j,k}}{\Delta X_{i-1}} \right) = K \left(\frac{T_{i,j+1,k} - 2T_{i,j,k} + T_{i,j-1,k}}{\Delta Z^2} \right) + \Phi_{i,j,k-1} \tag{4.19}$$

Equation (4.19) is implicit in temperature, and U and Φ are evaluated at an earlier time. Using the boundary conditions specified in equation (4.12), equation (4.19) can be solved successively to obtain a temperature field at any instant (assuming that at the beginning of the first time step, the temperature of the melt is equal to the temperature T_e everywhere). With the temperature known at the new time step, S can be obtained using equation (4.15) and h at an earlier time. Λ is obtained using equation (4.13). η, $\dot{\gamma}$, U and Φ can be determined using equations (4.11), (4.16), (4.17) and (4.18) respectively. Based on these results, the temperature field at the next time step can be determined. Using this approach, calculation is continued until the cavity is fully filled.

4.4.2 Two-dimensional flow analysis

When the 1-D flow analysis scheme, such as disk or strip cavity filling, has been established, 2-D flow analysis can be performed using a composite of 1-D primitive segments.

In this method, a 2-D cavity/runner system can be considered as a combination of flow paths, each consisting of a concatenation of primitive segments. As shown in Figure 4.11, there are four types of flow segments, *viz.*, (1) circular tube, (2) sector of a centre-gated disk of zero inner radius, (3) end-gated rectangle, and (4) sector of a disk having non-zero inner radius. An example of a cavity and runner represented with these primitive segments is illustrated in Figure 4.12.

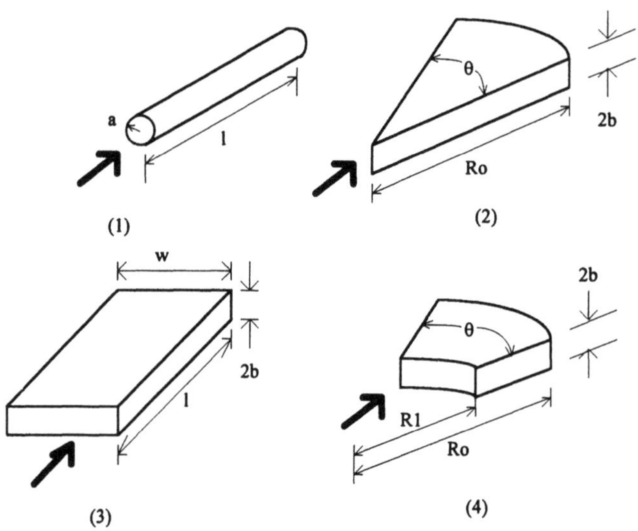

Figure 4.11: 1-D flow segment configurations.

There are two paths in this case. Both paths start at segment 1 and have common flow segments 1 and 2. The first flow path comprises segments 1, 2, 3, 4 and 5. The other flow path comprises segments 1, 2, 6, 7, 8 and 9. Segments 1 and 2 are circular tubes. Segments 3 and 6 are sectors of a centre-gated disk of zero inner radius. Segments 4 and 7 are sectors of a disk having non-zero inner radius, and segments 5, 8 and 9 belong to the end-gated rectangle type.

Based on the 1-D flow analysis, the procedures for the 2-D flow analysis at any time step are as follows.

1. Updating the temperature field in each non-empty segment where the flow has not stopped;
2. Updating the volumetric flow rate along each flow path using a Newton-Raphson procedure to satisfy the condition of equal overall pressure drop along each flow path; and
3. Advancing the melt-front in each partially filled segment based on the updated volumetric flow rates.

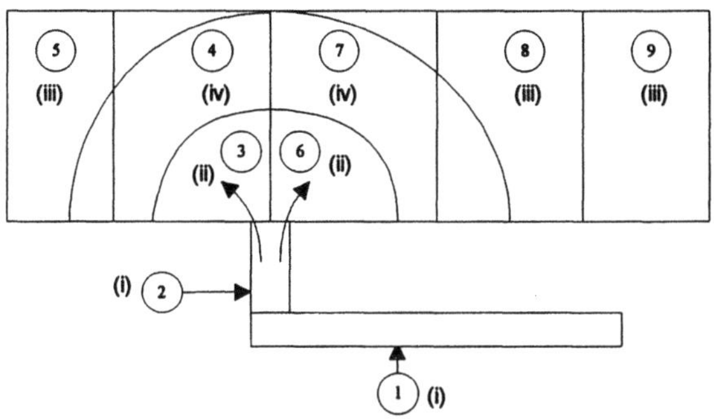

Figure 4.12: An example of a cavity and runner represented by segments.

This analysis is carried out with a nominal prescribed time step unless one of the segments is about to become filled. In this case, the time increment is appropriately changed for that time step.

Although more general 2-D flow analyses have been developed using the finite element method or the finite difference method, the 2-D scheme based on the 1-D flow analysis has many advantages. Practice has shown that predictions based on this method agree well with the corresponding predictions based on a more accurate but lengthy 2-D flow programs, especially in multi-gated cases.

4.4.3 Three-dimensional flow analysis

The 1-D flow analysis scheme can further be used for the three-dimensional (3-D) flow analysis. Using this method, the 3-D shape of a cavity must be flattened to a 2-D shape to decompose the cavity and runner geometry to 1-D flow paths based on a conjectured flow pattern. Each flow path is further decomposed into a series of 1-D segments such as strip, disk or tube. Figure 4.13 shows the flat pattern and its flow paths and segments of an injection moulded part.

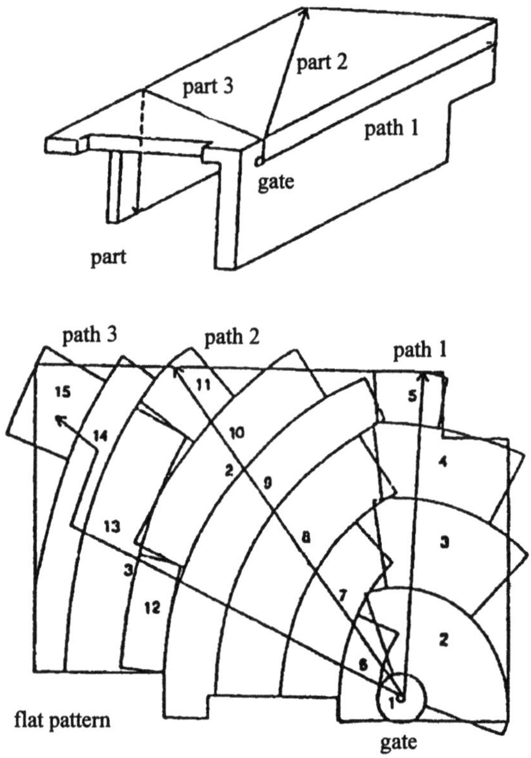

Figure 4.13: A flat pattern and its flow paths and segments.

The reason for using the 1-D flow approximation method instead of a 3-D flow simulation is based on economical considerations. Due to the substantial increase in the CPU time required to run a 3-D flow simulation program and the limited improvement in the predictive accuracy in some cases, it is appropriate to use the 1-D flow simulation program in multi-gated cases or simpler cavities. However, the 1-D analysis program depends strongly on the experience of mould designers. It is not easy for mould designers to flatten the 3-D cavities, divide the flow paths and decompose the segments, especially when there are complex cavities.

Using 3-D flow simulation programs that are based on the finite element method and the finite difference method, it is not necessary to flatten the cavities. This method is largely independent of the experience of designers, and is very accurate in predicting the resulting pressure and temperature of the melts during the filling process. One of the most successful flow analysis packages makes use of a fixed-mesh representation of the 3-D mould geometry. Thin-shell triangular elements and tubular elements are employed for geometric discretization of the cavity and runner system. A control volume concept is introduced to obtain the governing equations.

As shown in Figure 4.14, the centroids of the triangular elements are joined to the mid-points of the three corresponding sides to create polygonal control volumes that surround each vertex node in the calculation domain. In Figure 4.14, the solid lines denote the element edges, the dash lines represent the control volume boundaries, and the shaded area shows the control volume associated with a node N.

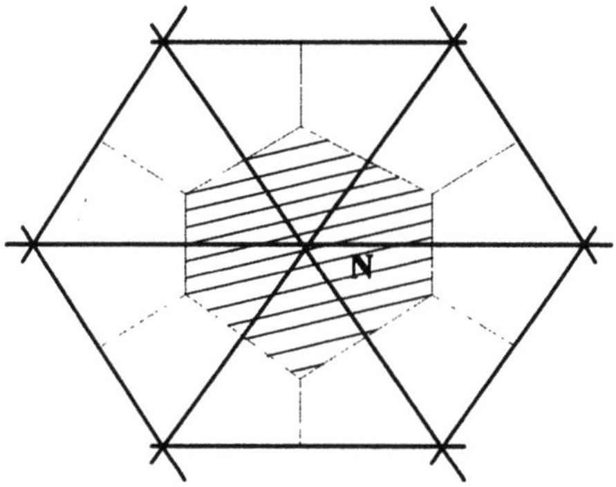

Figure 4.14: A polygonal control volume.

Four types of control volumes have been defined, as shown in Figure 4.15. The first type is the entrance control volume, through which the polymer melt enters the cavity. The second type is the interior control volume, which is completely filled with polymer melt (f = 1). The third type is the melt-front control volume, which is partially filled with polymer melt (0 < f < 1). The parameter f gives the ratio of the filled volume to the total control volume. The fourth type is an empty control volume that has not been filled by the melt (f = 0).

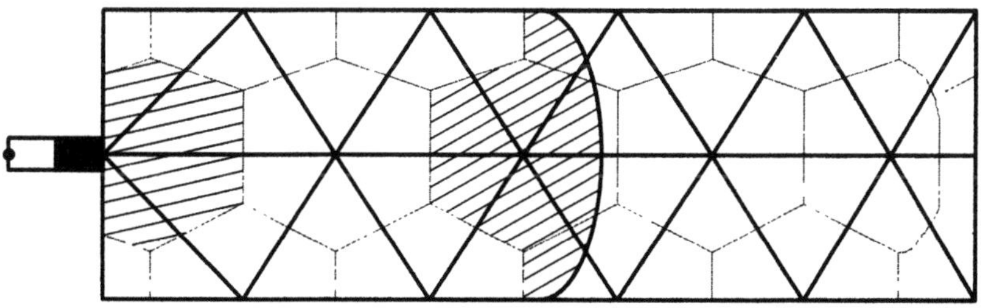

Figure 4.15: Control volume definition for melt-front advancement.

In this analysis, the time step is chosen such that only one melt-front control volume is fully filled per step. Each of its adjacent empty volumes becomes the new melt-front control volumes subsequently.

The flowchart for the 3-D flow simulation program is shown in Figure 4.16. The input data includes the geometry of the cavity, thin-shell triangular and tubular elements, and parameters of the polymer and flow conditions.

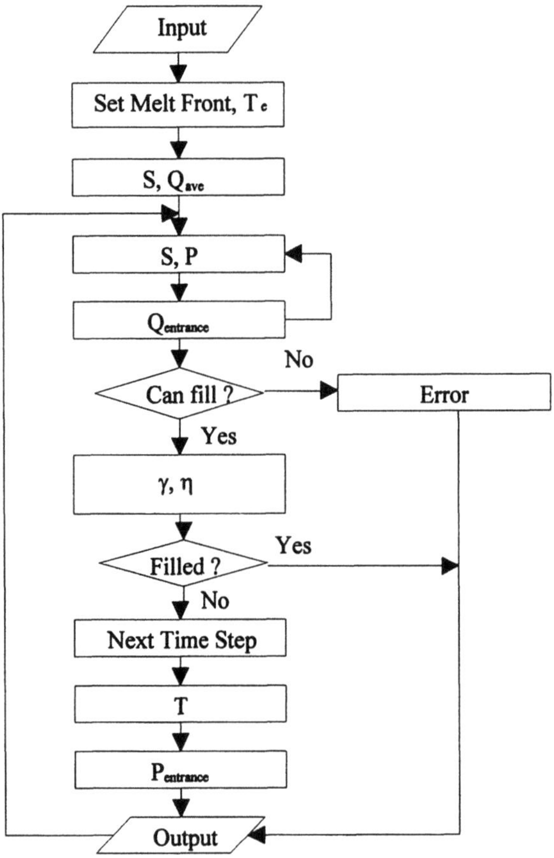

Figure 4.16: Flowchart of 3-D flow simulation program.

An assumption made here is that the entrance control volume has been totally filled at the first time step and the temperature distribution is uniform and equal to the melt temperature T_e. Therefore, the melt-front and the temperature field at the first time step are known. Thus, the shear viscosity η, flow conductance S and average volumetric flow rate Q_{ave} can be calculated. The pressure field is obtained through iteration. Based on these results, the temperature field at the next time step can be determined. After updating the pressure of the entrance control volume, the program iterates to calculate the new pressure field until the whole cavity is completely filled.

4.5 Cooling simulation of plastic injection moulding

4.5.1 One-dimensional cooling analysis

The cooling simulation for plastic injection moulding is of great importance as it significantly affects the productivity and quality of the final part. It is well-known that more than two-thirds of the cycle time in an injection moulding process is spent on cooling the hot polymer melt sufficiently, so that the part can be ejected without any significant deformation. An efficient cooling line design can considerably reduce the cooling time. Severe warpage and thermal residual stresses in the product may result from non-uniform cooling.

The use of cooling simulation program for determining the locations of cooling channels and process conditions to achieve uniform cooling and minimum cooling time would be of great help to mould designers.

The 1-D cooling analysis models the heat transfer problem with infinite number of parallel cooling channels with equal spacing pitch b and depth a from an infinite straight cavity surface, as shown in Figure 4.17.

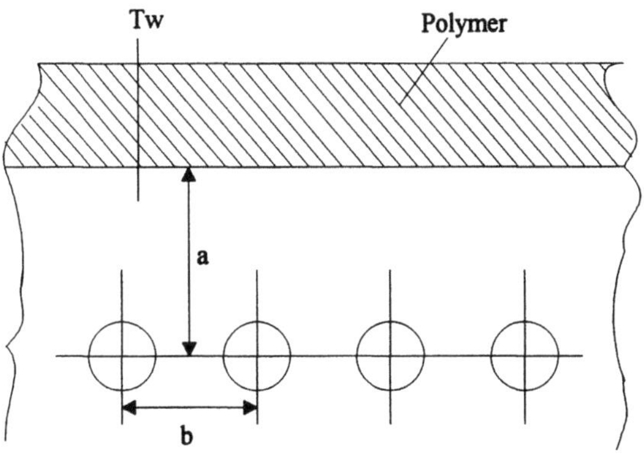

Figure 4.17: The layout of 1-D cooling channels.

By considering the heat balance in one cycle of the injection moulding operation and the uniformity of heat flux over the cavity surface, the 1-D cooling program determines several important design parameters such as the pitch b, depth a, diameter d, cooling time t_c, flow rate and the heat loss of coolant in a cooling channel.

Based on the assumption that all the heat dissipated from the plastic part is absorbed by the coolant, the rate of heat dissipation from the polymer to the mould wall is given by equation (4.20).

$$Q_P = \frac{\Delta H m_P}{t_c + t_i} \tag{4.20}$$

ΔH is the change in the enthalpy of the polymer between the time it enters the cavity to the time when the part is ejected from the mould. m_p is the polymer mass density multiplied by the cavity thickness. t_c is the cooling time, and t_i is the idling time of the cycle during part ejection.

The cooling time can be calculated using equation (4.21).

$$t_c = \frac{S^2}{\pi^2 \alpha_c} \ln\left(\frac{8}{\pi^2} \times \frac{T_M - T_W}{T_E - T_W}\right) \tag{4.21}$$

S is the thickness of the mould cavity. α_c is the effective thermal diffusivity of the polymer. T_M is the inlet melt temperature. T_E is the average temperature of the plastic part at ejection, and T_W is the average temperature of the cavity wall. Equation (4.21) is obtained by taking only the first term in the Fourier-series solution for the 1-D transient heat conduction equation.

The volumetric flow rate V_c of the coolant required to absorb the heat dissipated from the polymer is given in equation (4.22). ρ_c is the density of the coolant. ΔT_c is the increment of the coolant temperature. C_c is the specific heat of the coolant.

$$V_c = \frac{Q_P}{\rho_c C_c \Delta T_c} \tag{4.22}$$

The heat transfer coefficient h_c at the wall of the cooling channel can be estimated by the empirical expression in equation (4.23).

$$h_c = 0.023 R_e^{0.8} P_r^{0.4} K_c / d \tag{4.23}$$

R_e and P_r are the Reynolds and Prandtl numbers respectively. d is the diameter of the cooling channel that is assumed by mould designers at the beginning of the cooling line design. K_c is the thermal conductivity of the coolant.

The pitch b and depth a are determined by equations (4.24) and (4.25).

$$j = 2.4\left(h_c d / K_W\right)^{0.22} \left(b/a\right)^{2.8} \ln\left(b/a\right) \tag{4.24}$$

$$Q_c = \frac{0.87(T_W - T_C)K_W h_c d}{0.87bK_W + h_C d[a - 0.13(\pi d - b)]} \tag{4.25}$$

j is the percentage of temperature fluctuation in the mould at the plane of the cooling channels. The value of j is set to be less than 2.5% for amorphous materials. K_w is the heat conductivity of the mould plate.

Parameters b and a are solved using an iterative procedure, by first assuming a large b/a ratio of 5.0 in equation (4.24). Next, this value is reduced successively in increments of 0.1 until the required value of j is obtained. Equation (4.25) is then used with different values of a. The initial value of a is equal to the diameter of the cooling channels. a is incrementally increased by one millimetre until Q_c is approximately equal to Q_p.

Although the 1-D cooling analysis is not rigorous, and is only applicable to relatively flat mould cavities, it can help mould designers by providing a rational basis for obtaining quick estimates of the initial design parameters involved in the cooling system.

4.5.2 Two-dimensional cooling analysis

Cooling line design based on infinite number of parallel cooling channels with equal spacing pitch and depth is impractical for plastic parts with complex shapes. The 2-D cooling analysis is a useful method where the complex mould geometry can be approximated by a representative cross-sectional mould boundary.

It is well-known that the temperature of an injection mould changes periodically during the cyclic injection moulding process. In particular, the temperature can be decomposed into a steady cycle-averaged temperature field superimposed on a cycle transient temperature, as shown in Figure 4.18. Since the cyclic transient is usually small, the steady non-uniform temperature field is mainly considered in the 2-D cooling analysis.

The steady non-uniform temperature field in the mould region is governed by the well-known Laplace's equation, with proper cycle-averaged boundary conditions superimposed over the boundary of the mould.

$$\frac{\partial^2 T}{\partial X_1^{\,2}} + \frac{\partial^2 T}{\partial X_2^{\,2}} = 0 \qquad \text{on B} \tag{4.26}$$

The boundary conditions can be formulated as follows.

$$T = \overline{T} \qquad \text{on} \qquad \partial B_1 \tag{4.27a}$$

$$\frac{\partial T}{\partial n} = \frac{\partial \overline{T}}{\partial n} \qquad \text{on} \quad \partial B_1 \qquad (4.27b)$$

$$-K_m \frac{\partial T}{\partial n} = h_a (T - T_a) \qquad \text{on} \quad \partial B_3 \qquad (4.27c)$$

$$T_1 = T_2, \left(\frac{\partial T}{\partial n}\right)_1 = -\left(\frac{\partial T}{\partial n}\right)_2 \qquad \text{on} \quad \partial B_4 \qquad (4.27d)$$

T represents the temperature field. " $\overline{\ }$ " indicates the given value. K_m is the thermal conductivity of the mould material. n is the normal to the surface. h_a is the heat-transfer coefficient between the mould and the ambient environment of temperature T_a. $\partial B_1, \partial B_2, \partial B_3, \partial B_4$ represent the different boundaries and B represents the individual sub-mould region under consideration.

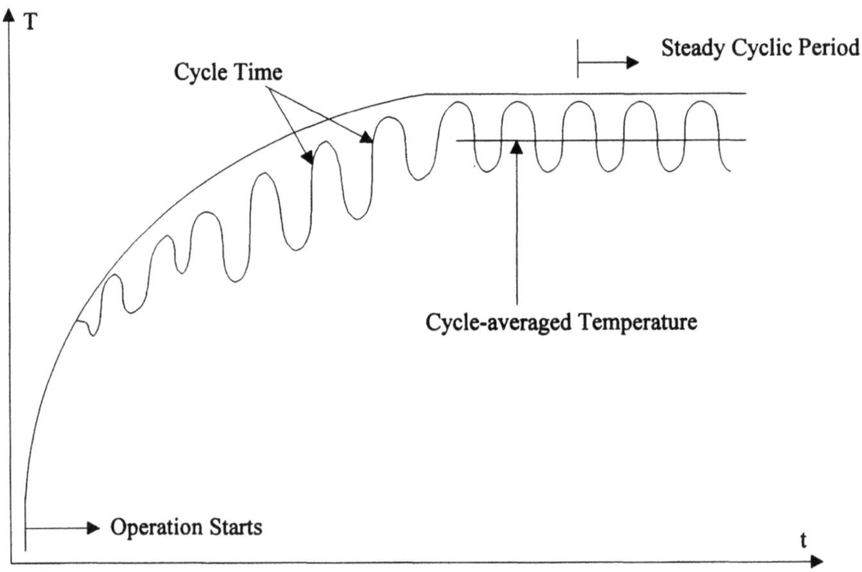

Figure 4.18: Schematic temperature transient at a certain point in the mould.

A mixed boundary condition is prescribed along the cavity wall as equation (4.28). T_o denotes the initial bulk temperature of the polymer melt. The cycle-averaged heat transfer coefficient h_{cv} is used to describe the heat transfer interaction between the mould and the polymer melt.

$$h_{cv}\left(T - T_o\right) = - K_m \frac{\partial T}{\partial n} \tag{4.28}$$

$$h_{cv}\left(t_c\right) = \frac{\int_0^{tc} q(t)dt}{t_c\left(T_o - T_W\right)} \tag{4.29}$$

t_c is the required cooling time. q(t) is the time dependent heat flux transferred from the melt with a uniform initial temperature T_o, to the mould wall held at constant temperature T_w. Since only heat conduction in the gap-wise direction is considered, the heat flux q(t) across the melt-mould interface can be readily obtained using the 1-D transient heat conduction analysis.

The heat transfer coefficient of the coolant h_{col} is used to describe the heat transfer interaction between the cooling channel surfaces and the mould.

$$h_{col} = 0.023\left(\frac{4\rho Q}{\pi \mu D}\right)^{0.8}\left(\frac{C_p \mu}{K_c}\right)^{0.4}\frac{K_c}{D} \tag{4.30}$$

ρ, μ and C_p represent the mass density, dynamic viscosity and specific heat of the coolant respectively. Q is the volumetric flow rate, which should be large enough to maintain a turbulent flow in the cooling channel. D is the diameter of the cooling channel, and K_c is the thermal conductivity of the coolant.

The boundary element method (BEM) has been employed for solving the temperature field along the mould boundaries for a 2-D cross-section perpendicular to the cooling channels, with the design parameters specified by a mould designer. The BEM is ideally suited for this particular problem because (1) the discretization associated with BEM does not extend over the interior of the mould region, which eliminates the need for extensive mesh regeneration when the cooling channels are re-arranged, and (2) discretization over only the boundary decreases the input data preparation time and reduces the dimension of the computational domain, and the memory space and computational cost required.

The discretized boundary integral equation, shown in equation (4.31), based on temperature and its normal derivative, can be derived from equation (4.26) using Green's second identity.

$$T_j R_{ij} = \left(\frac{\partial T}{\partial n}\right)_j L_{ij} \tag{4.31}$$

The numerical coefficients R_{ij} and L_{ij} depend only on the geometry of the boundary. They can be evaluated when the boundary elements are generated. If N is the total number of boundary elements, N linear algebraic equations, together with the boundary conditions for each element, can be used to solve for 2N unknowns, namely the temperature and its normal derivative at each boundary element.

The 2-D cooling analysis can display the cooling results in terms of temperature and its normal derivative (heat flux) along the boundary, together with many other important information such as the maximum and minimum temperatures along the cavity surface, heat flux through each cooling channel, and the cooling time based on the analysis. After examining the results on a graphic screen, mould designers may alter the design by changing a process condition or the cooling channel geometry, and re-run the cooling program for new results.

4.5.3 Three-dimensional cooling analysis

The 3-D cooling analysis is developed to handle precise geometry for more sophisticated problems. Three methods have been considered for the 3-D cooling analysis, *viz.*, the fully transient technique, periodic analysis and cycle-averaged approach. Based on the results obtained for several typical moulds, polymer melt materials and process conditions, it has been found that the fluctuating component of the mould temperature is small compared to the cycle-averaged component. Thus, it has been established that a cycle-averaged approach for the mould region is sufficient to accurately predict the transient and non-uniform melt and mould temperature distributions, and their effects on heat removal by the cooling field. In addition, a cycle-averaged approach is computationally more efficient. In a full numerical approach for transient analysis, the 3-D domain has to be discretized. In a periodic analysis, a Fourier decomposition is needed and detailed computation is necessary for each Fourier component, which requires a discretization of the volume domain to faithfully simulate the 3-D problem. Compared to these two approaches, the cycle-averaged approach does not require any meshing of the 3-D domain as only surfaces are meshed.

The governing equation for the cycle-averaged temperature field in the mould is given in equation (4.32). Ω is the mould region.

$$\frac{\partial^2 T}{\partial x^2}+\frac{\partial^2 T}{\partial y^2}+\frac{\partial^2 T}{\partial z^2}=0 \qquad \text{for} \qquad x\in\Omega \tag{4.32}$$

The boundary conditions are specified in the following sub-sections.

4.5.3.1 Mould cavity surface

The cycle-averaged heat flux on the mould cavity surface is given by equation (4.33).

$$-K_m \frac{\partial T}{\partial n} = \bar{q} \tag{4.33}$$

n is the normal to the surface. K_m is the thermal conductivity of the mould and $\bar{q}$ is the cycle-averaged heat flux.

During cooling, the polymer cools and solidifies due to conductive heat transfer. Plastic parts are usually thin, hence a local 1-D transient analysis is adequate for a 3-D plastic part. The governing equation for the melt temperature is given by equation (4.34).

$$\rho C \frac{\partial T}{\partial t} = \frac{\partial}{\partial S}\left\{K_p \frac{\partial T}{\partial S}\right\} \tag{4.34}$$

t denotes time. T is the melt temperature. ρ, K_p and C are the density, thermal conductivity and the specific heat. S is the local co-ordinate along the part thickness direction.

4.5.3.2 *Channel surface*

On the cooling channel surface, the heat transfer coefficient is defined in equation (4.35).

$$K_m \frac{\partial T}{\partial n} = h(T - T_\infty) \tag{4.35}$$

K_m is the thermal conductivity of the mould. h represents the heat transfer coefficient between the mould and the coolant at a temperature of T_∞. h is calculated using equation (4.36).

$$h = 0.023 \frac{K}{D} R_e^{0.8} P_r^{0.4} \tag{4.36}$$

R_e is the Reynolds number ($= 4Q/\pi D\gamma$), and P_r is the Prandtl number ($= \gamma/\alpha$). Q denotes the volumetric flow rate. D is the diameter of the channel, and γ is the kinematics viscosity of the cooling, with α and K being its thermal diffusivity and conductivity. This equation is valid for $10^4 < R_e < 1.2 \times 10^5$ and $0.7 < P_r < 120$. This range of parameters is adequate for injection moulding applications, as it is a common practice to maintain turbulent flow conditions in the cooling channels.

4.5.3.3 Mould exterior surface

Practice has shown that in most injection moulding applications, heat loss through the exterior surface is very small (typically less than 5%). In such cases, the exterior surface of the mould may be treated as an infinite adiabatic surface. This approximation does not require modelling of the exterior surface of a mould, and leads to substantial savings in computer memory requirements. For a typical model of a mould using the 3-D cooling analysis, the area of the mould exterior surface is about 50-75% of the total surface area.

A more accurate alternative to the above infinite adiabatic exterior formulation is to impose a heat transfer coefficient on the mould exterior surface, which is approximated as a sphere with an equivalent radius to preserve the surface area of the actual exterior surface. In this case, the heat transfer coefficient between mould surfaces and the ambient air is similar to equation (4.27c) discussed in the 2-D cooling analysis for cycle-averaged mould temperature. All the boundary conditions have to be known in order to obtain the cycle-averaged mould surface temperature distribution. The simulation procedure needs to be iterative. The convergence should be based on the consistency of solutions, i.e., to maintain compatibility of the temperature and the heat flux at the mould-melt interface. The following iterative procedure is recommended.

1. To begin the iteration, assume a mould-melt interface temperature distribution.
2. Carry out the part melt analysis to determine the heat flux variation with time along the mould-melt interface. Using the heat flux distribution, determine the cycle-averaged flux values along the mould-melt interface.
3. Using the resulting cycle-averaged flux distribution along the mould-melt interface, perform mould analysis to obtain the temperature distribution along the interface.
4. If the calculated mould-melt interface temperature distribution from the mould analysis is in close agreement with that assumed for the melt analysis, the iterative procedure is stopped. Otherwise, proceed to step (2) and repeat.

4.6 CAM for plastic injection moulds

Computer-aided manufacturing, along with CAD, have been described as "the wave of the future in manufacturing" and the new industrial revolution. CAM can be defined as the use of computer systems to plan, manage and control the operations of a manufacturing plant, through either a direct or indirect computer interface with the production resources. Generally, CAM includes the following fields:

1. Computer monitoring and control application.
2. Numerical control part programming.

3. Computer-aided process planning (CAPP), which is computer-aided preparation of a list of the operation sequence required to process a product.
4. Production scheduling, which is computer-aided preparation of an appropriate schedule for meeting production requirements.
5. Material requirements planning, where computers are used to determine when to order raw material and purchase components, and how many should be ordered to achieve the production schedule.

In this section, the discussion of CAM is concentrated on part programming for numerically controlled machines such as the NC machining centre and the NC wire-cutting machine.

4.6.1 Integrated CAD/CAM system

Experience has shown that the benefits that can be obtained from an integrated CAD/CAM system are much greater than those that can be realised from applying CAD and CAM separately. There is a significant overlap in the databases required for design and manufacturing. Since the data of the geometry of a part has already been created during design using the CAD/CAM system, a part programmer does not have to re-define the geometry of the part. An integrated CAD/CAM system usually has the two following features, *viz.*, (1) the CAM and CAD modules share a common database, and (2) several functions of the CAD modules, such as graphics display and geometry editing can be used within the CAM module. The functions of the CAM module of an integrated CAD/CAM system are as follows:

Parts Display
Both the finished part and the rough stock can be displayed.

Tool Definition
The programmer could either select one of the tools that have been stored in the tool library, or create a new tool by specifying the parameters and dimensions of the new tool (diameter, corner radius, cutter length, etc.).

Tool Path Generation
Once the geometric model of a part and the tools to cut this part have been defined, the tool path can be automatically generated using certain common machining routines. These automatic routines may include profile milling, end milling of pockets and surface contouring.

Tool Path Display and Verification
The graphic display provides a view of both the model and the path in different colours. This permits an easy visual differentiation of the tool path from the outline

of the part. Another feature that aids in the visualisation of the machining sequence is the dynamic simulation of the tool path on the graphic screen.

Automatic NC Output

After the generation of the tool path and the setting up of the machine information file, the software can output the APT program or the actual cutter location file (CL file). Figure 4.19 is a block diagram of the CAM module.

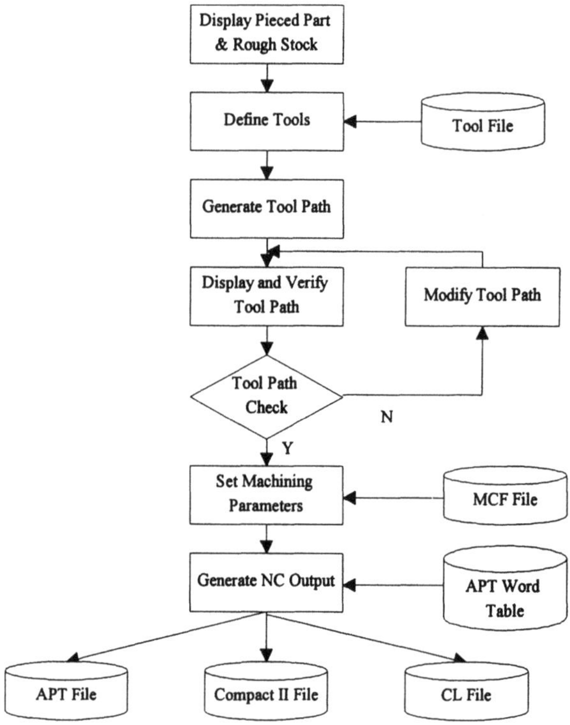

Figure 4.19: The block diagram of the CAM module.

4.6.2 Information transfer from CAD to CAM

Historically, computers have been used independently in the design and manufacturing processes. This has resulted in "separate islands of automation." There are several methods of transferring information from CAD to CAM.

In the early development of CAD and CAM, the conversion between CAD and CAM was completely accomplished manually. For both manual and language-based computer-aided programming, NC programmers would receive a drawing, which is an output of a CAD file, and interpret the drawing to re-define the geometry according to the machining requirements of the part. As can be seen from Figure 4.20, a drawing, which is an output from CAD, has to be entered into a part program using the APT language. The definition of free-form surfaces in CAD software is usually different from those in the CAM software. For example, in many CAD software, a complex surface is defined as a B-spline or a Bezier surface. The programmer has to define the same surface as a Coons surface using APT geometric statements. Thus, a new database for manufacturing has evolved. It is obvious that there is much duplication effort by the design and manufacturing personnel to integrate CAD and CAM. Engineering changes that have been made on a drawing during design have to be painstakingly communicated to all those who need them.

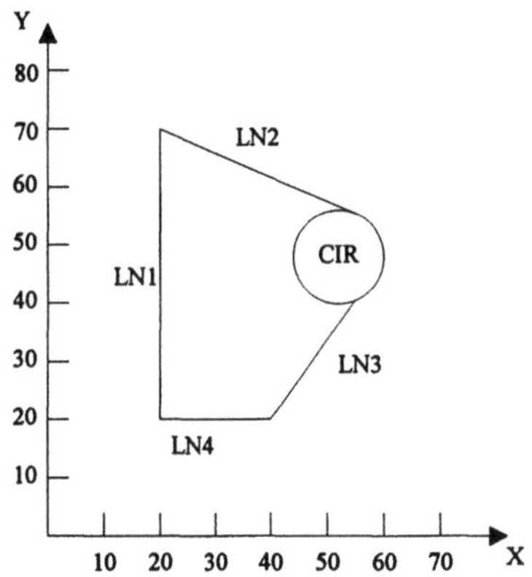

(a)

LN1=LINE/20,20,20,70
LN2=LINE/(POINT/20,70),
ATANGL,75, LN1
LN3=LINE/(POINT/40,70),
ATANGL,45
LN4=LINE/20,20,40,20
CIR=CIRCLE/YSMALL,LN2,
YLARGE, LN3,RADIUS, 10
XYPL=PLANE/0,0,10

(b)

Figure 4.20: Drawing to part program in APT language, (a) drawing (b) geometric statements describing the drawing.

With the development of interactive computer graphics, a more effective method to integrate CAD and CAM has been developed. When the geometric model of a part has been created using the CAD software, the NC programmer can proceed to label the various surfaces and elements of the part that need to be machined, on the graphic terminal. When the labelling has been completed, APT geometry statements can be generated automatically by the interface software. The efficiency of the information transformation in this method is higher than the previous methods. However, there is also a re-definition of the geometric model.

During the mid 1980s, CAPP was proposed and implemented as a more advanced method of linking CAD and CAM. CAPP provides a direct link between CAD and CAM by decomposing a workpiece into a set of machined surfaces (or extracting machined surfaces from a workpiece). A machined surface is a portion of the workpiece that can be processed by a certain metal removal operation. There is a standard machining routine to process each type of machined surface. Commonly defined machined surfaces in a non-rotational part are holes, pockets, slots, steps, 2D-contours, planes and sculptured surfaces.

Nowadays, the principal concern of the CAD/CAM community is the integration of CAM with CAD based on a single database. In an integrated CAD/CAM system, the CAD and CAM modules share a common geometric model. Based on this geometric model, common machining routines that can be automatically programmed are provided. These automatic routines might include profile milling around the outline of a part, end-milling a pocket and contouring a free-formed surface. After the part has been designed, the NC programmer can perform part programming by using one of these common routines.

4.6.3 Tool path generation in 2-D NC

Many components of an injection mould can be machined with wire EDM, or 2- or 3-axis NC milling machines. To generate the tool path for 2-axis NC machining, the user must first establish the geometric model of the part. NC programming of wire EDM has already been discussed in Section 2.8. In this section, tool path generation for CNC milling machines is discussed.

A common procedure to generate the tool path for 2-axis NC machining is as follows. Beginning with the engineering design or drawing, the part is displayed together with the rough stock. The necessary tools are selected. After the machining routines have been selected, tool paths are generated, displayed, verified and modified. Next, the machining parameters are set, and the desired NC output is generated.

The most common machining routines for generating tool paths in 2-axis machining are profiling and pocketing. The profiling routine is used to generate tool path sequences for machining around a series of geometric elements that define the outline of a workpiece. Figure 4.21 shows the profiling of a workpiece. The following parameters must be defined in order to automatically generate the tool path using the profiling routine.

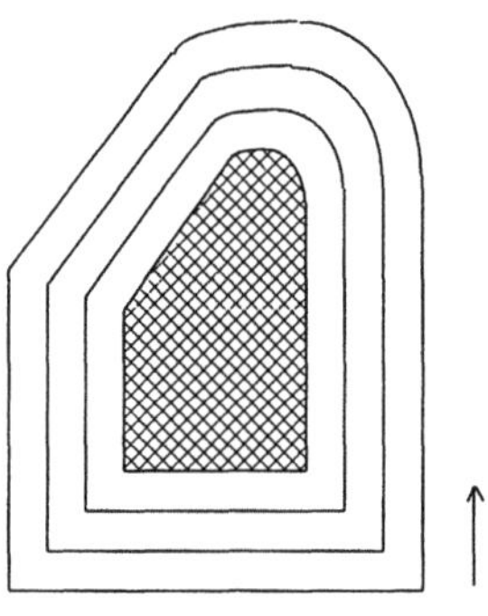

Figure 4.21: Profiling.

1. Position Planes
   Position planes include the cutting plane and the clearance plane. The cutting plane is defined as the level of the tool tip for the finish cut of the profile path. The clearance plane is defined as the level of the tool tip at which the tool can safely move over the entire model without interference.

2. Approach and Retract Type
   The tool can approach or retract the workpiece in three ways, *viz.*, straight, canted and depth.

3. Side and Base Offsets of Rough Cut

4. Tolerances
   Tolerances are used to control how closely the splines and conic arcs are being approximated by the tool path.

Figure 4.22 shows the position plane, approach and retract type of profiling.

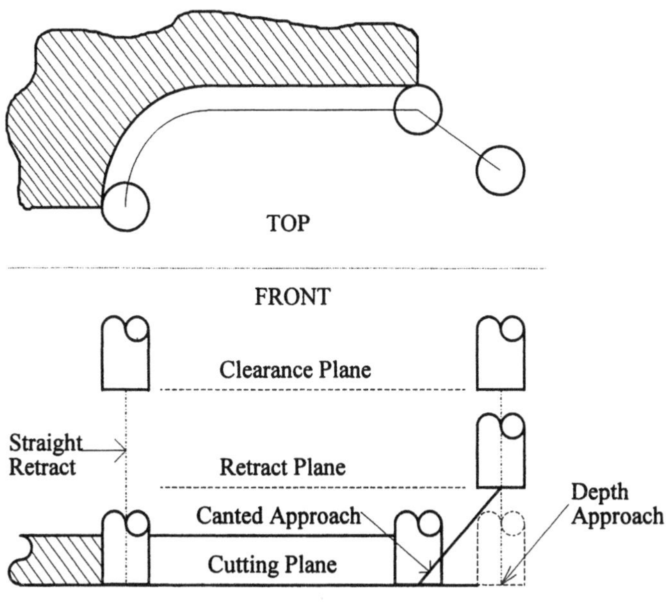

Figure 4.22: Parameter of profiling.

Pocketing consists of helical pocketing and lace pocketing as shown in Figure 4.23. The parameters to accomplish the tool path generation in the pocketing routine are similar to those in the profiling routine.

4.6.4 Manufacturing for 3-D core and cavity

Several cores and cavities of injection moulds have complicated geometric surfaces that cannot be accomplished using 2-axis NC machining. In this case, the 3-axis NC milling machine (or centre) is needed. The machining of cores, which usually have no limiting boundary, is easy as there is an open space that is wide enough for the approach and retract of the cutter. The most common method used to machine a core with complicated geometric surfaces is contouring using the cross-cut or parametric routines as shown in Figure 4.24. To prevent the tool from gouging the part surface, the CAM module should have the function of gouge avoidance, as shown in Figure 4.25. The tool would be automatically lifted at points that would gouge the part. In these cases, there will be material left on the part, which could be machined using a smaller tool.

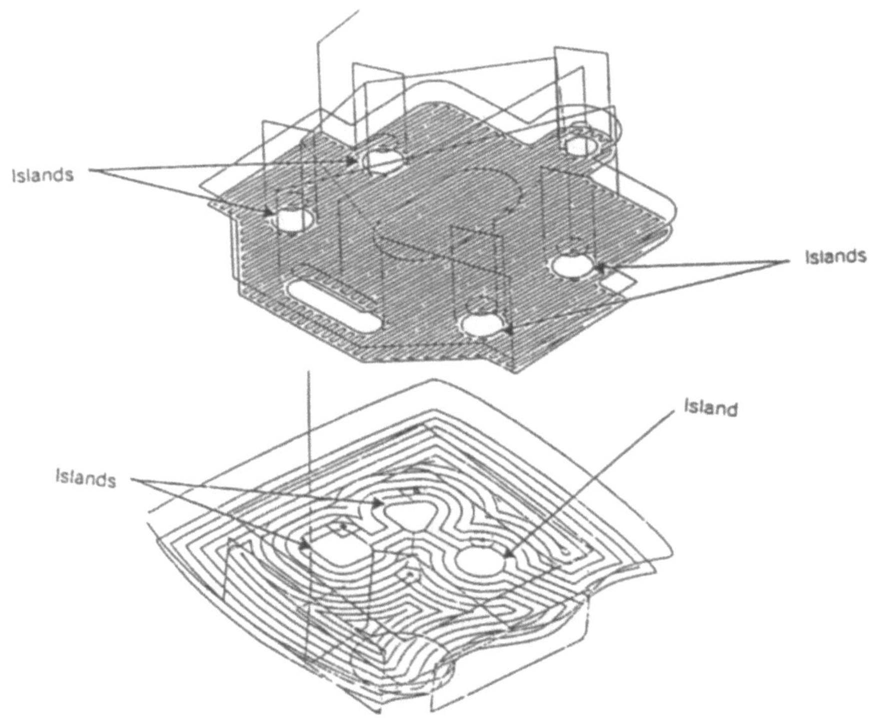

Figure 4.23: Pocketing (a) helical pocketing and (b) lace pocketing.

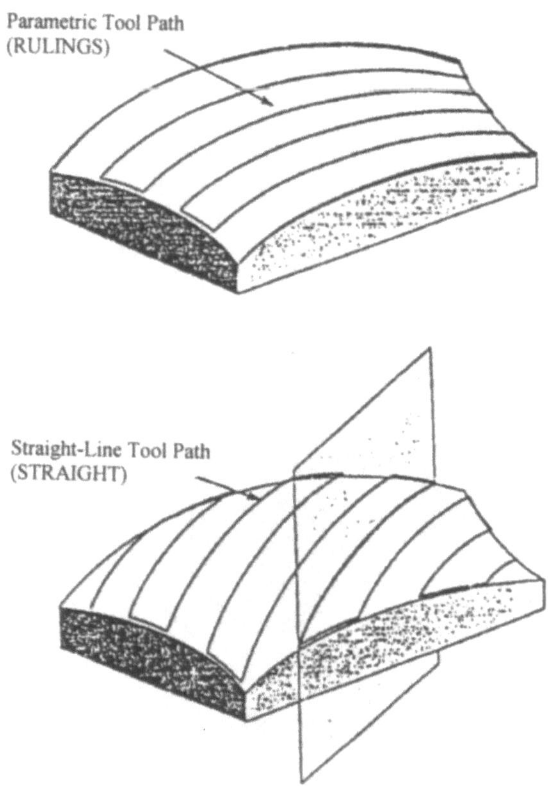

Figure 4.24: Contouring in cross-cut or parametric routines: (a) parametric routine and (b) cross-cut routine.

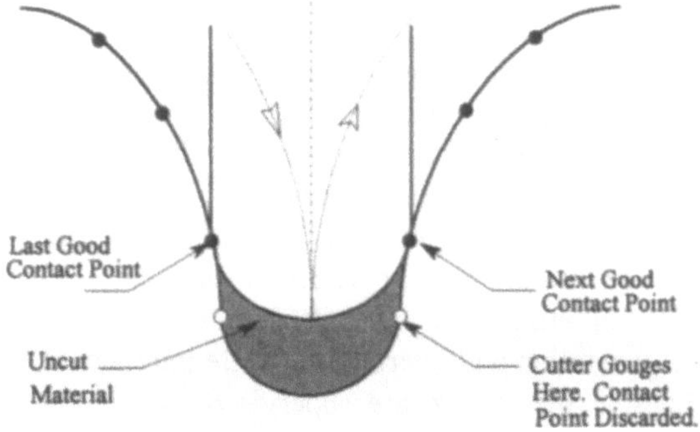

Figure 4.25: Gouge avoidance.

It is more difficult to machine a cavity than a core because of the containment boundary limits of the motion of the cutter. Tool motion is determined by the surface of a part, containment boundaries, and cutting parameters such as machining direction, step-over direction and scallop height. The machining directions consist of cross-cut and parametric directions. In the cross-cut direction, the tool remains tangent to part surfaces that are parallel to the cross-cut planes. When the tool reaches a containment boundary or a part surface boundary, it moves to the next cut (in the step-over direction) and reverses its direction. Figure 4.26 shows the machining of a cavity in the cross-cut direction. In the parametric machining direction, the tool remains tangent to the part surfaces along constant parametric contour lines.

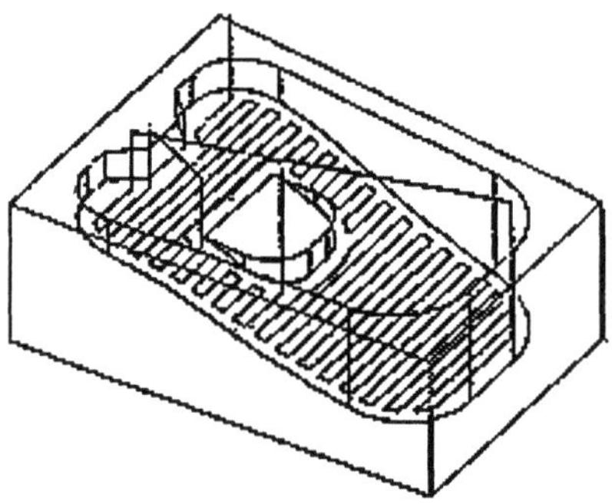

Figure 4.26: Machining of a cavity in cross-cut direction.

4.7 CAD/CAE/CAM system for plastic injection moulding

4.7.1 System configuration

The HUST has developed an integrated CAD/CAE/CAM system for plastic injection moulding. It has been used successfully in many factories in P.R. China. This system is developed on microcomputers, under the OS2 and MS-DOS environments. Figure 4.27 shows the flowchart of the system.

There are ten modules in this integrated system, *viz.*, geometrical input, core and cavity creation, standard mould base selection, mould construction design, 3D-flow simulation, 3D-cool analysis, runner balance calculation, mechanical check of the mould, NC tape generation and interface software interface. As shown in Figure 4.27, all the modules in the system are supervised by a main control program. Mould designers can invoke any module by using the menu displayed on the screen. The data exchange from one module to another is automatic in the form of data files.

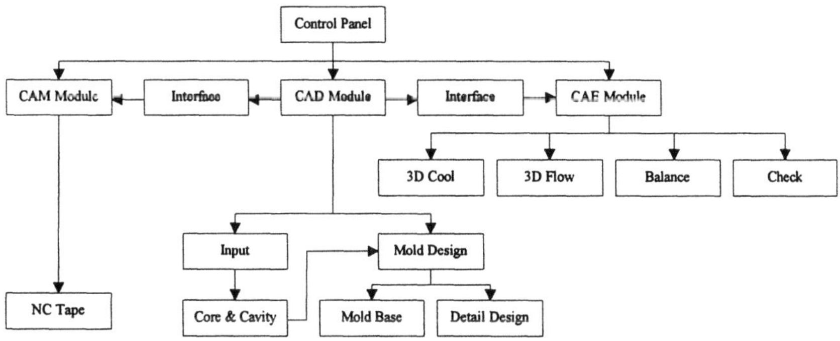

Figure 4.27: The flowchart of the system.

4.7.2 CAD software functions

The task of the CAD software is to transform the drawings of plastic parts into the drawings of the mould parts efficiently, and provide the necessary data for the simulation and NC modules.

Due to the complex shapes of the cavities, surface modelling is used for graphic input. The co-ordinates of the points on the surfaces can be calculated by using the dimensions of the part drawing, and the co-ordinates of the points that have been input earlier. While the part drawing is being input, the dimensions of the part are transformed into the dimensions of the cavity and core interactively. The data of the cavity and core are recorded for both mould design and simulation.

A database for standard mould bases has been set up. It contains standard mould bases, such as the D.M.E of USA, Futoba of Japan and E.M.C of China.

The software provides a group of functions for mould design, runner system design, editing of the construction of cavity and core, and arrangement of ejection pins and cooling lines. The design results include all the mould part drawings and mould assembly drawing.

4.7.3 CAE software functions

The CAE software includes mechanical check for mould plates, runner balance analysis, flow and cooling simulation. With the help of these functions, mould design can be improved and possible defects in the plastic parts such as warpage tendency, short shots and improper location of weld-lines, can be addressed before mould making.

A 2-D finite element method is used to analyse the strength and rigidity of a mould on a typical mould cross-section.

To ensure a uniform quality of the plastic parts produced with multi-cavities moulds, each cavity must be filled simultaneously with the same pressure and temperature. This requires a balanced runner system. In this software, this can be achieved by correcting the dimensions of the runners and gates designed by mould designers at the preliminary design stage.

Flow simulation is one of the most basic and useful analyses in the CAE software. This simulation makes use of a generalised Hele-Shaw model and a fixed-mesh representation of the surfaces of a cavity.

A control volume method is adopted to obtain the governing equations. These equations are based on a hybrid of finite element and finite difference methods by solving the flow and the heat transfer equations for an inelastic non-Newtonian fluid under the non-isothermal conditions. With this software, the user can simulate the filling of the cavity to obtain an optimum flow pattern by changing the number of gates and their locations.

Cooling simulation includes 3-D steady and transient cooling analyses. The 3-D steady cooling analysis uses the boundary element method. Formulae for modelling the cavity surfaces, cooling lines and exterior surfaces have been established and proved to be reliable and effective.

All the results of the CAE software can be displayed dynamically with colourful graphics to aid the users in improving their design to obtain optimum forming parameters.

4.7.4 CAM software functions

The cutter location files can be created based on the geometry of the cavity and the core, which are modelled in the previous CAD stage. NC codes can be generated using the post-processors for both NC wire cutting and NC milling machine tools.

Bibliography

Gosztyla J 1985 *CAD/CAM for Plastic Part and Mould Design.* Injection Moulding Handbook

Wang K K et al 1975 - 1988 *Research Report (1) - (14).* Cornell University.

Li D Q 1990 *CAD of Injection Moulding.* HUST Publishing House, China (In Chinese)

Himase K et al 1992 CAE of Mould Cooling in Injection Moulding Using a 3-D Numerical Simulation. *Journal of Engineering for Industry*, 114(5): 213-221

Wang K K and Wang V W 1987 Computer Aided Mould Design and Manufacturing. *Injection Moulding Fundamentals*, Marcel Dekker Inc., New York

Chapter 5

FEM applications in near net-shape operations

J.C. Xia, S.J. Li and Y.X. Ding

5.1 Introduction

The use of advanced numerical methods to design forming processes has found wide applications in the metal forming industry. Specifically, the Finite Element Method (FEM) is being used extensively to generate information on issues ranging from metal flow prediction to forming load calculations. The use of FEM to obtain detailed analytical solutions to highly complicated problems has provided a rational process design methodology for metal forming processes. Typical process design activities include pre-form design, die design and forming load prediction. The traditional process design approach in the forming industry is based on empirical rules, guidelines and actual production trials. Various metal forming processes can be simulated and designed accurately using the FEM methodologies to obtain a detailed description of the metal flow, load requirements and variations of important process variables such as the accumulated strain, strain rate and temperature. The common trial-and-error approach in the industry can therefore be avoided.

The potential of advancing the FEM-based methodology in the metal forming industry has not been fully realised. The reason is primarily due to the time required to develop the finite element models, especially mesh generation. Although mesh generation has been widely researched, it continues to be a major bottleneck due to the fundamental difficulties that are encountered during the numerical simulation of the forming processes. To simulate the metal flow that occurs during forming operations by FEM, it is necessary to update the mesh continuously as the deformation progresses. As a result, the finite element mesh becomes gradually distorted. This affects the predicting capability of the elements. The analysis is stopped altogether when one of the elements becomes highly distorted. It is then necessary to re-mesh the arbitrarily shaped intermediate configuration of the workpiece, and transfer the field variables onto the new mesh using accurate interpolation algorithms. Typically, the re-meshing of the intermediate shapes is a time-consuming process, and this makes the simulation exercise an arduous task. During the simulation of certain complicated forming processes, a number of re-meshings may be required. Several approaches are currently being pursued in the research community to alleviate the problems associated with FEM.

In this chapter, new algorithms are presented for the automatic triangular mesh generation of arbitrarily multi-connected planar regions, generations of new meshes and transfers of field variables when mesh distortions occur, and the generation of isograms in the FEM simulation of the metal forming processes. They can be extended to 3-D regions. Experiments and rigid-plastic FEM simulations of the radial extrusion and upsetting-backward extrusion processes have been conducted to compare the simulated results with the actual experimental data.

Besides the FEM simulation of near net-shape processes, the FEM methodology can also be applied in the optimisation of the die design process, and the analysis of the hydraulic presses for near net-shape operations. In the design and manufacture of dies, FEM can be used to calculate parameters such as the stress, and strain distributions, and the optimised construction and dimensions of the die components. This would greatly reduce the lead-time required to design and manufacture the dies. In this research, mathematical models have been developed for analysing the pre-stressing force and working stresses for the design of combined dies for cold extrusion, backward extrusion, forward extrusion and the forging processes. For the hydraulic presses, FEM is used to analyse the fluid transient in the pipelines of presses for a better design of these machines.

5.2 FEM applications and developments in near net-shape operations

5.2.1 New algorithms for automatic triangular mesh generation

With the rapid development of computer technology, complex engineering problems can be handled using the FEM. However, researchers are hindered by the enormous amount of input required. Automatic generation of FEM mesh is an issue in the development of FEM.

During the recent years, research on the algorithms for automatic generation of FEM mesh is in great demand. A number of algorithms with varying degrees of automation have been developed. In the early 1970s, Cavendish [1] described an algorithm for semi-automatic triangulation of arbitrary, multi-connected planar domains. Compared to the other types of elements, triangular elements have overwhelming advantages in handling domains with very irregular boundaries and openings. Presently, most research work on automatic mesh generation has been focused on triangular meshes.

A major characteristic of the 2-D Delaunay triangulation is that it satisfies the optimal maximum-minimum angle criterion [2], i.e., the sum of the minimum angles of all the triangles is maximised. Therefore, Delaunay triangulation is extremely useful for the automatic generation of triangular meshes. Watson [3] proposed an algorithm for computing the n-dimensional Delaunay triangulation, and its dual Voronoi diagram of a given arbitrary set of random points in an n-dimensional Euclidean space ($n \geq 2$). Cavendish [4] implemented the automatic generation of 3-D FEM meshes using Watson's algorithm. However, his algorithm requires a solid modeller to supply a list of points to be triangulated, and tetrahedrons that are poorly shaped cannot be avoided easily. Frey [5] proposed a new node placement algorithm using an improved version of Watson's algorithm. This method could control the mesh size based on a "spacing function" provided by the users. However, the algorithm cannot solve the triangulation problem of arbitrary multi-connected regions. The algorithm proposed below is not only useful in the automatic mesh generation of arbitrary multi-connected planar regions, it can also be extended to 3-D regions.

5.2.1.1 Discretization of the boundary curves

The geometric boundary of a planar domain is usually represented as a disjoint union of several closed loops which are composed of straight line segments, circles, arcs and spline curves. These boundary curves must be discretized into smaller straight line segments during automatic mesh generation. The algorithm that is being presented in this section requires the boundary curves to be discretized before the new nodes are inserted. The approach to discretizing the boundary curves is the same as that for discretizing the boundary straight line segments.

The straight line segment is used as an example of a general case. Let L be the length of the line segment, and I be the number of intervals that this line segment should be divided. The number I can be calculated as follows:

$$I \geq \int_0^L \frac{dx}{S_x} = \int_0^L \frac{dx}{\frac{x}{L}(S_2 - S_1) + S_1} \tag{5.1}$$

S_1 and S_2 are the spacing values at the two end nodes. The variable x is the distance between node 1 and the integral point. S_x is the spacing function that is assumed to vary linearly on this line segment. The geometric interpretation of equation (5.1) is that the node density function along the edge is a reciprocal of the spacing function.

When the intervals I is known, the positions of (I - 1) new nodes can be determined. Equation (5.2) is applied iteratively to calculate the ultimate positions of the new nodes. The iteration starts with a uniform distribution of (I - 1) new nodes, and the number of iterations is set to be one or twice the number of new nodes.

$$Z_i = \frac{(Z_{i+1} + Z_{i-1})}{2} + \frac{(Z_{i+1} - Z_{i-1})}{2}\left[\frac{S_{m_i} - S_{m_{i+1}}}{S_{m_i} + S_{m_{i+1}}}\right] \tag{5.2}$$

$i = 1, 2, ..., I-1$. Z_i is the co-ordinates of the ith new node. Z_0 and Z_I are the co-ordinates of the two end nodes respectively. m_i is the mid-point of the ith interval which spacing value is S_{m_i}. The first term on the right hand side of equation (5.2) is the mid-point of the interval between nodes i-1 and i+1. The second term can be interpreted as a correction due to the gradient of the spacing function. This term makes the new node approaches the end node, which has a smaller spacing value. The approach is larger when the gradient is larger.

5.2.1.2 *Initial Delaunay triangulation of arbitrary polygons*

After the boundary curves have been discretized, the geometric boundaries form a polygon consisting entirely of straight line segments. Next, the arbitrary polygons are triangulated.

In practical applications such as CAD/CAM, a complete planar graphics usually consists of an outer loop and a few inner loops. In general, the nodes on the exterior loop are entered in the counter-clockwise order and the nodes on inner loops are entered in the clockwise order.

There is a unique way of generating a valid triangulation for an arbitrary polygon. The triangulation algorithm for an arbitrary polygon proposed here is essentially a preparation for FEM mesh generation. It makes full use of an important property of Delaunay triangulation, i.e., the circum-circle of any Delaunay triangle does not contain the nodes in its interior [2], [3], [4]. Although the computational efficiency of this algorithm is not high, the resulting triangulation satisfies the Delaunay properties. It is useful for FEM mesh generation. Figure 5.1 shows the triangulation of a polygon that consists of two loops and seven straight line segments.

Line segment 12 is first considered to search for the node P_k, among those nodes which are visible from edge 12, where the dihedral angle of the lines P_k1 and P_k2 is maximum. In Figure 5.1, only nodes 5 and 6 are visible from edge 12, and $\angle 152 > \angle 162$. Therefore, $P_k = 5$. Given a point P_i, any other point P_j is said to be visible from P_i if and only if segment $\overline{P_i P_j}$ lies completely inside the domain or on the boundary, and does not intersect any of the edges of the current polygon. The three nodes 1, 2 and 5 must form a Delaunay triangle. After triangle 125 has been removed from the current polygon, edge 15 is next considered and a Delaunay triangle 156 is formed. The procedure is repeated until only one triangle remains.

Note that the thick straight line segments in Figure 5.1 show the geometric boundary of the graphical region, and the thin line segments are the edges of the Delaunay triangles.

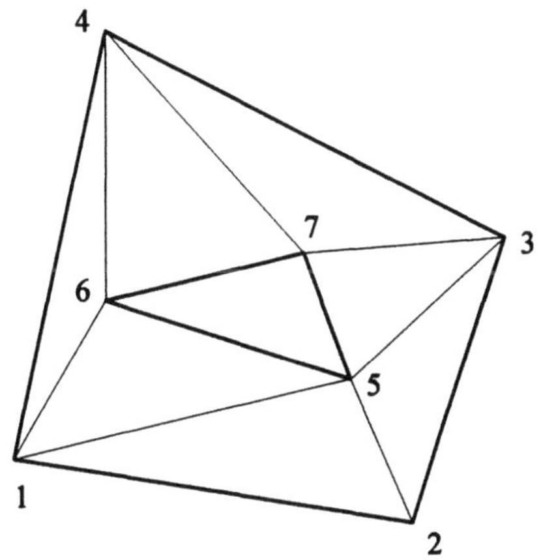

Figure 5.1: Delaunay triangulation of arbitrary polygons.

5.2.1.3 Algorithm for new nodes placement

In general, the resulting Delaunay triangulation above provides a polygonal representation of the domain. However, it is not useful for FEM analysis. For a complete mesh generation, several boundary and interior nodes must be added into the graphical region to improve the quality of the mesh of the current Delaunay triangulation.

A. New boundary node placement

The spacing function on the edges connecting the adjacent boundary nodes is assumed to vary linearly. In general, the length of the straight line segment on the boundary is larger than the spacing values at its end nodes. Thus, a few new nodes must be added. The positions of these new nodes can be calculated using the approach described above to discretize the boundary curves.

The approach to insert new nodes on the straight line segments of the geometric boundaries is the same as the approach to insert new interior nodes, which will be explained in the following section. However, it should be noted that some triangles, which are invalid when new nodes are inserted on the geometric boundaries, should be removed from the new Delaunay triangulation. Figure 5.2 shows a Delaunay triangulation of a polygon consisting of two loops when the boundary node placement process ends.

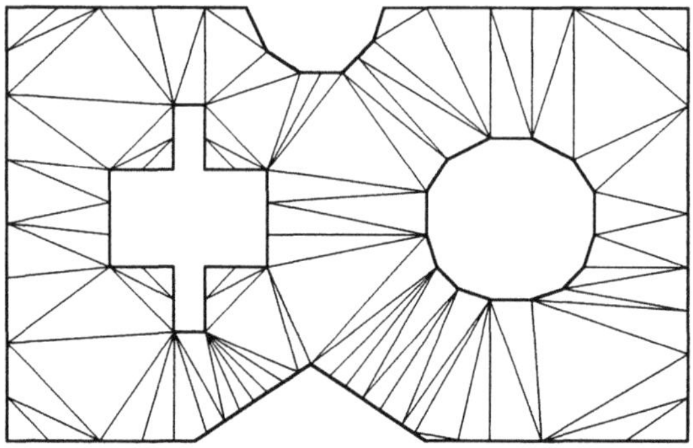

Figure 5.2: Delaunay triangulation of boundary node placement.

B. Automatic local refinement

An important characteristic of the algorithm proposed in this chapter is its automatic local refinement capability which is not available in many of the algorithms. If a user knows in advance that the mesh density should be larger (or smaller), or is not satisfied with the resulting triangulation, he can locally refine the mesh by placing some appropriate control nodes in those regions to be refined, and assigning the corresponding spacing values at these control nodes. The procedure for control node placement is the same as that of the interior nodes. However, it should be noted that the spacing values of the control nodes are assigned by the user while that of the interior nodes are calculated using an interpolation approach. Figure 5.3 shows a Delaunay triangulation after inserting a control node into Figure 5.2. The black point is the control node for local refinement.

C. New interior node placement

From Figures 5.2 and 5.3, it can be seen that the mesh quality is not suitable for FEM analysis after the Delaunay triangulation of boundary node placement and control node placement have been completed. To obtain a desired mesh quality, sufficient new interior nodes must be added one at a time into the current Delaunay triangulation.

For most FEM analysis processes, an equilateral triangle is the most desirable. The circum-centre of a triangle is usually the ideal position to insert a new node. The generation of a new node at the circum-centre of a triangle requires little additional computational effort. However, local gradation of the mesh is somewhat inhibited because the circum-centre of a triangle is equidistant from the three nodes defining the triangle, and more distant from any other nodes. An illegal node outside the problem domain may result when a prospective node is placed at the circum-centre of an obtuse triangle because the circum-centre of the obtuse triangle lies outside itself. Therefore, all obtuse triangles are not considered in the new node placement algorithm. This algorithm only considers acute triangles when inserting the new nodes.

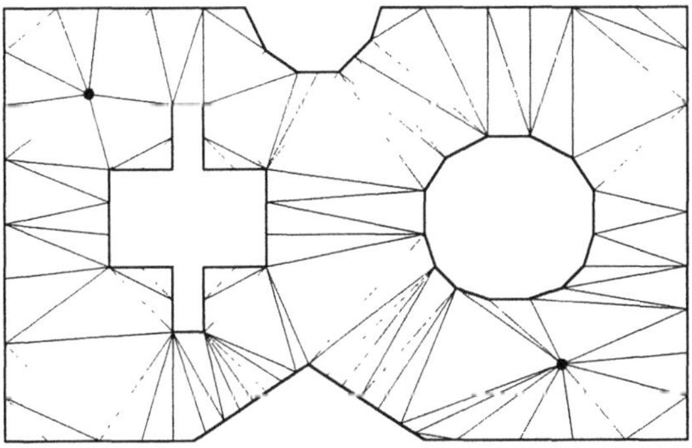

Figure 5.3: Delaunay triangulation of control node placement.

First, the acute triangle which area is maximum is taken as the ideal triangle to insert a node. A weighted average method is adopted to calculate the position of the new node to be inserted. Let $\rho = 2r/R$ represent how closely an arbitrary triangle approaches this ideal triangle. r is the in-radius of the triangle while R is the circum-radius. The value of ρ ranges from 0 to 1. Let the in-centre of the ideal triangle be I, and the circum-centre be C. The position of the new node P can be determined as follows:

$$(x_P, y_P) = (1 - \rho)(x_I, y_I) + \rho(x_C, y_C) \tag{5.3}$$

(x_I, y_I) and (x_C, y_C) are the in-centre and circum-centre of the selected triangle respectively. (x_P, y_P) is the co-ordinate of the new node to be determined.

In this chapter, the spacing function is assumed to be linearly distributed within each triangle. The spacing value at the new node can be obtained based on the concept of the area co-ordinates in FEM.

$$S_P = \sum_{i=1}^{3} L_i S_i \tag{5.4}$$

S_P is the spacing value at the new node P. S_i is the spacing value at the ith node of the selected triangle, and L_i is the ith area co-ordinate value of the new node P.

If the distance from the new node to the nearest of the three nodes of the selected triangle is smaller than the spacing value at this new node, the node is rejected. Another new triangle in terms of the area values of all the acute triangles is selected. Otherwise, this new node is inserted into the current Delaunay triangulation. This insertion procedure is stated as follows.

From Figures 5.4 and 5.5, it can be seen that most of the obtuse triangles would have been removed when the new nodes are added one at a time. The weighted average approach in equation (5.3) is more feasible than the approach of inserting new nodes only at the circum-centres of the selected triangles.

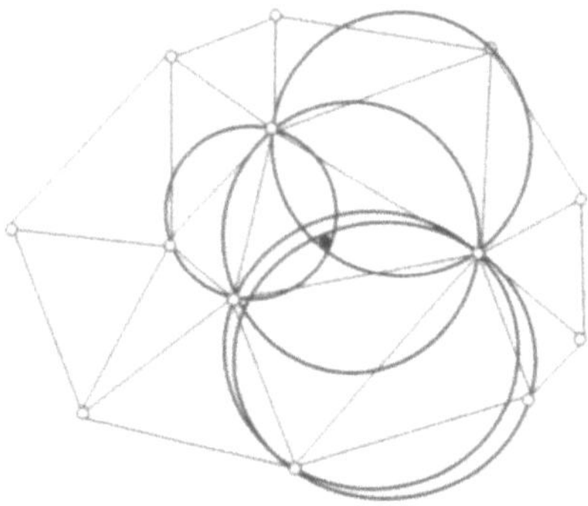

Figure 5.4: Formation of "Insertion Polygon".

Let the selected triangle be T_o. The procedures are extended gradually outwards from the selected triangle T_o. Triangles which circum-circles contain the new node to be inserted are found to form a list $\{T_k\}$. According to the optimal properties of Delaunay triangulation, the union of all these triangles in the list $\{T_k\}$ forms an "Insertion Polygon" containing the new node [3]. It can be shown that no previously inserted node is contained in the interior of this insertion polygon, as shown in Figure 5.4. A new triangulation of the insertion polygon can be formed when each boundary node of the insertion polygon is connected to the new node by a straight line. When this local Delaunay triangulation in the insertion polygon is combined with the triangles outside the polygon, a new Delaunay triangulation which includes the new node is formed, as shown in Figure 5.5.

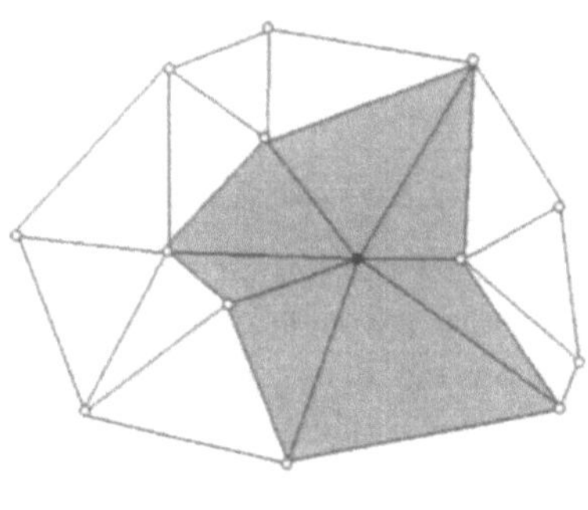

Figure 5.5: Delaunay triangulation of new node.

The interior node insertion process stops when no acute triangle can be found in which acceptable new nodes can be inserted. The number of interior nodes generated is largely controlled by the spacing function. The resulting triangulation is suitable for FEM analysis. However, a further improvement to the mesh quality can be realised using an appropriate smoothing algorithm on the triangulation. Figure 5.6 shows the resulting mesh after post-processing the triangulation using the Laplacian smoothing technique [6]. The Laplacian smoothing technique improves the triangulation so that elements are approximately equilateral triangles. Each interior node is moved successively to the centre of its connected neighbours. More than twenty passes are required for the entire set of interior nodes.

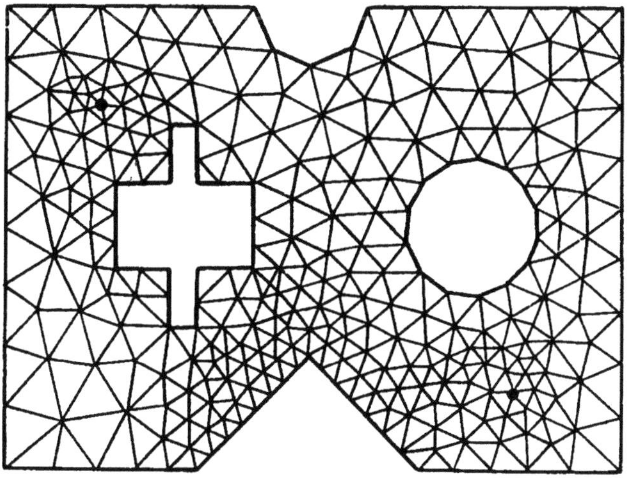

Figure 5.6: The application of the Laplacian smoothing technique.

5.2.2 New algorithms for mesh rezoning in FEM simulation

In practical metal forming processes, deformation is usually very large. It is common to encounter 60% to 70% deformation. Moreover, the relative motion between the die surface and the deforming material is also large. Such large deformations and displacements are liable to cause serious mesh distortions, which may result in a drop of the computing accuracy and a break off in the simulation procedure.

To completely and accurately simulate large deformation processes, it is necessary to establish some rules on mesh distortion. When a mesh distorts to a certain degree, the simulation procedure must be suspended to re-define a new mesh system.

The mesh rezoning process consists of two procedures. The first procedure is the assignment of a new system to the workpiece. Generation of the new mesh system is essentially the same as the initial mesh generation process. It can be performed using a commercially available mesh generator or using procedures described in Section 5.2.1. The zone of the new meshes should be entirely the same as that of the old meshes although the number of elements and nodes of the new meshes are not equal to that of the old meshes. The second procedure is the transfer of field variables (for example, effective strain, temperature, etc.) which depend on the deformation history, from the old mesh system to the new one using the interpolation method. The interpolation should be ideally performed in such a way that the distribution of the field variables in the old mesh system is the same as that in the new one.

During the derivation of the rigid-plastic FEM model, a four-node quadrilateral iso-parametric element is introduced to establish the shape function. The determinant of the Jacobian matrix is used as a criterion to recognise the degree of deformation when mesh rezoning is necessary.

The area differential is defined as $dA = |J| d\zeta d\eta$ in iso-parametric transformation, where $|J|$ is the well-known Jacobian determinant. The iso-parametric transformation requires each elemental inner angle to be less than 180°, i.e., each element must be strictly convex. In this algorithm, the formulation $|J| \leq \delta$ is used as a suitable criterion for mesh rezoning, which value changes in the various processes.

The least square fit method is usually used to interpolate the field variables from the old system to the new mesh system [7], [8]. Among the various interpolation methods available, the area-weighted average method is the most convenient. It can provide sufficient computing accuracy for mesh rezoning during metal forming simulation [7]. This method often fails at the boundary of the workpiece when the field variables have large gradients and the boundary nodes do not have sufficient number of surrounding elements. For four-node quadrilateral iso-parametric elements, the effective strain is only known at the centroid of each element. The effective strains at all the old nodes must be known during the transfer of the field variables. These effective strains can be obtained using the area-weighted average method.

A node N surrounded by several adjacent elements shown in Figure 5.7 is considered. The effective strain value at node N can be obtained as follows:

$$\bar{\varepsilon}_N = \frac{\sum_{i=1}^{4} \bar{\varepsilon}_i A_{iN}}{\sum_{i=1}^{4} A_{iN}} \tag{5.5}$$

$\bar{\varepsilon}_N$ is the effective strain at node N, and $\bar{\varepsilon}_i$ is the effective strain at the centroid of the ith element surrounding node N. A_{iN} is the area contribution of the ith element to node N. The summation in equation (5.5) is performed on all the elements that surround node N.

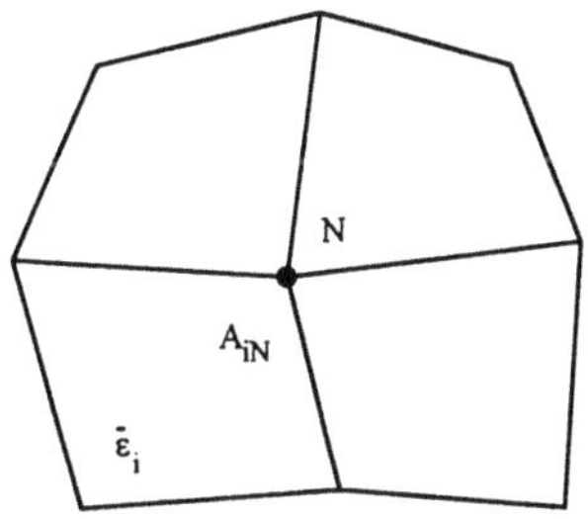

Figure 5.7: Node surrounded by the adjacent elements.

When the effective strains at all the old nodes have been obtained, the effective strains at all the new nodes can be calculated in terms of the new node co-ordinates which are known, using equation (5.6).

$$\bar{\varepsilon} = \frac{1}{4}\sum_{i=1}^{4}(1+\zeta_i\zeta)(1+\eta_i\eta)\bar{\varepsilon}_i \tag{5.6}$$

(ζ_i, η_i) are the local co-ordinates of the ith old nodes. $\overline{\varepsilon_i}$ is the effective strain at the old node. (ζ, η) is the local co-ordinates of the new node in the old mesh system. $\bar{\varepsilon}$ is the effective strain at the new node. In equation (5.6), (ζ, η) is unknown. The effective strains at all new nodes can be calculated after (ζ, η) has been determined.

To calculate the value of the local co-ordinates (ζ, η), the elemental order of the new node in the old mesh system must be known. There are three methods to determine the old elemental order of a new node. The area method is the best among the intersection-point method and the angle method. In general, most new nodes are located in the workpiece. Therefore, they must be within certain old elements. In Figure 5.8, the new node is P and the quadrilateral IJKL is in the old mesh system. If the sum of the four triangular areas that consist of node P and the four nodes of the old element IJKL is equal to the area of the old element IJKL, the node P is within the old element IJKL. Otherwise, the new node P is outside the old element IJKL.

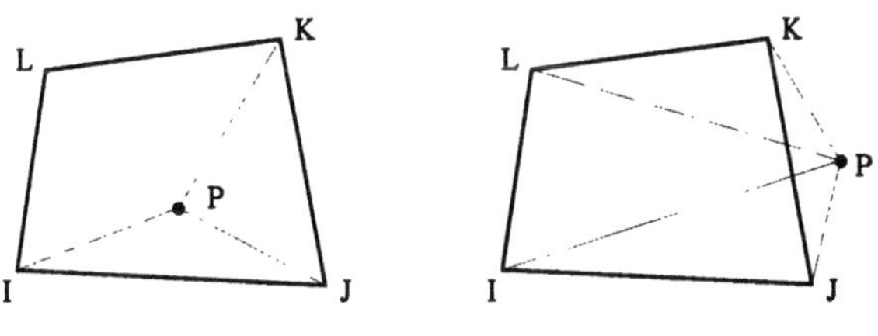

Figure 5.8: The relation of new node and old element.

After the order of the old element, where the new node is located, has been determined, the local co-ordinates (ζ, η) of the new node in the old element can be calculated by the following equation.

$$\left.\begin{aligned} x &= \frac{1}{4}\sum_{i=1}^{4}(1+\zeta_i\zeta)(1+\eta_i\eta)x_i \\ y &= \frac{1}{4}\sum_{i=1}^{4}(1+\zeta_i\zeta)(1+\eta_i\eta)y_i \end{aligned}\right\} \tag{5.7}$$

The overall co-ordinate (x, y) is known in the new mesh system. (x_i, y_i) is the overall co-ordinates of the ith node of the old element. (ζ_i, η_i) is the local co-ordinate of the ith node of the old element. The elimination method is adopted to solve equation (5.7). Equation (5.7) can be simplified into equation (5.8).

$$\left.\begin{aligned} a_{11}\zeta\eta + a_{12}\zeta + a_{13}\eta &= b_1 \\ a_{21}\zeta\eta + a_{22}\zeta + a_{23}\eta &= b_2 \end{aligned}\right\} \tag{5.8}$$

where

$$\begin{cases} a_{11} = \sum_{i=1}^{4}\zeta_i\eta_i x_i, & a_{21} = \sum_{i=1}^{4}\zeta_i\eta_i y_i \\ a_{12} = \sum_{i=1}^{4}\zeta_i x_i, & a_{22} = \sum_{i=1}^{4}\zeta_i y_i \\ a_{13} = \sum_{i=1}^{4}\zeta_i x_i, & a_{23} = \sum_{i=1}^{4}\zeta_i y_i \\ b_1 = 4x - \sum_{i=1}^{4}x_i, & b_2 = 4y - \sum_{i=1}^{4}y_i \end{cases}$$

In iso-parametric transformation, the definition region of the local co-ordinates (ζ, η) is [-1, 1] and the equation has two solutions. However, only one solution is legal and the other solution is illegal.

To reduce the computing time for solving equation (5.8), the location of the new node is examined to determine if it lies in the circumscribed rectangle of the old element. If the new node lies outside the old element, equation (5.8) need not be solved.

When the old element where the new node is located, and the local co-ordinates (ζ, η) of the new node have been obtained, the field variables at all the new nodes can be calculated using equation (5.6).

The nodal values of the new mesh system should be transferred to the elemental values for the simulation to proceed. For four-node quadrilateral iso-parametric elements, the average value of the four nodal values is taken as the elemental value.

$$\bar{\varepsilon} = \frac{1}{4}\sum_{i=1}^{4}\bar{\varepsilon}_i \tag{5.9}$$

The transfer of the effective stresses from the old to the new mesh system is similar to the transfer of the effective strains. Since the nodal values of the temperature and velocity fields are known from FEM analysis, the transfer is simpler.

5.2.3 Algorithms for generating isogram in FEM

Automatic generation of isogram is an important part of an FEM analysis system. With the development of FEM, research on the algorithm for generating isogram in FEM is gradually evolving. Isograms have an extensive application in engineering analysis. Many algorithms for generating isogram have been developed. However, until now, there is no general-purpose algorithms for the automatic generation of isogram that is suitable for FEM analysis. Gray [9] adopted the concept of standard iso-parametric interpolation function. The computation accuracy of his method is high, but the efficiency is low. His method is suitable for analyses that have high-order elements. It is not suitable for rigid-plastic analysis systems that only have low-order triangular or four-node quadrilateral iso-parametric elements. A fast algorithm based on the linear interpolation principle is proposed here. It uses the lists of element and node orders, and stress-strain field values from the FEM analysis results. It has an acceptable computing accuracy.

In general, an isogram has the following important properties:

1. The contour lines are smooth and continuous.
2. For an assigned height value, the number of contour lines may be more than one.
3. Since the defining regions are bounded, the contour lines may be closed or intersected with the boundary.
4. The contour lines do not intersect each other.

Based on the proposed linear assumption, the nodal field values should be known before the generation of the isogram. However, only the elemental field values of stresses and strains are available from the rigid-plastic FEM analysis. Therefore, field values must be transferred from the elements to the nodes. The area-weighted average method described in Section 5.2.2 is the most suitable transfer algorithm.

5.2.3.1 Determination of contour points

Theoretically, a 3-D smooth curve surface z = f(x, y) can be fitted to the known nodal field values. If the curve surface is cut by a plane which height is Z_c, the projection of all the intersecting lines forms the isogram which field value is Z_c. However, this fitting method is too time-consuming.

Based on the linear assumption, the stress-strain fields are linearly distributed within all the elements. For each element lateral, the field values of the two end nodes determine whether there are any contour points on it. If the height value Z_c is between the two nodal field values, there exists a contour point within the element lateral. Otherwise, there are no contour points within it.

In general, an isogram can consist of contour lines with different height values. Since the procedure for the generation of contour lines with different height values is completely the same, only the algorithm for generating an isogram of a given height value Z_c is discussed here.

Let two points i and j be the nodes of an element lateral. The following formulation is used to determine if the lateral ij intersects the contour lines.

$$f = (Z_i - Z_c)(Z_j - Z_c) \tag{5.10}$$

Generally, the height values of nodes i and j are assumed to be unequal. If $f < 0$, the lateral ij intersects the contour lines at only one point. If $f > 0$, they do not intersect each other. If $Z_c = Z_i$ (or $Z_c = Z_j$), $f = 0$, the contour line passes through node i (or node j). These are illustrated in Figure 5.9. There may be some difficulties in following the track of the contour lines at this time. A remedy for this problem, which is very useful in actual applications, is to slightly perturb the field value of node i (or node j). The perturbation is only slightly large enough to make the contour line deviate from node i (or node j). The graphs do not distort and the contour line passes through the node i (or node j) in the graphic display. After all contour lines of height value Z_c have been generated, the field value of node i (or node j) can be returned to its original value.

It is obvious that there is only one contour point on each lateral of an element. It can be proved that only one contour line passes through a triangular element, and not more than two contour lines pass through an quadrilateral element, i.e., the contour points always occur in pairs.

A list of contour points can be established by determining whether there are contour points on the laterals of all the elements. This list, which structure is a chain-list, records the information of the elements on which there are contour points. This data domain has two data terms that record the number of contour points and the element order. The point domain has a point that records the next element on which there are contour points.

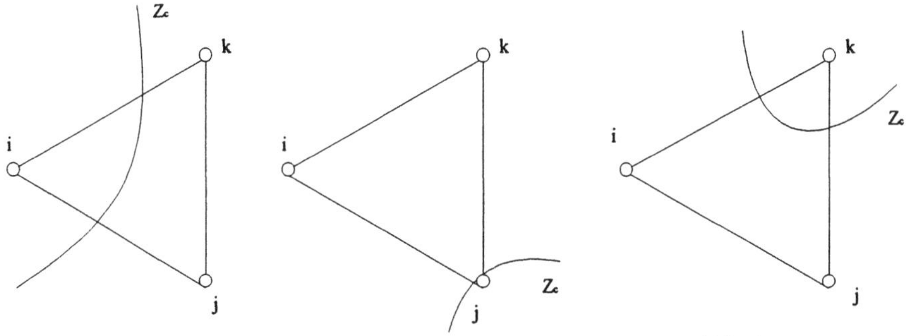

Figure 5.9: The relationship between contour line and element lateral.

5.2.3.2 *Following the track of contour lines*

Based on the linear distribution of the field values within all the elements, if there is a contour point on the lateral ij, its co-ordinates can be calculated as follows:

$$\left.\begin{aligned} x_c &= x_i + \frac{Z_c - Z_i}{Z_j - Z_i}(x_j - x_i) \\ y_c &= y_i + \frac{Z_c - Z_i}{Z_j - Z_i}(y_j - y_i) \end{aligned}\right\} \qquad (5.11)$$

(x_c, y_c) is the co-ordinate of the contour point to be obtained. Z_c is the given height value. (x_i, y_i) and (x_j, y_j) are the known co-ordinates of nodes i and j respectively. Z_i and Z_j are the corresponding field values.

For an assigned height value, there may be more than one contour line in the defining domain. The determination of the alignment of the contour lines and the recording of the data concerned is the key in the automatic generation of an isogram. To reduce the computer memory required and increase the computing speed, contour lines are generated one at a time. A dynamic array is allocated to record the co-ordinates of all the contour points of the contour line. When a contour line has been generated, the dynamic array is cleared to record the information of the next contour line. Let FE be the first element in the chain list were there are contour points. FN is the first contour point in FE. In Figure 5.10, FE = 2 and FN = S_1. Since the contour points always occur in pairs, there must be another contour point on FE. Let this point be S_2. The co-ordinates of the two contour points S_1 and S_2 are recorded in the dynamic array. Since there are only two contour points on the second element, and they have been tracked, the number of contour points in the corresponding node of the chain-list is set to zero. At the same time, this node is deleted from the chain-list before generating the next contour line.

Since the contour point S_2 is not on the boundary of the mesh, this contour line must have an outlet. The second element shares a lateral and the contour point S_2 with the third element. The third element has another contour point S_3, besides the contour point S_2. S_3 is recorded in the dynamic array. In this way, tracking of contour line is followed repeatedly until the element FE and the node FN are encountered again and the contour line has closed. The contour line is open and the procedure reaches the boundary of the mesh if it has no outlet. When this happens, the record of the chain-list is reversed and the contour line is tracked in the opposite direction from the element FE until this contour line goes into a dead end. For example, the other contour line in Figure 5.10 is open.

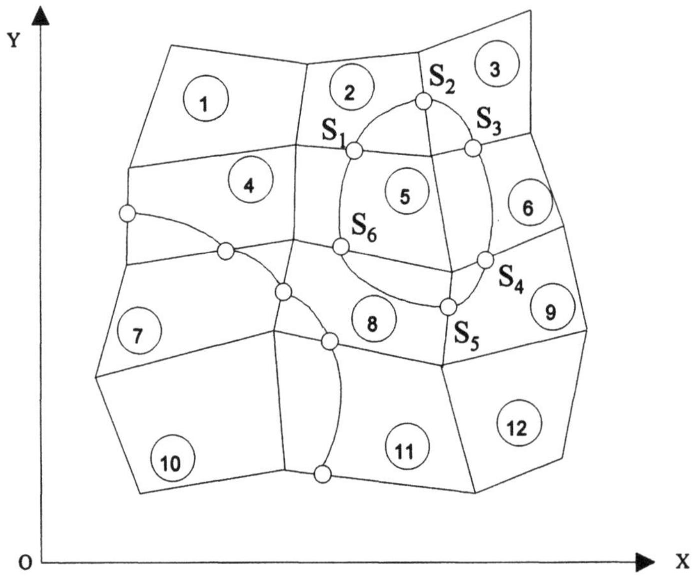

Figure 5.10: The alignment and tracking of contour lines.

Based on the linear interpolation principle, there are only two contour points in each triangular element. If the quadrilateral element IJKL has four contour points, as shown in Figure 5.11, a wrong trend of the contour line will result in an illegal isogram. The correct trend of the contour lines is determined based on linear interpolation. P_1, P_2, P_3 and P_4 are the four contour points in Figure 5.11. P is the intersection point of the two diagonals. Its field value Z_P is the average of the field values of the four nodes.

$$(Z_I - Z_c)(Z_P - Z_c) \leq 0 \tag{5.12}$$

If the above inequality is true, points P_1 and P_4 are on the same contour line. Otherwise, points P_1 and P_2 are on the same contour line. The contour line has an outlet when it passes through one of the four contour points.

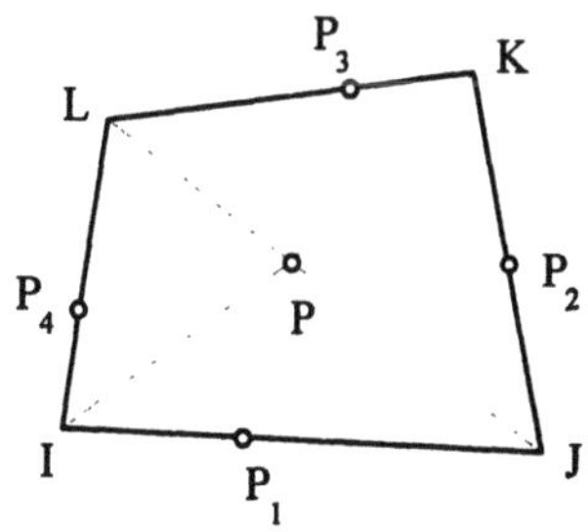

Figure 5.11: Tracking of the contour lines in the quadrilateral elements.

5.2.3.3 Connection of contour lines

When the information of all the contour points of a contour line has been recorded in the chain-list, all the contour points can be connected using a smoothing algorithm. Among the curve-fitting methods available, the parabolic spline fitting method ensures the continuity of the first derivatives of the curve at all the given points. This method requires less computing time and memory.

For a given height value Z_c, only one contour line is generated in the determination, tracking and connection of contour points process. If there are more than one contour line for a given height value Z_c, the last two steps are repeated until the list of contour points is cleared. An isogram with different heights can be easily generated by repeating the above procedure.

5.2.4 Calculation of rigid regions using rigid-plastic FEM

During the rigid-plastic FEM simulation, a Lagrange multiplier or a penalty term is usually introduced into the function to overcome the difficulty in the stress calculation due to the volumetric constraint $\dot{\varepsilon}_V$.

In general, the deforming body is divided into m elements inter-connected at n nodes. The surface in element $V^{(k)}$, where the external force is applied is $S_F^{(k)}$. The die-workpiece interface is $S_C^{(k)}$. The variable **u** is assigned to represent the elemental velocity array. The shape function matrix is **N**, and the strain rate matrix is **B**. **F** is the traction matrix over surface S_F. σ_{ij} is distribution of stress, and f_i is frictional stress. The body is composed of a rigid-plastic material that obeys the von Mises yield criterion and its associated flow rule. Body forces are assumed to be absent. By introducing a Cartesian co-ordinate system $\{n, t_1, t_2\}^T$ on the interface, the modified penalty function matrix ($\pi^{(k)}$) can be expressed as follows:

$$\pi^{(k)} = \int_{V^{(k)}} E(\mathbf{Bu})dV + \frac{\lambda_k}{2V_k}\left[\int_{V^{(k)}} \mathbf{C}^T\mathbf{Bu}dV\right]^2 - \int_{S_F^{(k)}} \mathbf{F}^T\mathbf{Nu}dS - \int_{S_C^{(k)}} f_i t_i^T\mathbf{Nu}dS + \int_{S_C^{(k)}} hdS \tag{5.13}$$

λ_k is the coefficient of the penalty term (e.g., $\lambda_k = 10^6$). V_k stands for the element volume. $C = \{\delta_{ij}\}$ is the matrix representation of Kronecker delta, which is $\{1, 1, 0\}^T$ for plane-strain deformation, and $\{1, 1, 1, 0\}^T$ for axi-symmetric deformation. For most plastic-deformation metal forming processes, $S_F^{(k)}$ is a force-free surface, i.e., $\mathbf{F} = 0$.

The work function E (function of $\dot{\varepsilon}_{ij}$) is defined as follows:

$$\frac{\partial E}{\partial \dot{\varepsilon}_{ij}} = \frac{2\bar{\sigma}}{3\dot{\bar{\varepsilon}}}\dot{\varepsilon}_{ij} = \frac{2Y}{3\dot{\bar{\varepsilon}}}\dot{\varepsilon}_{ij} \quad (i, j = 1, 2, 3) \tag{5.14}$$

$Y = Y(\bar{\varepsilon})$ is the effective stress of the rigid-plastic deforming material.

The physical interpretations of the individual terms on the right hand side of equation (5.13) can be made based on the distortion, dilatation, input energy, frictional work, and frictional work increment that have resulted from the change of the relative velocities due to the change of velocity fields.

$$h = \int_0^{V_t} \frac{\partial f_j}{\partial v_{ti}} v_{tj} dv_{ti} \tag{5.15}$$

h has a unique determinate solution. f is the frictional stress.

$$\frac{\partial h}{\partial v_{ti}} = \frac{\partial f_j}{\partial v_{ti}} v_{ij} \quad (i = 1, 2)$$

It can be proved that the true solution, among all the admissible velocity fields that satisfy the compatibility equations and the velocity boundary conditions, optimises the minimum value of equation (5.13).

The principles described above apply only when the entire body is being deformed plastically. However, in practical metal-forming processes, there are situations when the rigid zones exist and unloading occurs. These rigid zones are characterised by an extremely small effective strain rate compared to that of the deforming body. If these rigid zones are included within the control volume, the value of the first derivatives of equation (5.13) cannot be uniquely determined as the effective stress is undefined when the effective strain rate approaches zero. Since these derivatives cannot be determined, iterative solutions cannot be obtained, except for cases where the rigid zones have been eliminated from the actual calculation. However, the rigid zone elimination technique is virtually inapplicable in most problems due to the difficulties involved in determining the rigid-plastic boundaries.

In an effort to eliminate the problem of the indeterminate functional derivatives, Kobayashi [10] assumed that the stress and strain rate relationship in the constitutive equations can be approximated by the following equation.

$$\sigma_{ij}' = \frac{2}{3}\frac{\bar{\sigma}_0}{\dot{\bar{\varepsilon}}_0}\dot{\varepsilon}_{ij} \qquad \text{for } \dot{\bar{\varepsilon}} \le \dot{\bar{\varepsilon}}_0 \tag{5.16}$$

$\bar{\sigma}_0 = \bar{\sigma}(\bar{\varepsilon}, \dot{\bar{\varepsilon}}_0)$. $\dot{\bar{\varepsilon}}_0$ takes an assigned limiting value for all the elements lying inside the rigid zones. $\bar{\sigma}$ is the effective stress. This stress and strain rate relationship is equivalent to the assumption of a Newtonian fluid-like material behaviour for the near-rigid regions. Elements with effective strain rates less than the assigned value $\dot{\bar{\varepsilon}}_0$ (e.g., 10^{-4}), are called near-rigid elements. These elements are not eliminated from the control volume during actual computation. The use of the limiting value $\dot{\bar{\varepsilon}}_0$ ensures a unique value of the deviatoric stress.

There exists an inflexion point in the constitutive equation where the element stiffness matrix is disconnected. For the rigid-perfectly plastic and work-hardening rigid-plastic material, the stress fields are disconnected. Kobayashi presented a few examples that have several solutions for the stress and strain fields. To overcome this difficulty, the constitutive equation for the rigid-plastic deforming body is modified. A new mathematical model is proposed as follows:

$$\sigma_{ij}' = \frac{2\bar{\sigma}}{3\dot{\bar{\varepsilon}}}\dot{\varepsilon}_{ij} = \frac{2Y}{3\dot{\bar{\varepsilon}}}\dot{\varepsilon}_{ij} \qquad (i, j = 1, 2, 3) \tag{5.17}$$

$\bar{\sigma} = Y = \varphi(\dot{\bar{\varepsilon}})Y_0$. $Y_0 = Y_0(\dot{\bar{\varepsilon}})$ is the former expression of yield stress, which is based on the assumption of the unique curve. $\varphi(\dot{\bar{\varepsilon}}) = \frac{2}{\pi}\tan^{-1}\left(\frac{\dot{\bar{\varepsilon}}}{a}\right)$, in which $a = 10^{-6}$

Figure 5.12 shows the stress and strain rate relationship curves for a rigid-plastic material. The effective stress of the rigid-perfectly plastic material is constant for different effective strain-rate values. It is shown as the line AB in Figure 5.12. When the effective strain-rate approaches zero, the effective stress cannot be uniquely determined as the value of the first derivative is undefined. The geometric meaning of equation (5.16) is shown as line OC in Figure 5.12. The new mathematical model in equation (5.17) is shown as the curve OD. It can be seen that there is a small difference between equations (5.16) and (5.17) in the plastic deforming zones. The new mathematical model continues into the entire deforming body such that it is not necessary to distinguish the rigid zones from the plastic zones. This new mathematical model simplifies the program structure considerably and reduces the computing time.

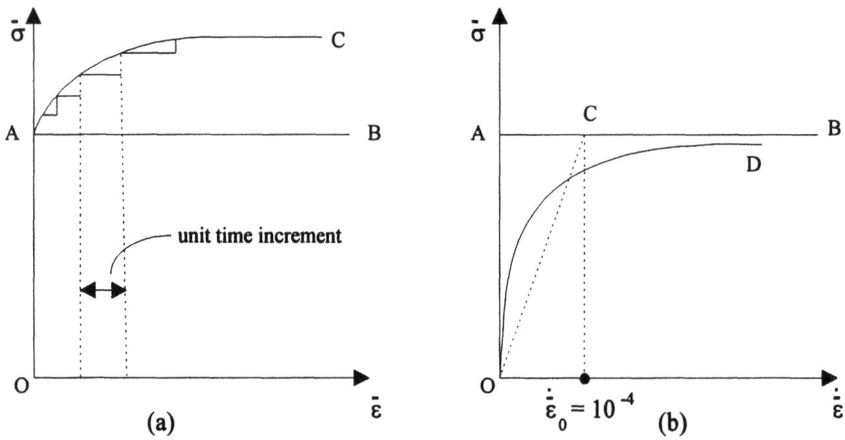

Figure 5.12: The constitutive equation for deforming-perfectly material.

5.3 FEM applications in massive (bulk) metal forming processes

5.3.1 Simulation of rigid-plastic finite element of radial extrusion process

Closed die forging with branches is an advanced die forging process. It can save materials considerably and improve the forging qualities. For forgings with branches, deformations of the billet in the cavity are mainly upsetting, forward and backward extrusion, and radial extrusion. These three deformations are widely applied in the industry. However, there is comparatively less research on radial extrusion. An experiment of a rigid-plastic FEM simulation on the radial extrusion of equal diameter three-way joint has been conducted in this research.

Figure 5.13 shows the mesh deformation of a multi-way die forging of a three-way joint. The experiment is carried out in a multi-way die forging set. The three-way joint is a typical forging part with branches which main deformation is radial extrusion.

The so-called equal diameter Tee joint is a part where the diameter of the main body is entirely the same as one of its branches. The experiment material is lead and the dimensions of the billets are 20 mm × 70 mm. The billets are divided into two groups. The first group of billets consists of six-layer lead plates that have equal thickness welded together. The second group consists of billets with six equal height closed lines etched along their profile. A 2000 kN hydraulic press was used in the experiment. For the first group of specimens, the deformation in each layer mesh was found to be approximately the same, and the thickness of the layers are almost constant. For the second group of specimens, the equal height closed lines remained almost unchanged. Therefore, the radial deformation of the equal diameter Tee joint can be considered to be a plane-strain deformation.

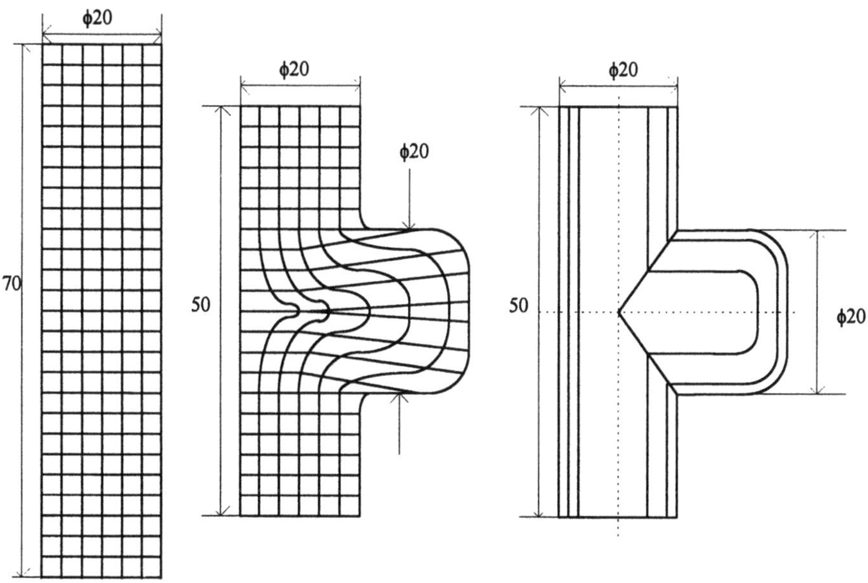

Figure 5.13: Radial extrusion of equal diameter tee joint, (a) original billet, (b) deformation of middle layer, and (c) deformation of closed lines.

Simulation of the deformation of the radial extrusion of the equal diameter three-way joint was performed to compare the simulation results with the experimental data. Figures 5.14(a), (b) and (c) show respectively the mesh deformation, velocity field and effective strain field when the punch stroke ends. It can be seen from Figure 5.14 that the deformation of the radial extrusion concentrates mainly at the passing portion between the major cavity and the side branch. Deformation degrees of the two ends of the body are very small. When the metal flows through the passing portion, its direction changes. At the same time, its velocity increases gradually and exceeds twice the velocity in the main body. The metal flow in radial extrusion is typically non-steady. The distributions of the stress and strain in the workpiece are extremely non-uniform as the punch stroke increases. When the punch stroke ends, the deformation in the part is mainly located within the plastic zones. The deformations around the passing portions are very large, and the maximum value of the effective strain is 1.60. When the metal flows through the side branches, the effective strain decreases gradually, and the minimum value is approximately 0.1. The effective strain of the main body is approximately 0.01, and its minimum value is approximately 0.3×10^{-3}.

Figure 5.15 shows the extrusion force displacement curves of the results obtained from the rigid-plastic FEM analysis and the actual experiment. From Figure 5.15, it can be seen that the two curves approach each other and have the same tendency of change. Since the frictional coefficient in the rigid-plastic FEM simulation is larger than the actual value in the experiment, the simulation value of the extrusion force is slightly higher than the experimental results.

5.3.2 Rigid-plastic finite element simulation of the upsetting-backward extrusion process

The upsetting-backward extrusion process of the cup head of a cup-rod part is very common in the automobile industry. As the deformation mechanism of this process is more complex than the common forward and backward extrusion processes, little research has been done on analysing the upset-backward processes using numerical techniques such as the rigid-plastic FEM. With the rapid development of the precision die-forging processes, there is an urgent need to find the deformation mechanisms of complex metal forming processes such as the upsetting-backward extrusion. As an efficient numerical technique, the rigid-plastic FEM simulation can provide the detailed information on the deformation for engineering designers. Thus, it has been applied extensively in the metal forming problems.

The upsetting-backward extrusion process of the cup head of a cup-rod part in the universal coupling spline shaft of an automobile transmission shaft is used as an actual example in this section. The stress-strain equation for lead is $Y = 1.13 + 3.35\bar{\varepsilon}^{0.5}$ MPa, which has been established using the least square fit method based on experimental data [11]. The specification of the lead rod in this experiment is 52 mm × 75 mm based on the head volume and rod diameter of the spline shaft.

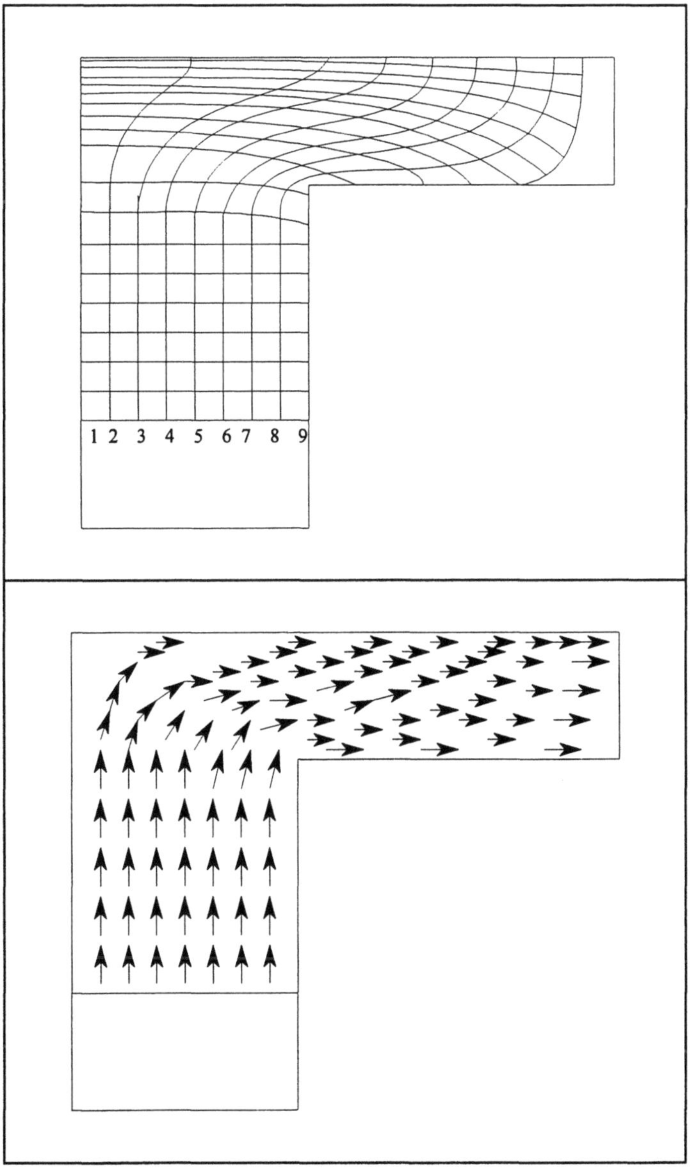

Figure 5.14: Simulation results of equal diameter tee joint, (a) mesh deformation, (b) velocity field.

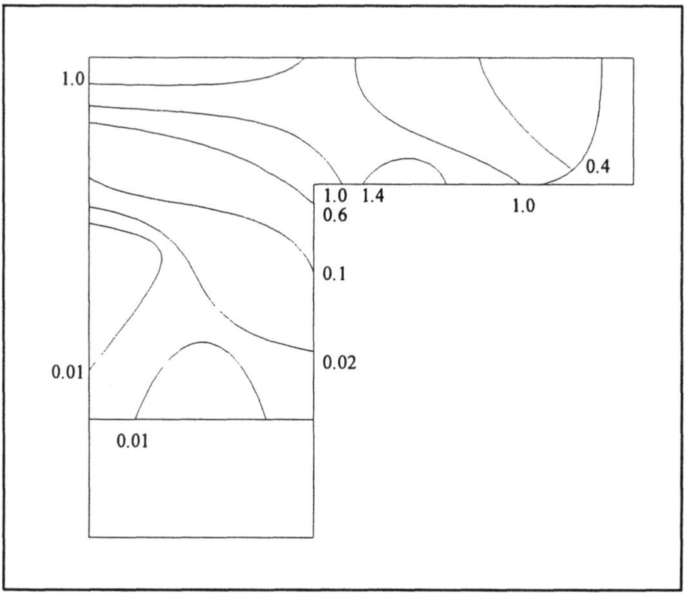

Figure 5.14: (c) isogram of effective strain field (continued).

Based on the rigid-plastic FEM method developed by Lee and Kobayashi [12], the upsetting-backward extrusion process has been analysed. Distribution values of strain and stress have been calculated and compared with the results from the rigid-plastic FEM analysis.

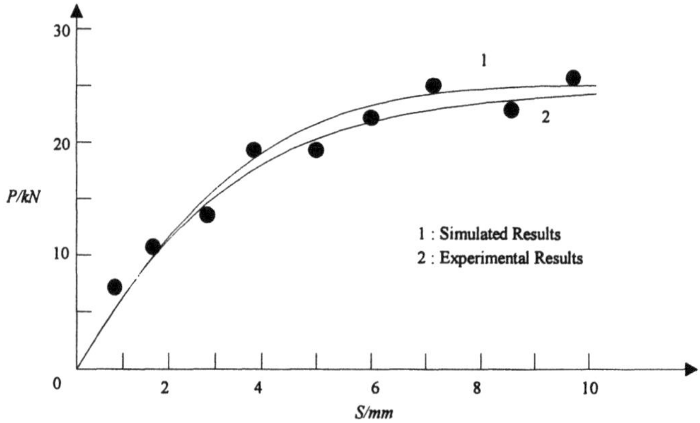

Figure 5.15: Extrusion force displacement curves.

Since the billet diameter is equal to the nominal diameter of the rod of the final part, the rod can be considered as a large rigid zone during metal forming. Deformation occurs only in the cup head. The upset-backward extrusion process of the cup head is analysed using the rigid-plastic FEM simulation. Based on the results obtained, the upset-backward extrusion process can be divided into three stages, *viz.*, (1) the upsetting stage, (2) the combined-upsetting backward extrusion stage, and (3) the backward extrusion stage.

Figure 5.16(a) shows schematically the mesh deformation at the end of the third stage, i.e., the backward extrusion stage, when the mesh system has been rezoned three times. From this diagram, it can be seen that the deformation around the circular angle of the punch is very large, and mesh distortions are extremely serious. This is because this portion is a major passageway, where the metal flows inwards into the circumferential portion, from the lower end of the punch.

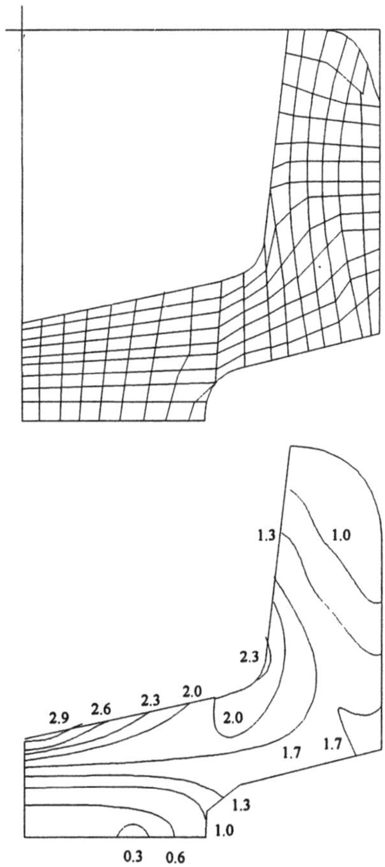

Figure 5.16: Deformation when the punch stroke has ended, (a) mesh deformation, and (b) isogram of effective strain fields.

Figure 5.16(b) shows the isogram of the effective strain fields when the punch stroke has ended. It can be seen that the deformation becomes greater with the increase of the punch stroke, and there is greater metal flow under the punch. The effective strain at the outside of the circular angle of the punch is small during the upsetting stage. However, it increases rapidly during the combined upsetting-backward extrusion stage, and increases drastically during the backward stage.

Figure 5.17 shows the extrusion force-displacement curves during the upsetting-backward process of the cup head of a cup-rod part. From the curves, it can be seen that the simulation results agree with the experimental data, although the former is less than the latter. The difference between them increases with the increase of the punch stroke and decreases when the punch stroke reaches the end. The main reason is that the frictional factor is constant, and the frictional resistance depends only on the work-hardening degrees and the changes of the relative velocities during the rigid-plastic FEM simulation. However, the actual frictional conditions worsen rapidly with the increase of the deformation in metal forming, and the actual frictional resistance increases rapidly.

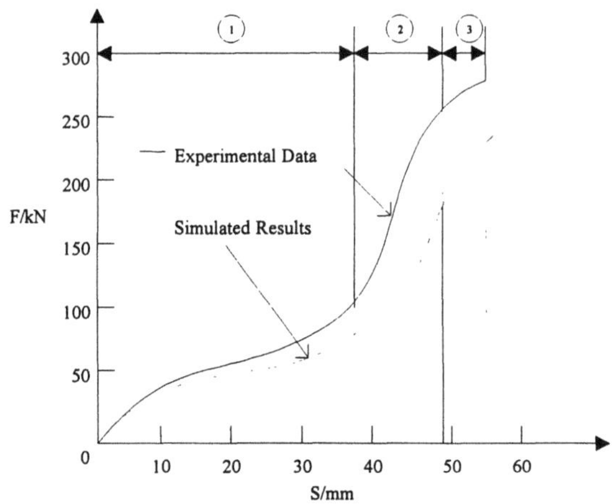

Figure 5.17: Extrusion force-displacement curves, (1) upsetting stage, (2) combined upsetting-backward extrusion stage, and (3) backward extrusion stage.

The extrusion force-displacement curves have two inflexion points when the upsetting stage ends. This is one of the deformation characteristics of the upsetting-backward extrusion process of the cup head. During the combined upsetting-backward extrusion stage, the metal in the cup, which has contacted the die surface, is forced to flow into the cup head along the circular angle of the punch. The frictional resistance increases rapidly and a serious strain-hardening case exists. The extrusion force increases rapidly at the same time. During the backward extrusion stage, the change of the frictional resistance is small. When the cup head has been filled up, the deformation degree decreases.

5.3.3 Simulation of forward extrusion and upsetting-backward extrusion process

5.3.3.1 Simulation conditions

The hot forging process can be regarded as an isothermal process by neglecting the effect of temperature changes during the process. The shapes and dimensions shown in Figure 5.18 are used in the simulation.

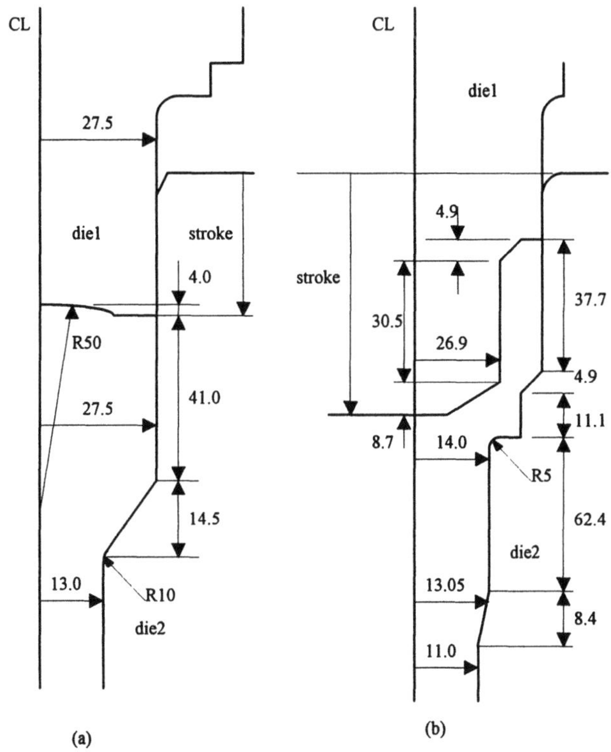

Figure 5.18: Dimensions of dies.

In the numerical computations, the following conditions were used: (1) flow stress: $K\bar{\varepsilon}^{0.23}\dot{\bar{\varepsilon}}^{0.026}$ (MPa), (2) friction factor: 0.5, (3) ram speed: 3.6 mm/sec, (4) billet size: 50.0 mm in diameter, 80.0 mm in height, (5) number of nodes: 351 (for initial meshes only), and (6) number of elements: 312 (for initial meshes only).

The mesh systems are remeshed whenever the elements have been distorted severely. The usual number of nodes is 350-360 after remeshing. Remeshing was done with the commercial pre/post-processing package SDRC's I-DEAS. To compare the flow of the numerical model with that of the experiments, the layer boundaries were traced after remeshing.

5.3.3.2 Result analysis

Figure 5.19 shows the deformation of the billet at various strokes. A small unfilled cavity remains around the outer corner of the cup in the final forging. This is because of the relatively lower sensitivity rate of the material property used in the numerical model. It should be noted that during finished forging, the top edge of the cup is flat, while it has declined in the experiments. The strains are high near the inner surface and outer neck of the cup and the rod. The flow patterns obtained through the isothermal simulation are illustrated in Figure 5.20. The patterns are similar to those of the experiments, except at the outer surface of the cup where the experiments showed an upward flow. Since the gap between die 1 and die 2 has been neglected in the simulation, the simulation results show no upward flow along the side wall of the lower die. The simulation results show little deformation in the container, while the experiments showed that the flow takes place in the same area below the upper die. This means that the sensitivity rate of the plastic is higher than the flow stress used in the simulation.

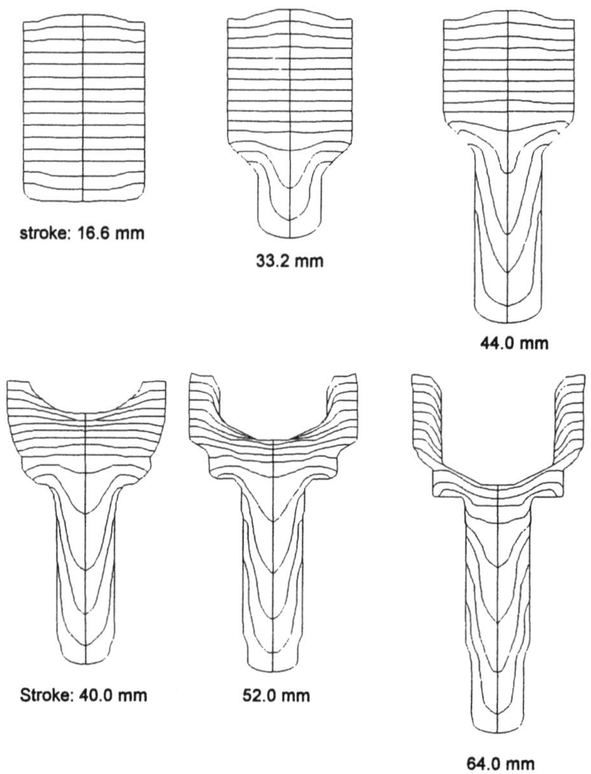

Figure 5.19: Deformation of billet in isothermal forging, (a): total effective strain, (b): deformed grid.

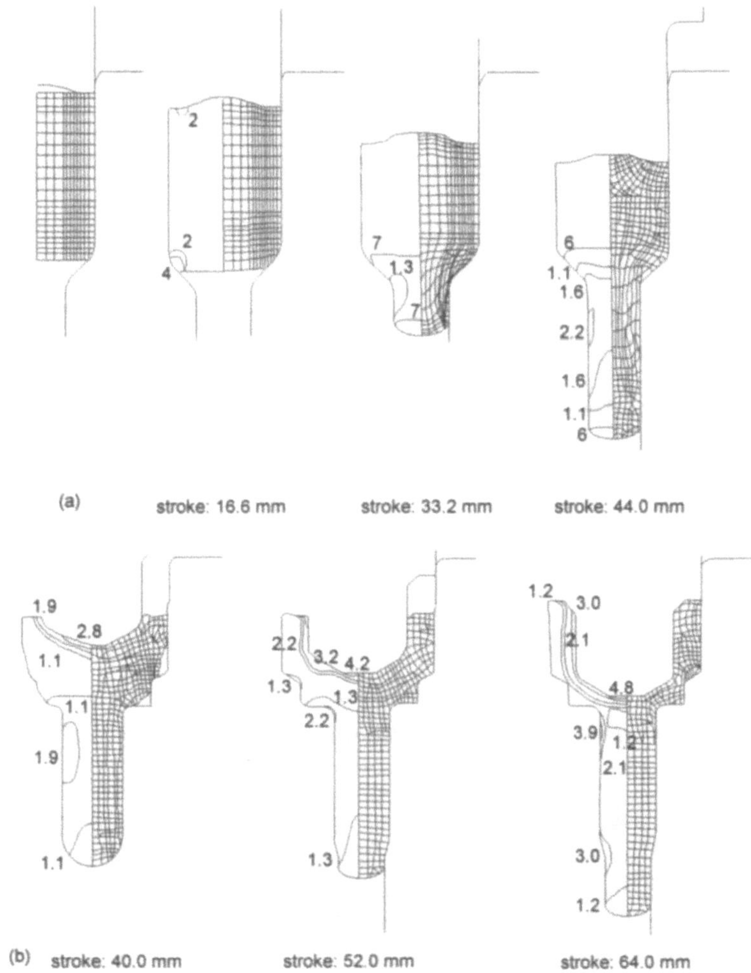

Figure 5.20: Simulation of flow pattern in isothermal forging.

5.4 FEM application in die design

In modern die design, FEM has been used for calculating the stress and strain distributions, checking the strength and stiffness, and optimising the die construction and dimensions. FEM is suitable for the analysis of extrusion and precision forging dies which are complex in geometric and loading conditions. As compared with the conventional design calculation method, the model is established based on the characteristics of the die construction and load using FEM. Therefore, the results are closer to practical situations.

5.4.1 Basic equations for FEM analysis of combined die and mathematical modelling of pre-stressing force

5.4.1.1 Basic equations for analysing combined die

The basic finite element equation for the hot forging die can be obtained from the general equation in the hot elasticity condition:

$$[K]\{\delta\}=\{Q\}+\{Q_T\} \tag{5.18}$$

[K] is the die stiffness and $\{K\}=\int[B]^T[D][B]dx$. $\{\delta\}$ is the knot displacement column matrix. {Q} is the equivalent knot force column matrix of the external applied force, and $\{Q_T\}$ is the equivalent knot force column matrix caused by the temperature.

For the cold extrusion or cold forging die, $\{Q_T\} = 0$. Therefore, the basic finite element equation is as follows:

$$[K]\{\delta\}=\{Q\} \tag{5.19}$$

5.4.1.2 Mathematical model of pre-stressing force in combined die for cold extrusion

The pre-stressing force in a combined die is caused by the interference fit. The calculation model is shown in Figure 5.21. The line ab represents the location of the contact surface of the internal and external rings after assembly. The dotted line on the right of line ab represents the location of the internal ring before assembly, and the dotted line on the left represents the external ring before assembly. $\Delta^{(o)}$ represents the amount of radial interference between the internal and external rings. The basic finite element equations for the internal and external rings are equations (5.20) and (5.21) respectively.

$$[K^{(1)}]\{\delta^{(1)}\}=\{Q^{(1)}\} \tag{5.20}$$

$$[K^{(2)}]\{\delta^{(2)}\}=\{Q^{(2)}\} \tag{5.21}$$

Besides a constrained opposite force, an equal and opposite unknown surface force in the radial direction acts on the contact surface of the internal and external rings. Both these forces are unknown. Therefore, it is essential to eliminate the unknown equivalent contact forces on the right side of equations (5.20) and (5.21) to solve the equations.

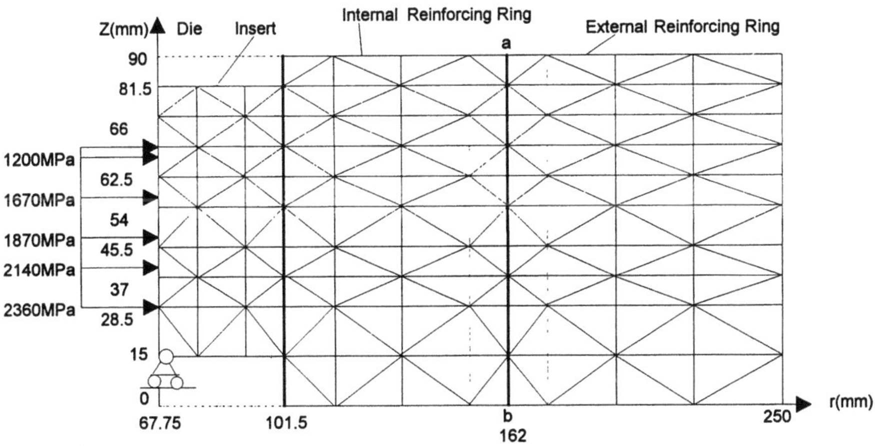

Figure 5.21: FEM modelling for combined die.

Let the internal ring be constricted uniformly during the assembly. For equations (5.20) and (5.21), the radial displacement of every node on the contact surface is as follows:

$$U_1 = \ldots = U_{i-1} = U_i = U_{i+1} = \ldots = U_b = \Delta^{(0)} \quad (5.22)$$

Equation (5.22) is substituted into equation (5.20) to obtain equation (5.23).

$$\left[K_r^{(1)}\right]\left\{\delta_r^{(10)}\right\} = \Delta^{(0)}\left\{J_r^{(1)}\right\} \qquad (r = 2i-1, 2i) \quad (5.23)$$

Similarly, equation (5.4.4) becomes as follows:

$$\left[K_r^{(2)}\right]\left\{\delta_r^{(20)}\right\} = \Delta^{(0)}\left\{J_r^{(2)}\right\} \qquad (r = 2i-1, 2i) \quad (5.24)$$

$\left\{J_r^{(1)}\right\}$ and $\left\{J_r^{(2)}\right\}$ represent the summation of the element stiffness matrices relating to knots that are on the contact surface of the internal and external rings respectively. Each knot displacement $\left\{\delta_r^{(10)}\right\}$ and $\left\{\delta_r^{(20)}\right\}$ in the internal and external rings can be found from equations (5.22) and (5.23) respectively. Each element strain $\left\{\varepsilon_{ij}^{(10)}\right\}$ and $\left\{\varepsilon_{ij}^{(20)}\right\}$, and stress $\left\{\sigma_{ij}^{(10)}\right\}$ and $\left\{\sigma_{ij}^{(20)}\right\}$ can be evaluated.

Subsequently, the average radial stress $\sigma_r^{(1)}$ and $\sigma_r^{(2)}$ acting on the contact surface can be determined.

Based on the characteristic of a linear element (Figure 5.22), the relationships of the amount of compression $U^{(1)}$ and $U^{(2)}$ of the internal and external rings to the average radial stress are shown in equation (5.25).

$$\left.\begin{aligned} U^{(1)} &= \frac{\sigma_r^{(1)}}{\sigma_r^{(1)} + \sigma_r^{(2)}} \Delta^{(0)} \\ U^{(2)} &= \frac{\sigma_r^{(2)}}{\sigma_r^{(1)} + \sigma_r^{(2)}} \Delta^{(0)} \end{aligned}\right\} \tag{5.25}$$

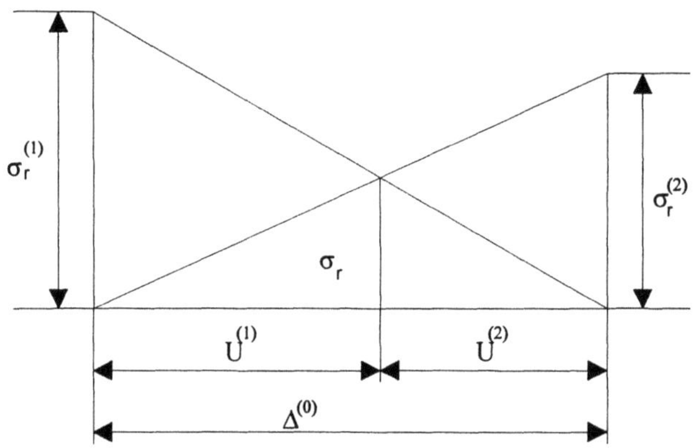

Figure 5.22: The co-ordinate of an element.

Based on the characteristics of the linear element, the effect of friction in the contact surface has been neglected. Thus, the formulae for calculating the displacement of each knot and the strain of each element in the internal and external rings can be obtained as follows:

$$\left.\begin{aligned} \{\delta^{(1)}\} &= \frac{U^{(1)}}{\Delta^{(0)}}\{\delta_r^{(10)}\}, & \{\delta^{(2)}\} &= \frac{U^{(2)}}{\Delta^{(0)}}\{\delta_r^{(20)}\} \\ \{\varepsilon_{ij}^{(1)}\} &= \frac{U^{(1)}}{\Delta^{(0)}}\{\varepsilon_{ij}^{(10)}\}, & \{\varepsilon_{ij}^{(2)}\} &= \frac{U^{(2)}}{\Delta^{(0)}}\{\varepsilon_{ij}^{(20)}\} \\ \{\sigma_{ij}^{(1)}\} &= \frac{U^{(1)}}{\Delta^{(0)}}\{\sigma_{ij}^{(10)}\}, & \{\sigma_{ij}^{(2)}\} &= \frac{U^{(2)}}{\Delta^{(0)}}\{\sigma_{ij}^{(20)}\} \end{aligned}\right\} \tag{5.26}$$

5.4.2 FEM solution and program flowchart

Figure 5.23 shows the flowchart of the FEM program for calculating the working stress in a die. For the calculation, the combined die is considered as a whole die, i.e., the difference in the material properties of each ring is neglected. The die is assumed to be completely elastic. Its elasticity modulus is E and Poisson ratio is γ.

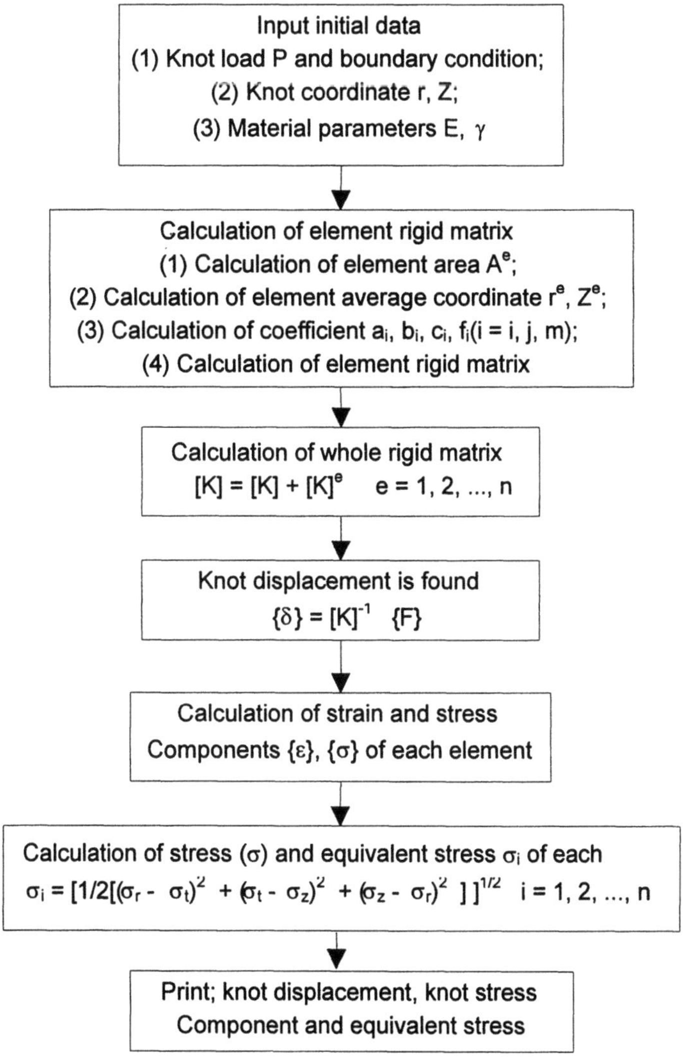

Figure 5.23: FEM program flowchart.

According to the order in the flowchart, the formula for each step is as follows. The element area:

$$A=\frac{1}{2}\left[r_i\left(Z_j-Z_m\right)+r_j\left(Z_m-Z_i\right)+r_m\left(Z_i-Z_j\right)\right]$$

The average co-ordinate of an element (Figure 5.24):

$$\left[r^e=\frac{1}{3}\left(r_i+r_j+r_m\right)\right]Z^e=\frac{1}{3}\left(Z_i+Z_j+Z_m\right)$$

The coefficient of each term in the displacement function, i.e., the co-ordinate difference:

$$a_i=r_jZ_m-r_mZ_j$$

$$b_i=Z_j-Z_m$$

$$C_i=r_m-r_j$$

$$f_i=\frac{a_i}{r^e}+b_i+\frac{c_iZ^e}{r^e}$$

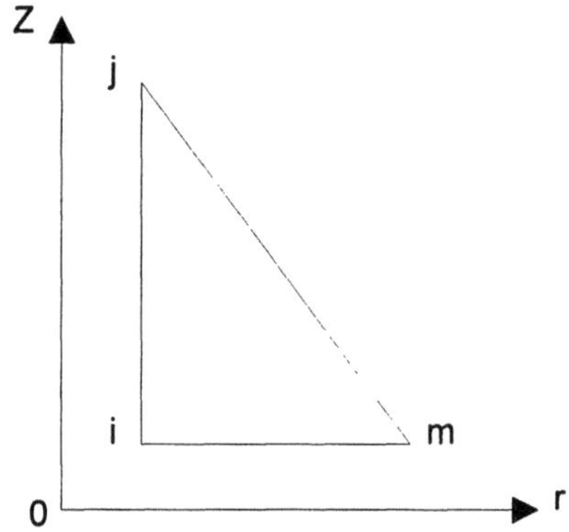

Figure 5.24: A typical linear triangular element.

The element rigid matrix:

$$\begin{bmatrix} K_{ii} & K_{ij} & K_{im} \\ K_{ji} & K_{jj} & K_{jm} \\ K_{mi} & K_{mj} & K_{mm} \end{bmatrix}$$

The expression of each sub-matrix is as follows:

$$[K_{st}] = \frac{2\pi r A_3}{A}\begin{bmatrix} b_s(b_i + A_i f_i) + f_3(f_i + A_1 b_i) + A_2 C_s C_i & A_1 C_i(b_s + f_s) + A_2 C_s C_i \\ A_1 C_i(b_s + f_s) + A_2 b_s b_i & C_s C_i + A_2 b_s b_i \end{bmatrix}$$

where $A_1 = \frac{\gamma}{1-\gamma}$, $A_2 = \frac{1-2\gamma}{2(1-\gamma)}$, $A_3 = \frac{(1-\gamma)E}{4(1+\gamma)(1-2\gamma)}$, and (s = i, j, m; t = i, j, m).

The whole rigid matrix:

$$[K] = \sum_{e=1}^{n} [K]^e$$

The equivalent knot force:

$$[F]^e = 2\pi r z P_i$$

z represents the distance between the mid-point of knots i-1 and i, and the mid-point of knots i and i+1, on the inside surface of a die.

The displacement of knots is found by solving the following linear equation set:

$$\{\delta\} = [K]^{-1}\{F\}$$

The strain and stress components of each element are as follows:

$$\{\varepsilon\} = \begin{Bmatrix} \varepsilon_r \\ \varepsilon_i \\ \varepsilon_z \\ \varepsilon_{rz} \end{Bmatrix} = \frac{1}{2A}\begin{bmatrix} b_i & 0 & b_j & 0 & b_m & 0 \\ f_i & 0 & f_j & 0 & f_m & 0 \\ 0 & c_i & 0 & c_j & 0 & c_m \\ c_i & b_i & c_j & b_j & c_m & b_m \end{bmatrix}\begin{bmatrix} U_i \\ W_i \\ U_j \\ W_j \\ U_m \\ W_m \end{bmatrix}$$

$$\{\sigma\}=\begin{Bmatrix}\sigma_r\\ \sigma_i\\ \sigma_z\\ \tau_{rz}\end{Bmatrix}=\begin{bmatrix}S_i, & S_j, & S_m\end{bmatrix}\begin{bmatrix}U_i\\ W_i\\ U_j\\ W_j\\ U_m\\ W_m\end{bmatrix}$$

where $[S_i]=\frac{2A_3}{A}\begin{bmatrix}b_i+A_if_i & A_1C_i\\ A_1b_i+f_i & A_1C_i\\ A_1(b_i+f_i) & C_i\\ A_2C_i & A_2b_i\end{bmatrix}$, and similarly for S_j and S_m.

The flowchart for the FEM program for calculating the pre-stressing force is similar to that for determining the working stress.

For a three-ply combined die, the evaluation of the pre-stressing force must be divided into three steps. In the first step, the pre-stressing force caused by the interference fit between the internal and external rings is evaluated. In the second step, the pre-stressing force caused by the interference fit between the internal ring and the assembly of the middle and external rings is determined. The total pre-stressing force is obtained in the third step by superimposing the pre-stressing forces obtained from the first and second steps.

The corresponding pre-stressing force is superimposed on the working stress of each knot, on the inside surface of the die to obtain the final total stress distribution. Next, the equivalent stress of each knot is calculated and checked against the third strength theory.

5.4.3 FEM analysis for combined backward extrusion die

5.4.3.1 FEM analysis for working stress

Three types of construction for a combined die for backward extrusion of the same part are represented as A, B and C. Figure 5.21 shows the axi-symmetric section of die B, which is a step construction. The height ratio of the internal ring to the workpiece is 1.8. The height ratio of the assembly of the middle and external rings to the internal ring is 1.4. The workpiece is located at the middle section of the internal ring, i.e., the location of the radial pressure shown in Figure 5.21. The ratios for construction A are equal to that for construction B. However, the top of the internal and external rings are on the same level, and the upper end of the workpiece and the top of the internal ring are also on the same level. Therefore, the radial extrusion pressure acts on the upper part of the internal surface of the die. The height of the internal ring, and the middle and external reinforcing rings for construction C are equal, i.e., for the common level mouth reinforcing die, the workpiece is located at the middle section inside the internal ring. The height ratio of the workpiece to the combined die is the same, and equals to 2.4 for all the three types of construction.

For the ease of calculation, the difference in the material properties of the internal ring, and the middle and external reinforcing rings is neglected. The elastic modulus E is 2.1×10^5 kN/mm^2 and the Poisson's ratio γ is 0.3. The mesh is shown in Figure 5.21.

The basic equation is solved using the method for large sparse symmetric positive definite equations. Comparing with the Gaussian elimination method, this method only requires half the total computing time needed. The computer storage requirement can be reduced by 80%. The key to this method lies in building the correct format of the 1-D contraction store for the matrix [k].

The evaluated results are plotted as the curves shown in Figures 5.25 to 5.28. As can be readily seen from Figures 5.25 to 5.27, during the extrusion pressure, there are both radial and axial displacements of the die. However, the axial displacement is the main displacement. The radial displacements of the knots on the inside surface of die A increase gradually from the bottom to the top of the internal ring. The maximum displacement occurs at the die mouth. The knot displacement curve of die B is similar to that of die C, and the maximum displacement occurs at the middle part of the internal ring. The stress curves in Figures 5.25 to 5.27 show that the tangential stress for die A, caused by the extrusion pressure increases from the bottom to the top of the internal ring. The maximum tangential stress occurs at the die mouth. For dies B and C, the change of the tangential stress distribution curves is not obvious.

The maximum radial and tangential stresses in the three types of combined dies A, B and C are: -1900 MPa, 2383 MPa, -2100 MPa, 1090 MPa, -2155 MPa and 830 MPa respectively. The maximum radial stress in die A is slightly smaller than that in dies B and C. However, the maximum tangential stress in die A is 2.2 and 2.9 times larger than the stresses in dies B and C. Generally, the internal ring, i.e., the die insert, is made of high strength hardened steel or hard alloy. It is suitable for bearing compressive but not tensile stress. Due to the great tangential tensile stress at the mouth of die A, the die insert is likely to break during extrusion, making die A unsuitable. Both dies B and C are more suitable. The curve shown in Figure 5.28 shows that the axial stress changes from compressive at the middle part to tensile at the two ends for die C. When the insert material is a hard brittle material and the tensile stresses at the ends are very large, the die will break in the axial direction. For die B, the axial stress is completely compressive. Since the insert is shorter than the pre-stressing ring in die B, less material is needed. The machining and assembling quality is easily ensured. Die B is therefore of better construction.

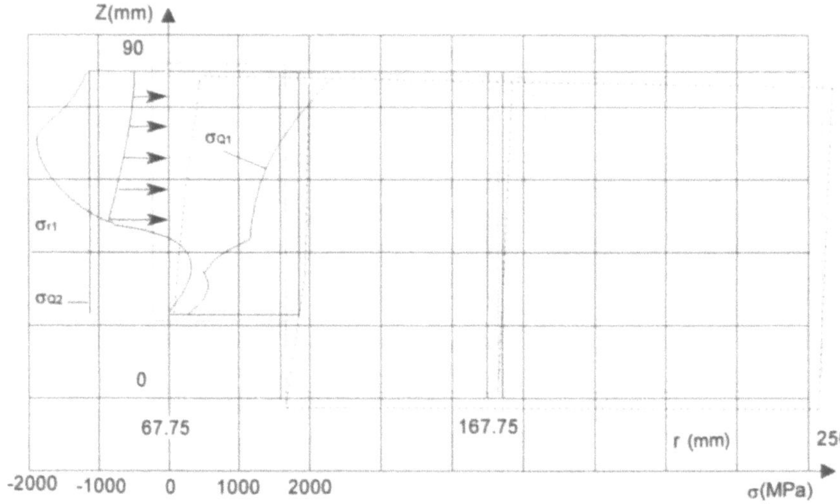

Figure 5.25: Stress distribution curve on the inside surface of die and deformation of combined die A, σ_{r1} and σ_{Q1} - radial and tangential stresses caused by working load, σ_{Q2} - tangential stress caused by interference fit, ... deformation (radial displacement: 0.2 mm/mm, axial displacement: 0.1 mm/mm) caused by working load; deformation (radial displacement: 0.2 mm/mm; axial displacement: 0.02 mm/mm) caused by interference fit.

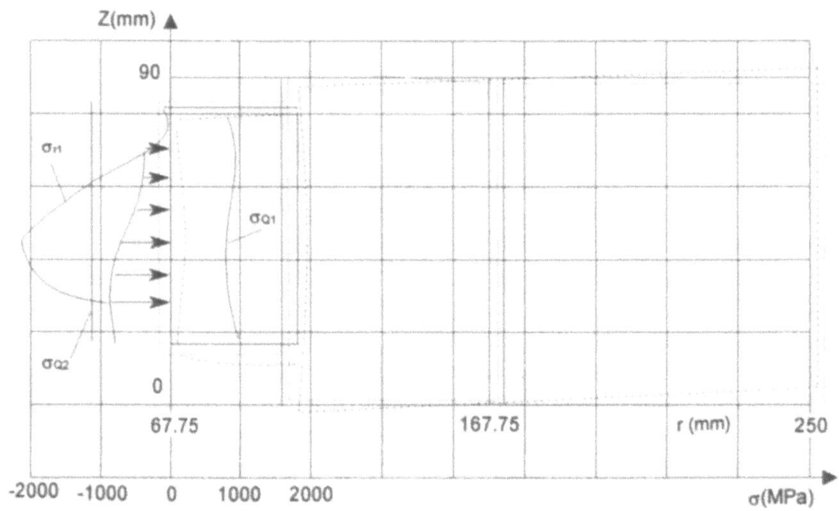

Figure 5.26: Stress distribution curve on inside surface of die and deformation of combined die B (the meanings of the symbols are the same as in figure 5.25).

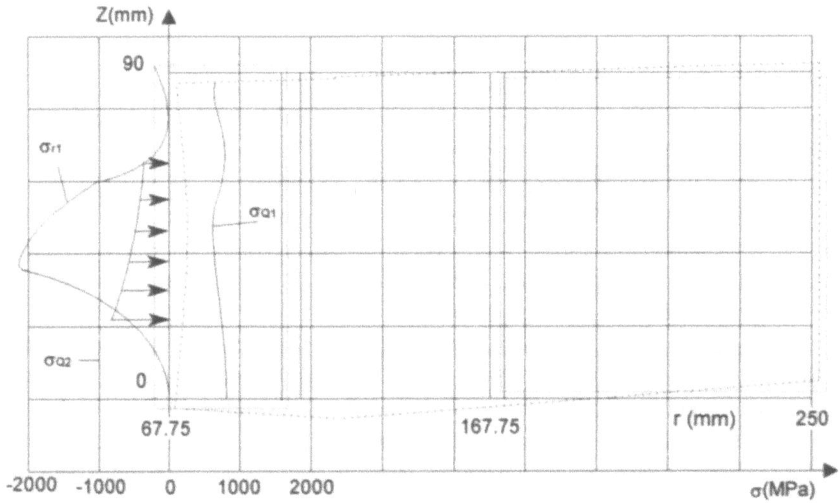

Figure 5.27: Stress distribution curve on inside surface of die and deformation of combined die C (the meanings of the symbols are the same as in figure 5.25).

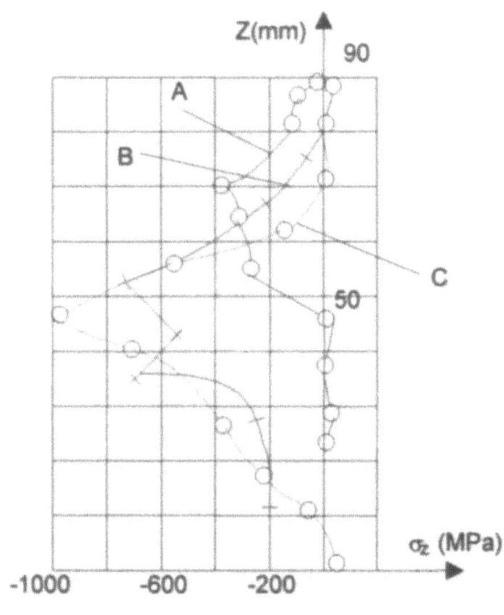

Figure 5.28: Axial stress distribution curve caused by extrusion pressure in the inside surface of three types of combined die.

5.4.3.2 FEM analysis for pre-stressing force

For the above three types of combined dies, the amount of radial interference between the middle and external rings, and between the internal ring and the assembly of the middle and external rings are 0.3 mm and 0.6 mm respectively. Since the middle and external rings of these three types of combined dies are of equal size, the amount of compression after the assembly are $\Delta^{(1)} = 0.122$ mm and $\Delta^{(2)} = 0.178$ mm. After the internal ring, the middle and external rings have been assembled, the amount of radial compression $\Delta^{(1)}$ and $\Delta^{(2)}$ of the internal ring, and the assembly for dies A, B and C are 0.365 mm and 0.235 mm; 0.362 mm and 0.238 mm; 0.344 mm and 0.256 mm respectively.

The broken lines in Figures 5.25 to 5.27 show the deformed conditions of the combined die caused by the interference fit. In the middle section (z = 45 mm), the axial displacement is zero and there is only radial displacement. Therefore, there are only radial and tangential strains ε_r and ε_θ in the middle section, i.e., the strain at the middle section is plane strain. The strains in other sections are both radial and axial displacements, which are similar to the deformed condition caused by the working load. Therefore, the main displacement is radial. After the internal, middle and external rings have been assembled, the radius increases and the height decreases for the external ring. For the middle ring, its external radius decreases, internal radius increases, and its height increases (For dies A and B, the upper end expands and the low end contracts. For die C, the outside reduces and the inside expands. For all the three dies, the amount of expansion is larger than the amount of contraction.). For the internal ring, its radius decreases and its height increases.

The maximum tangential stress acting on the internal surface of the three dies caused by the interference fit are: -1140 MPa, -1130 MPa and -1070 MPa respectively. These results are very close to the result obtained from the Lame formula, -1120 MPa. The stress distribution curve approximates a straight line, as shown in Figures 5.25 to 5.27. This is the reason for the uniform compression of the contact interface of the internal and external rings, which results in the external surface of the internal ring bearing the uniform pressure.

Using the combined die B as an example, the maximum tangential pre-stressing pressure and the working stress at the internal surface of the die are superimposed. The total is -40 MPa, which is relatively insignificant and can be equated to zero. Using the Lame formula, a tangential pre-stressing pressure of -2080 MPa is needed to meet the superimposed stress at the internal surface of the die. The example shows that under the condition of local working pressure, the radial pre-stressing force can be set at a smaller value than that calculated using the Lame formula. For dies of the same internal diameter, a larger extruded workpiece height requires a larger pre-stressing force.

5.4.4 FEM analysis for the combined forward extrusion die

The cavities of all the combined forward extrusion dies are of the step type. The thickness of the die wall is considerably non-uniform. Therefore, a large stress convergence in the transition corner of the die is present. The finite element analysis indicates that for the combined die, the radial non-uniform interference fit or the axial pre-stressing construction can be applied to decrease the effect of stress convergence.

5.4.4.1 FEM analysis for radial non-uniform interference fit

A. The uniform interference fit combined die

To explain the combined forward extrusion die with the radial non-uniform interference fit, it is essential to first introduce the combined forward extrusion die with uniform interference fit. Using the forward extrusion die shown in Figure 5.29 as an example, $d_1 = 20$ mm, $d_2 = 30$ mm, $d_3 = 50$ mm, $d_4 = 100$ mm, $H = 40$ mm, $h = 20$ mm, $\alpha = 60°$ and $r = 3$ mm. This die is the two layers combined construction type. The fit surface shown as the broken line in Figure 5.29 denotes the amount of radial interference. A fit, that minimises the maximum equivalent stress at the internal surface of the die, and raises the bearing capacity of the die, has to be selected. During evaluation, it is assumed that the load distribution, q is uniform. $q = 1000$ MPa acts on the internal surface of the die, and there is no axial displacement in the die during extrusion. The mesh element for the axial section is shown in Figure 5.30. From the evaluation, the maximum equivalent stress acting on the internal surface of the die is minimum when δ is the equivalent stress $\bar{\sigma}_{max} = 1700$ MPa. If the die is the single layer type, its maximum equivalent stress will increase by 11.2%.

B. Non-uniform interference fit combined die

The distinction between the construction and size of non-uniform and uniform interference fit combined dies lies in the different amounts of interference on the upper and low ends in the axial direction. Let δ_1 and δ_2 denote the amount of interference at the upper and low ends respectively. δ_1 is not equal to δ_2 as shown in Figure 5.29. Calculation shows that when δ_1 is positive and δ_2 is negative, i.e., the upper part of the internal and external rings is an interference fit, and the low part is a gap fit, the stress distribution on the internal surface of the die is considerably improved. When the non-uniform interference fit coefficient $K = \delta_1/\delta_2 = 0.4$ and $\delta_1 = 0.19$ mm, the maximal equivalent stress $\bar{\sigma}_{max}$ is reduced to 1464 MPa, the maximal axial tensile stress $\delta_Z = 580$ MPa and the tangential tensile stress $\delta_\theta = 370$ MPa at the transition corner. The maximal equivalent stress is reduced by 12.5% from that of the uniform combined die, and 23.7% from that of the whole die. The maximal equivalent stress distribution curve in the three types of dies for the forward extrusion is shown in Figure 5.30.

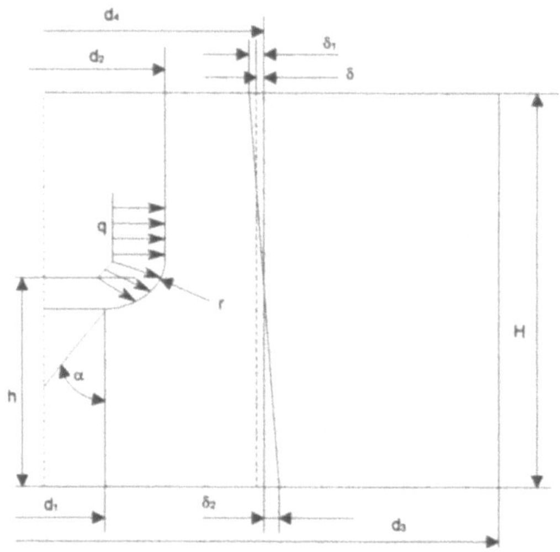

Figure 5.29: Modelling of the die.

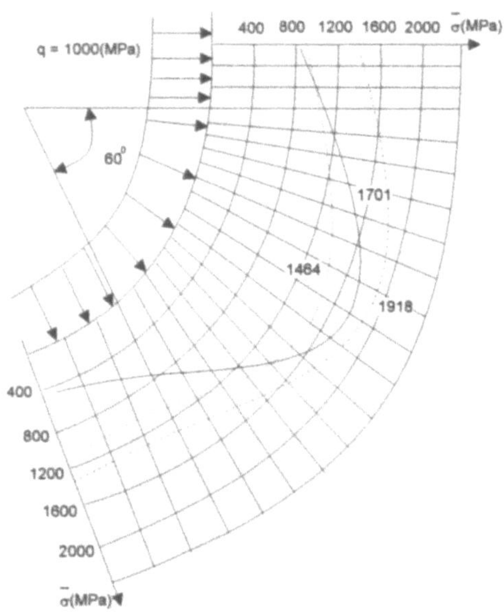

Figure 5.30: The maximum equivalent stress distribution curve in the die corner.

5.4.4.2 FEM analysis for axial reinforcing

For the forward extrusion combined die, its internal ring can be reinforced in the radial and axial directions simultaneously to considerably reduce the working stress spike acting on the die. Figure 5.31 shows the modelling of the extrusion die with axial reinforcing. It consists of the die cavity with the shoulder and the radial pre-stressing ring.

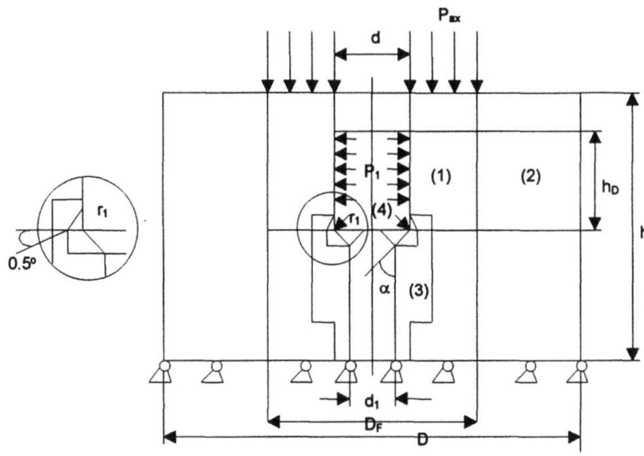

Figure 5.31: Extrusion die with axial and radial reinforcing.

Figure 5.32 shows the stress distribution curve on the internal surface of the die. The expression of the effective stress is as follows:

$$\bar{\sigma} = \sqrt{\frac{1}{2}\left[(\sigma_r - \sigma_z)^2 + (\sigma_z - \sigma_\theta)^2 + (\sigma_\theta - \sigma_r)^2 + 6\tau_{rz}^2\right]} \qquad (5.27)$$

σ_z, σ_θ, σ_r and τ_{rz} denote the axial, tangential, radial and shearing stresses respectively.

Curves 1 and 2 represent the stress distribution conditions on the internal surface of a die with radial reinforcing, and radial and axial reinforcing respectively. The die is of the shoulder angle type. The pressure space height ratio is h_D/h_M = 0.425. The diameter ratios are d/D = 0.15 and D_F/D = 0.5. The amount of radial interference is δ = 0.06. The extrusion pressure P_i is 1800 MPa. The curve and the data indicate that the stress spike is produced at the transition corner between the internal surface and the shoulder. In practice, the transition corner often produces the break. To reduce the stress spike, the radial pre-stressing force or the transition fillet radius can be increased. This can be seen from Figure 5.32. Another method for reducing the stress spike is by applying an axial pre-stressing force on the upper end of the die. Curve 2 shows that when the axial pre-stressing pressure P_{ax} is equal to 1200 MPa, the stress spike will be reduced by 44%.

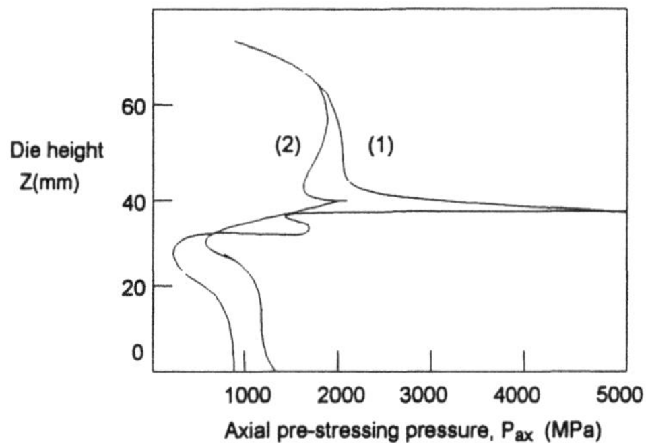

Figure 5.32: Stress distribution curve on the internal surface of die with axial and radial reinforcing, and the radial reinforcing

(1) radial reinforcing		(2) axial and radial reinforcing	
x	y	x	y
977.04	80	1221.4	80
703.17	75	938.45	75
850.72	70	937.98	70
2002.2	60	1943.8	60
2112.7	50	1998.6	50
2088.8	40	1773	40
2203.1	37.5	1654.2	37.5
4941.3	35.15	1738.8	35.5
3983.5	34.81	2224.6	34.997
2490.4	34.62	1648.7	34.56
2066.6	34.5	1577.6	32.69
1465	33	838.89	30
1649.4	32.26	759.7	29
697.4	30	920.28	25
400.97	28	1135.5	20
742.69	22	1249.1	16
912.23	16.34	1251.4	10
959.83	10	1335.6	0
927.23	0		

The relation between the die shoulder transition fillet radius r_1, stress spike in the middle part of the die, and the axial pre-stressing pressure is shown in Figure 5.33. The curve illustrates that the stress spike $\bar{\sigma}$ at the transition fillet radius and the middle part of the cavity decreases with the increase of the axial pre-stressing pressure. When P_{ax} is larger than 800 MPa, $\bar{\sigma}$ increases. The optimum axial pre-stressing pressure depends on the geometry of the die and the load. For the analysed die, the optimum axial pre-stressing pressure is equal to 800 MPa.

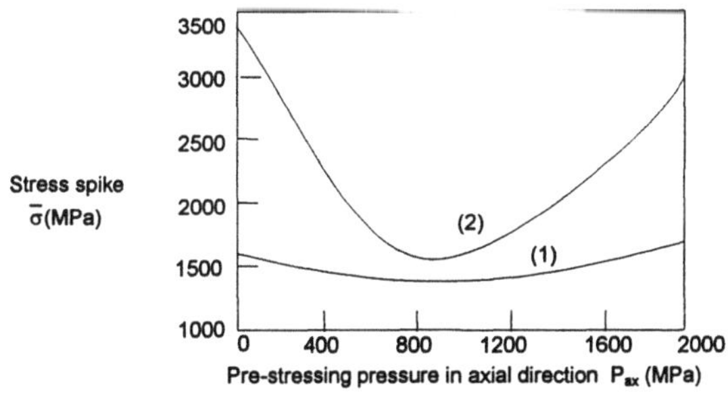

Figure 5.33: Relation of fillet radius r_1, stress spike, and axial pre-stressing pressure.

(1) middle point of die cavity		(2) transition fillet radius	
X	Y	X	Y
0	1595.1	0	3347.4
400	1463	400	2429.5
800	1419.5	800	1631.6
1200	1451.5	1200	1921.7
1600	1575.5	1600	2376.7
2000	1772.3	2000	2920.4

5.4.5 FEM analysis for precision forging die for blades

In the precision die forging and re-shaping operations for blades, the forging shape differs from the cavity shape due to the considerable elastic deformation of the die. This results in a large and non-uniform forging allowance, which causes difficulties during subsequent machining of the forging.

It is necessary to consider the elastic deformation of a die construction to reduce the non-uniform degree of allowance. Factors that affect the elastic deformation are the temperature and the load for the metal deformation. The load is of the non-uniform diffusion type and causes complex deformation of the cavity.

The stiffness of a die can usually be increased by magnifying the dimension of the die, or by applying a pre-stressing ring, to reduce the deformation of the cavity. Practice has shown that these measures do not have obvious optimising effects.

The sizes of the die are corrected based on the predicted results in the design phase. The established predicting model is shown in Figure 5.34. During loading, the determination of the geometry of the cavity can be converted into the thermal-elastic problem. The displacement of each element knot in the die is obtained by solving equation (5.18). The knot force column matrix {Q} is converted from the distribution contact pressure, and the contact pressure is obtained from the output in module 1.

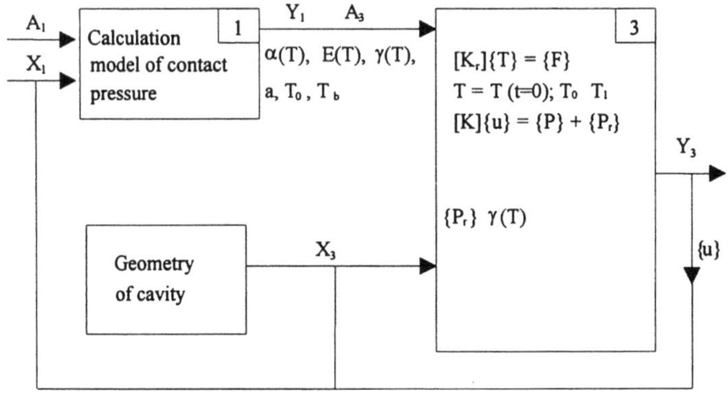

(a)

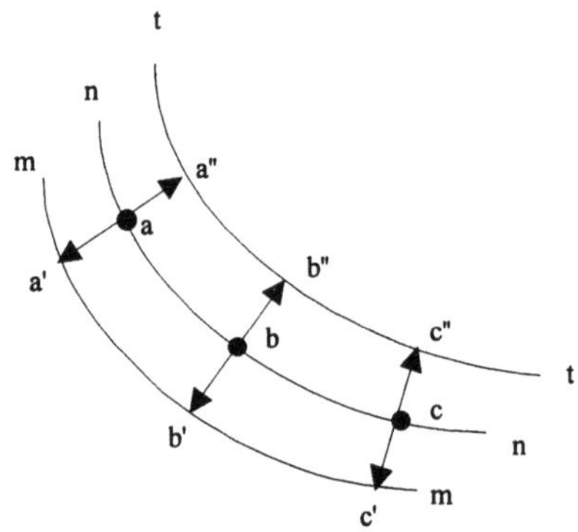

(b)

Figure 5.34: Model for predicting die size.

Input parameters for the model are the contact pressures, $\rho(x)$ and $\tau(x)$, corresponding with the finished forging, the die geometry, the fixed conditions, the physical constants of the die material related to temperature $\alpha(T)$, $E(T)$, a, the starting temperatures T_0 and T_b in the die and forging, the contact time t_1 of the block with die, the parameters of the die construction, and the displacement vector $\{\delta(x)\}$ of the element knots in the cavity section.

The correction of the cavity size is made according to Figure 5.34. The original size n-n is first determined based on the size of the hot forging. Next, the knot displacement vector caused by the contact pressure in some assigned sections of the forging die is evaluated. The knots are distributed on the contact surface aa', bb' and cc' of the cavity corresponding to the evaluation model. The surface outline of the cavity is m-m under the acting contact pressure. To obtain the cavity surface outline n-n under the application of the contact pressure, the designed cavity t-t is determined on the basis of aa" = aa', bb" = bb' and cc" = cc'.

Figure 5.35 shows a type of precision forging die for the compressor blade. FEM evaluation modelling for predicting the elastic deformation is shown in Figure 5.36. The die section is divided into 600 ~ 700 triangular elements. The contact pressure, which is distributed non-uniformly in the cavity, is converted into the knot load, and applied to 26 knots to ensure their repeated accuracy. The evaluation indicates that for a guiding blade with a length of 240 mm, the elastic deformation of the cavity, u_y is 0.3 mm. The cavity knot displacement is considerable for both the vertical and horizontal components. u_x approximates to 0.2 mm in the most abrupt part of the die cavity. The co-ordinate of each point in the cavity is corrected to ensure uniform allowance of the forging.

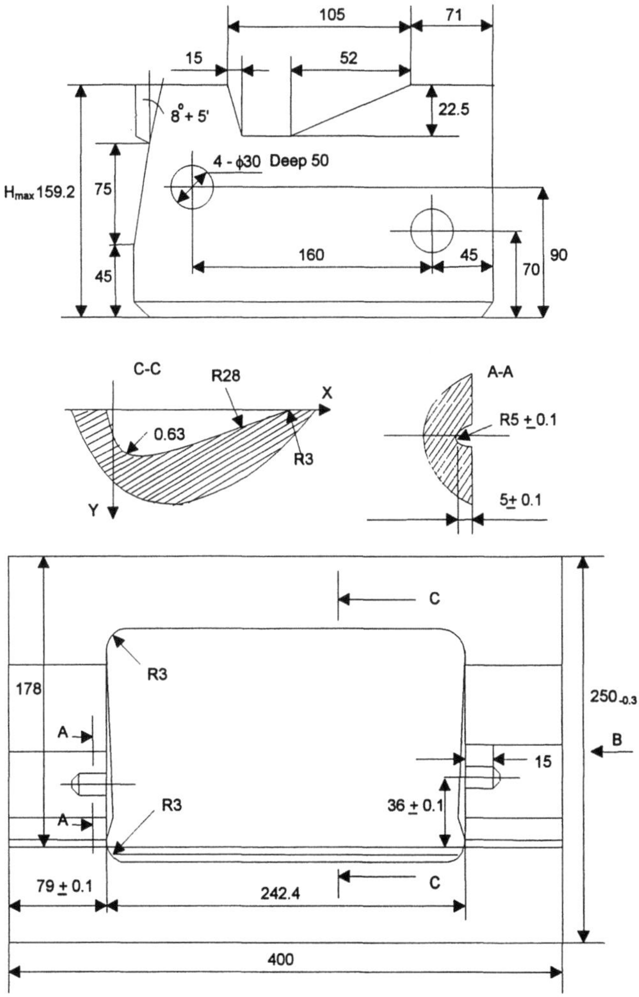

Figure 5.35: Precision forging die for blades.

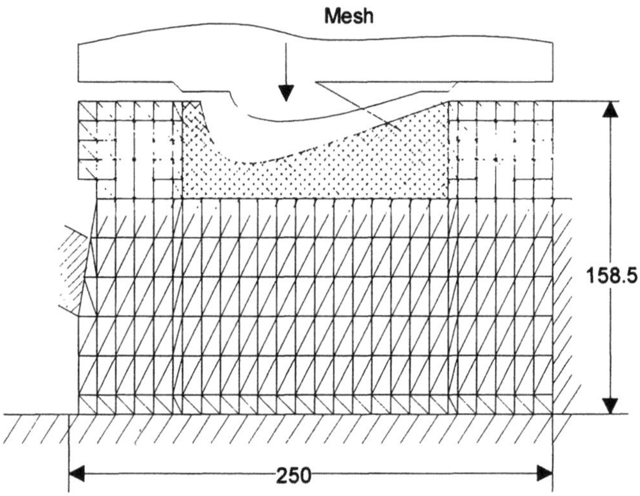

Figure 5.36: Corrected precision forging die construction.

5.4.6 Optimisation design of combined dies

5.4.6.1 Mathematical modelling of optimisation design

For a combined die, especially for the step type, the optimum ratio of the height of the internal and external rings, and the optimum ratio of the height of the internal ring and the internal surface height acted upon by an extrusion pressure can be obtained using the FEM. The optimum modelling of the radial construction of a combined die can be established using the thick-wall cylinder theory and the third strength theory. The wall thickness of the internal and reinforcing rings, and the amount of radial interference can be found by applying the Sequential Unconstrained Minimisation Technique (SUMT) optimisation method. The designed combined die is likely to be an optimum construction.

As shown in Figure 5.37, the internal surface of a combined die is assumed to have N layers. No tensile stress is allowed on these layers. The radius of each layer from inside to outside is $R_{(1)}$, $R_{(2)}$ ..., and $R_{(N+1)}$ respectively. The yield stresses of the material from the inner layer to the external layer are respectively $\sigma_{(1)}$, $\sigma_{(2)}$, ..., and $\sigma_{(N)}$. The elastic moduli are respectively $E_{(1)}$, $E_{(2)}$, ..., and $E_{(N)}$. The linear expansion coefficient are respectively $\alpha_{(1)}$, $\alpha_{(2)}$, ..., and $\alpha_{(N)}$. The Poisson's ratios are $\gamma_{(1)}$, $\gamma_{(2)}$, ..., and $\gamma_{(N)}$ respectively.

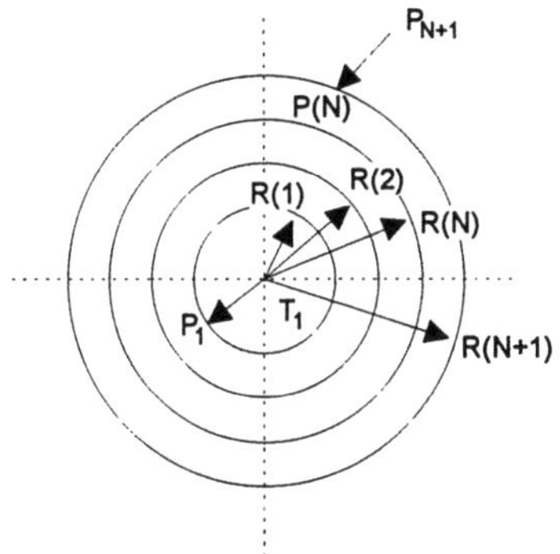

Figure 5.37: N layers of the internal surface of a combined die.

Based on the thick-wall cylinder theory, the third strength theory and the thermo-stress calculation formula, the target function is derived as follows:

$$F_{(R_{(2)}\dots R_{(N+1)})} = \left\{ P_1 - \frac{R_{(2)}^2 - R_{(1)}^2}{R_{(2)}^2 + R_{(1)}^2}\left[\frac{2R_{(2)}^2}{R_{(2)}^2 - R_{(1)}^2}\left[P_{(N+1)} + \sum_{i=2}^{N}\frac{R_{(i+1)}^2 - R_{(i)}^2}{2R_{(i+1)}^2}\right.\right.\right.$$
$$\left.\left.\left[\sigma_{(i)} - \frac{\alpha_{(1)}E_{(i)}}{1-U_{(i)}}\left(\frac{2}{R_{(i+1)}^2 - R_{(i)}^2}\int_{R_{(i)}}^{R_{(i+1)}} T_{(R)}RdR - T_{(R_{(i)})}\right)\right]\right]\right.$$
$$\left.\left. - \frac{\alpha_{(1)}E_{(i)}}{1-\gamma_{(i)}}\left(\frac{2}{R_{(2)}^2 + R_{(1)}^2}\int_{R_{(1)}}^{R_{(2)}} T_{(R)}RdR - T_{(R_{(1)})}\right)\right]\right\} \tag{5.28}$$

The combined die radii $R_{(2)}$, ..., and $R_{(N+1)}$ can be obtained using SUMT to obtain the minimum of this target function. $T_{R(i)}$ is the thickness of ring $R_{(i)}$.

Based on the relation of the amount of radial interference with the pre-stressing force $P_{(i)}$, i = 2, ..., N, under the heated state, the amount of radial interference can be determined using the following equation.

$$\sigma_{(i)}^{h} = P_{(i)}\left[\frac{2R_{(i)}}{E_{(i)}}\left(\frac{R_{(N+1)}^2 + R_{(i)}^2}{R_{(N+1)}^2 - R_{(i)}^2} + \gamma_{(i)}\right) + \frac{2R_{(i)}}{E_{(i-1)}}\left(\frac{R_{(i)}^2 + R_{(i-1)}^2}{R_{(i)}^2 - R_{(i-1)}^2} - \gamma_{(i-1)}\right)\right] \tag{5.29}$$

For the ease of design and manufacture, it is necessary to transform the radial interference amount from the hot state δ_i^h into the cold state δ_i^c:

$$\delta_{(i)}^{c} = \delta_{(i)}^{h} + 2\left(\delta_{r(i)}^{h'} - \delta_{r(i)}^{h}\right) \qquad (i = 1, \dots, N) \tag{5.30}$$

5.4.6.2 *Program flowchart of the optimisation design module*

The program flowchart of the optimisation design module for combined dies is shown in Figure 5.38. The optimum wall thickness and the amount of interference of combined dies in cold or hot state can be determined automatically using this program. In the optimisation of the design of a combined die, the original conditions include the existence of tensile stresses in the internal ring, the consideration of the effect of temperature, determination of whether the diameter of the external ring is constant, the use of experience, etc.

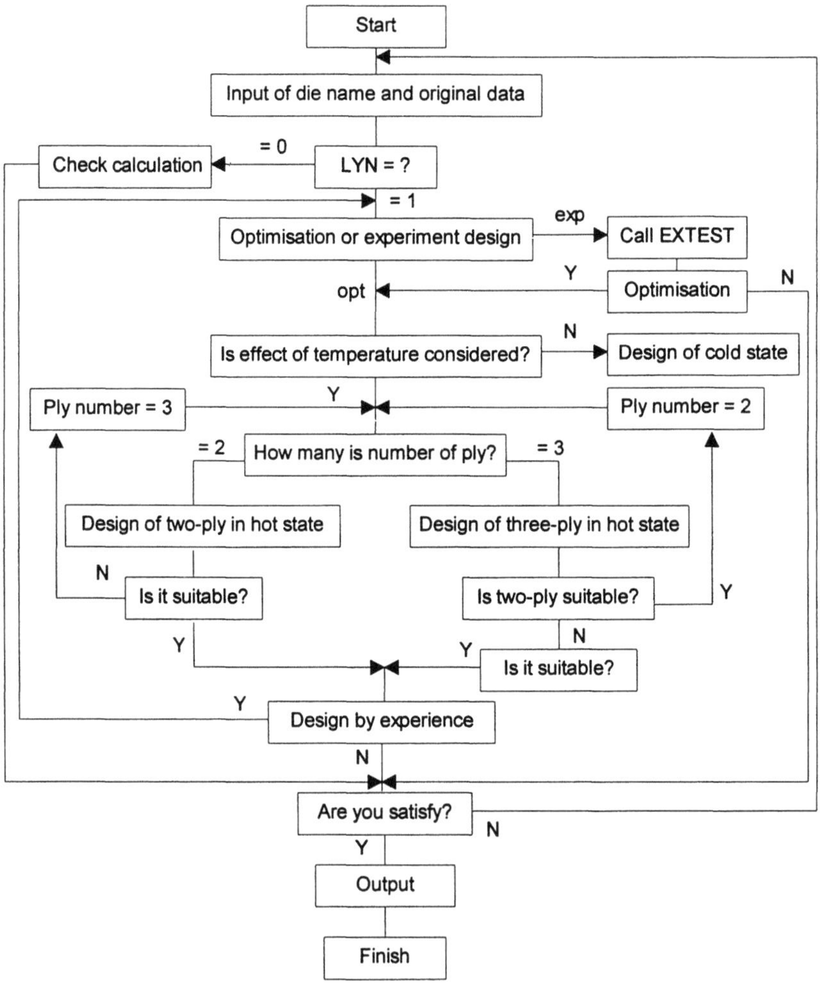

Figure 5.38: Program flowchart of the optimisation design module.

5.5 FEM application in analysis of hydraulic presses

5.5.1 Finite element analysis of fluid transients

Fluid transients have a strong influence on the operational quality of hydraulic presses. Since fluid transient is a multi-dimensional non-stationary flow process, and fluid has compressibility, moving inertia and viscosity, the main equations describing the process are often non-linear-partial differential equations coupled with field variables (velocity and pressure). It is difficult to solve the equations using conventional methods. The following is a transient analysis using the finite element method [13].

Let Ω be the solving field of flow in a pipe, $\Gamma\sigma$ is the given border of the stress, Γu is the given border of the velocity and Γs is the border of the pipe wall. The equation for fluid transients can be described as follows:

Continuous equation,

$$\frac{\partial p}{\partial t}+u\frac{\partial p}{\partial x}+\rho a^2\frac{\partial u}{\partial x}=0\text{, in }\Omega \tag{5.31}$$

Movement equation,

$$\frac{\partial u}{\partial t}+u\frac{\partial u}{\partial x}+\frac{1}{\rho}\frac{\partial p}{\partial x}-\frac{1}{r}\frac{\partial}{\partial r}(r\nu\frac{\partial u}{\partial r})=0\text{, in }\Omega \tag{5.32}$$

where $u = \hat{u}(x, r, t)$, on Γu

$u = 0$, on Γs

$u = u_0(r)$ and $p = p_0(x)$, when $t = 0$

$\nu = (ky + m)\, \nu_L$, in Ω

$\pi_{ij}n_j = f_i$, on $\Gamma\sigma$

p and u are the pressure and velocity of the fluid respectively. x and r are the axial and radial co-ordinates respectively. t is time, and ν_L and ν are the fluid viscous coefficients of layered flow and equivalence respectively. k and m are the coefficients for calculating ν. y is the dimension-less radial length to wall ratio. ρ is the fluid density. a is the transmission velocity of the pressure wave. π_{ij} is a component of the stress tensor for moving viscous fluid. n_j is the outer nominal line component on $\Gamma\sigma$. f_i is the given stress component on $\Gamma\sigma$. $\hat{u}$ is the given axial component of the velocity. u_0 and p_0 are the initial distributions of the velocity and pressure respectively. The field Ω is divided into N unit sub-fields with regular geometric shape, and an interpolating function is constructed in units. Next, the

Galerkin weak solution integral expressions of equations (5.33) and (5.34) based on the Galerkin weighted residual method can be obtained as follows:

$$\int_\Omega \left[\left(\frac{\partial p}{\partial t} + u\frac{\partial p}{\partial x} \right)\delta p - \rho a^2 u \frac{\partial(\delta p)}{\partial x} \right] r d\Omega = -\int_u \rho a^2 \hat{u} \delta p n_x r d\Gamma \qquad (5.33)$$

$$\int_\Omega \left[\left(\frac{\partial u}{\partial t} + u\frac{\partial u}{\partial x} \right)\delta u - \frac{1}{\rho} p \frac{\partial(\delta u)}{\partial x} + v\frac{\partial u}{\partial r}\frac{\partial(\delta u)}{\partial r} \right] r d\Omega =$$

$$\int_\sigma \left(-\frac{1}{\rho} f_x n_x + v\frac{\partial u}{\partial r} n_r \right) \delta u r d\Gamma \qquad (5.34)$$

f_x is the known stress component in the x-direction on the forced border. n_x and n_r are the directional cosines on the face of the border. When the considered face of the border is perpendicular to the pipe wall, $n_x \equiv \pm 1$ and $n_r \equiv 0$.

The rectangular unit is selected for discretization of the moving area. The linear function is chosen as the pressure interpolating function ψ_k and the second order function is selected as the velocity interpolating function φ_i. The plane (x, r) is mapped onto the plane (ξ, η) with equivalent parameter co-ordinate transformation. FEM equations of a typical unit for analysing fluid transients can be derived from equations (5.33) and (5.34):

$$A_{ij}\dot{u}_j + B_{ijl}u_j u_l + C_{ir}p_r + D_{ij}u_j = E_i \qquad (5.35)$$

$$F_{kr}\dot{p}_r + G_{krl}u_l p_r + H_{kj}u_j = I_k \qquad (5.36)$$

Mass matrix:

$$A_{ij} = \int_\Omega \phi_i \phi_{jr} d\Omega = \int_{-1}^{1}\int_{-1}^{1} \phi_i \phi_{jr} |J| d\xi d\eta$$

Matrix of velocity convection:

$$B_{ijl} = \int_\Omega \phi_i \frac{\partial \phi_i}{\partial x} \phi_{lr} d\Omega = \int_{-1}^{1}\int_{-1}^{1} \phi_i \frac{\partial \phi_i}{\partial x} \phi_{lr} |J| d\xi d\eta$$

Pressure matrix:

$$C_{ir} = \int_\Omega -\frac{1}{\rho}\frac{\partial \phi_i}{\partial x} \psi_r r d\Omega = \int_{-1}^{1}\int_{-1}^{1} -\frac{1}{\rho}\frac{\partial \phi_i}{\partial x} \psi_r |J| d\xi d\eta$$

Dissipation matrix:

$$D_{ij} = \int_\Omega \nu \frac{\partial \phi_i}{\partial r}\frac{\partial \phi_j}{\partial r} r d\Omega = \int_{-1}^{1}\int_{-1}^{1} \nu \frac{\partial \phi_i}{\partial r}\frac{\partial \phi_j}{\partial r} r|J| d\xi d\eta$$

Outer force vector:

$$E_i = \int_\sigma -(1/\rho) f_x n_x \phi_i r d\Gamma$$

Compression matrix:

$$F_{kr} = \int_\Omega \psi_k \psi_r r d\Omega = \int_{-1}^{1}\int_{-1}^{1} \psi_k \psi_r r|J| d\xi d\eta$$

Matrix of pressure convection:

$$G_{krl} = \int_\Omega \psi_k \frac{\partial \psi_r}{\partial x}\phi_l r d\Omega = \int_{-1}^{1}\int_{-1}^{1} \psi_k \frac{\partial \psi_r}{\partial x}\phi_l r|J| d\xi d\eta$$

Continuous matrix:

$$H_{kj} = \int_\Omega -\rho a^2 \frac{\partial \psi_k}{\partial x}\phi_j r d\Omega = \int_{-1}^{1}\int_{-1}^{1} -\rho a^2 \frac{\partial \psi_k}{\partial x}\phi_j r|J| d\xi d\eta$$

Velocity vector of border:

$$I_k = \int_u -\rho a^2 \hat{u} n_x \psi_k r d\Gamma$$

In the above coefficient matrices, $i, j, l = 1, 2, \ldots, 8$, and $k, r = 1, 2, 3, 4$. J is the .cobian matrix,

$$J = \begin{bmatrix} \frac{\partial x}{\partial \xi} & \frac{\partial r}{\partial \xi} \\ \frac{\partial x}{\partial \eta} & \frac{\partial r}{\partial \eta} \end{bmatrix}$$

The FEM equations can be solved using the non-symmetrical frontal solution ethod (see Section 5.5.2). Figure 5.29 is a velocity distribution diagram in the pipe at a time when fluid transients have occurred. This diagram is plotted using the FEM method stated above. Figure 5.30 shows the velocity-time curves at different nodes.

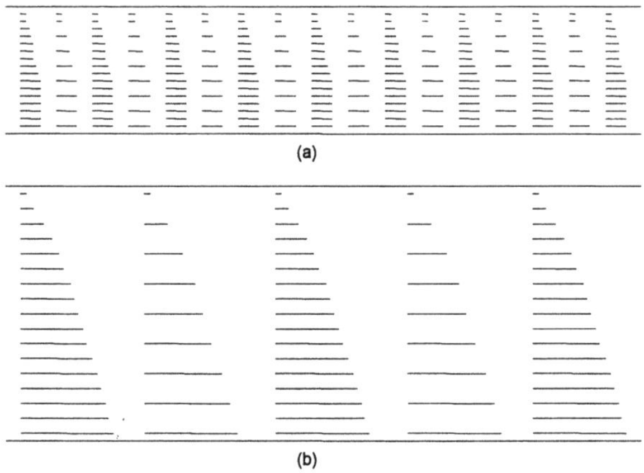

Figure 5.29: A velocity distribution diagram in the pipe at a time when fluid transients have occurred.

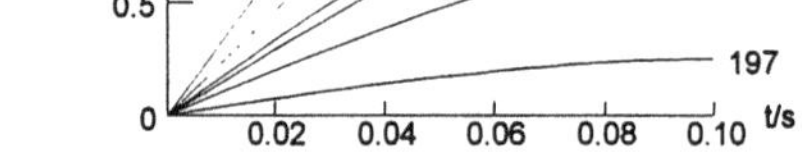

Figure 5.30: Velocity-time curves at different nodes.

5.5.2 Non-symmetrical frontal solution method for fluid FEM

Methods for solving finite element algebraic equations are mainly divided into two categories, namely the direct solution and the iterated solution methods. The frontal solution method is the only direct solution method that has now generally been recognised as a better evaluation technique for solving finite element equations with sparse coefficient matrices, effectively on a micro-computer, which usually has a smaller storage capacity. The method has been widely proven to be very effective for solving positive definite matrix equations arising from the finite element method when an external memory with a large storage capacity is provided. However, for some numerical calculations for practical engineering problems, such as the fluid transient simulations of hydraulic presses, the finite element matrices are non-symmetrical. In these cases, there will be no guarantee of the stability of the solution procedure if the pivoting search is still being made along the leading diagonal line of a matrix as in the conventional (or symmetric) frontal methods. In fact, a smaller or zero entry may occur on the diagonal and cause the computation to stop or fail. Therefore, when finite element equations with non-symmetrical coefficient matrices are to be evaluated, it is necessary to consider other forms of pivoting choice.

5.5.2.1 Symmetrical frontal method

The frontal method was originally devised by Irons [14]. Considering the nature of the elements in FEM, Irous constructed a systematic method to handle sub-divisions, and combined the sectional technique with the assembly and elimination of these sub-divisions, to produce the pithy frontal solution algorithm. The frontal technique mainly uses the properties in forming the finite element equations. The global coefficient matrix and the right-hand side vector are formed by summing up the element matrices and the right-hand terms one-by-one. Thus, it is not necessary to eliminate the variables only after the global equations have been fully formed. Once an unknown has been fully summed, it can be eliminated immediately. Hence, in the frontal method, the equations are being assembled while the unknowns are being eliminated. A four element mesh shown in Figure 5.31 will be used to explain the alternation of the two procedures. It is assumed that there is only a single degree of freedom, i.e., the number of unknowns is one, at each node.

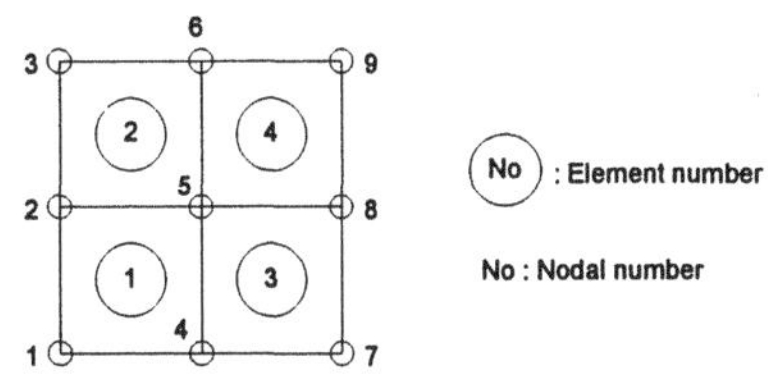

Figure 5.31: A four-element mesh.

The totality of the finite element equations can be written as:

$$\{A\}\boldsymbol{X} = \boldsymbol{B} \tag{5.37}$$

$\{A\}$ is a linear matrix of the coefficients of the vector $\boldsymbol{X}$, and $\boldsymbol{B}$ is the right-hand side vector. After the assembly of the first element, the above matrix equation becomes as follows:

$$\begin{bmatrix} a_{11}^{(1)} & a_{14}^{(1)} & a_{15}^{(1)} & a_{12}^{(1)} \\ a_{41}^{(1)} & a_{44}^{(1)} & a_{45}^{(1)} & a_{42}^{(1)} \\ a_{51}^{(1)} & a_{54}^{(1)} & a_{55}^{(1)} & a_{52}^{(1)} \\ a_{21}^{(1)} & a_{24}^{(1)} & a_{25}^{(1)} & a_{22}^{(1)} \end{bmatrix} \begin{bmatrix} X_1 \\ X_4 \\ X_5 \\ X_2 \end{bmatrix} = \begin{bmatrix} b_1^{(1)} \\ b_4^{(1)} \\ b_5^{(1)} \\ b_2^{(1)} \end{bmatrix} \tag{5.38}$$

From equation (5.38), it is obvious that the equations associated with the unknown X_1 have already been fully summed. Therefore, these equations may be employed as a pivot to eliminate X_1. The following equations can be obtained after X_1 has been eliminated.

$$\begin{bmatrix} a_{44}^{(1)} - \frac{a_{41}^{(1)}}{a_{11}^{(1)}} a_{14}^{(1)} & a_{45}^{(1)} - \frac{a_{41}^{(1)}}{a_{11}^{(1)}} a_{15}^{(1)} & a_{42}^{(1)} - \frac{a_{41}^{(1)}}{a_{11}^{(1)}} a_{12}^{(1)} \\ a_{54}^{(1)} - \frac{a_{51}^{(1)}}{a_{11}^{(1)}} a_{14}^{(1)} & a_{55}^{(1)} - \frac{a_{51}^{(1)}}{a_{11}^{(1)}} a_{15}^{(1)} & a_{52}^{(1)} - \frac{a_{51}^{(1)}}{a_{11}^{(1)}} a_{12}^{(1)} \\ a_{24}^{(1)} - \frac{a_{21}^{(1)}}{a_{11}^{(1)}} a_{14}^{(1)} & a_{25}^{(1)} - \frac{a_{21}^{(1)}}{a_{11}^{(1)}} a_{15}^{(1)} & a_{22}^{(1)} - \frac{a_{21}^{(1)}}{a_{11}^{(1)}} a_{12}^{(1)} \end{bmatrix} \begin{bmatrix} X_4 \\ X_5 \\ X_2 \end{bmatrix} = \begin{bmatrix} b_4^{(1)} - \frac{a_{41}^{(1)}}{a_{11}^{(1)}} b_1^{(1)} \\ b_5^{(1)} - \frac{a_{51}^{(1)}}{a_{11}^{(1)}} b_1^{(1)} \\ b_2^{(1)} - \frac{a_{21}^{(1)}}{a_{11}^{(1)}} b_1^{(1)} \end{bmatrix} \tag{5.39}$$

The equations from the second element are continually assembled to obtain the following equation.

$$\begin{bmatrix} a_{22}^{(2)} + a_{22}^{(1)} - \frac{a_{21}^{(1)}}{a_{11}^{(1)}} a_{12}^{(1)} & a_{23}^{(2)} & a_{24}^{(1)} - \frac{a_{21}^{(1)}}{a_{11}^{(1)}} a_{14}^{(1)} & a_{25}^{(2)} + a_{25}^{(1)} - \frac{a_{21}^{(1)}}{a_{11}^{(1)}} a_{15}^{(1)} & a_{26}^{(2)} \\ a_{32}^{(2)} & a_{33}^{(2)} & 0 & a_{35(1)}^{(2)} & a_{36}^{(2)} \\ a_{42}^{(1)} - \frac{a_{41}^{(1)}}{a_{11}^{(1)}} a_{12}^{(1)} & 0 & a_{44}^{(1)} - \frac{a_{41}^{(1)}}{a_{11}^{(1)}} a_{14}^{(1)} & a_{45}^{(1)} - \frac{a_{41}^{(1)}}{a_{11}^{(1)}} a_{15}^{(1)} & 0 \\ a_{52}^{(2)} + a_{52}^{(1)} - \frac{a_{51}^{(1)}}{a_{11}^{(1)}} a_{12}^{(1)} & a_{53}^{(2)} & a_{54}^{(1)} - \frac{a_{51}^{(1)}}{a_{11}^{(1)}} a_{14}^{(1)} & a_{52}^{(2)} + a_{55}^{(1)} - \frac{a_{51}^{(1)}}{a_{11}^{(1)}} a_{15}^{(1)} & a_{56}^{(2)} \\ a_{62}^{(2)} & a_{63}^{(2)} & 0 & a_{65}^{(2)} & a_{66}^{(2)} \end{bmatrix} \begin{bmatrix} X_2 \\ X_3 \\ X_4 \\ X_5 \\ X_6 \end{bmatrix}$$

$$= \begin{bmatrix} b_2^{(2)} + b_2^{(1)} - \frac{a_{21}^{(1)}}{a_{11}^{(1)}} b_1^{(1)} \\ b_3^{(2)} \\ b_4^{(1)} - \frac{a_{41}^{(1)}}{a_{11}^{(1)}} b_1^{(1)} \\ b_5^{(2)} + b_5^{(1)} - \frac{a_{51}^{(1)}}{a_{11}^{(1)}} b_1^{(1)} \\ b_6^{(2)} \end{bmatrix} \tag{5.40}$$

The procedure above can be summarised as follows. Firstly, the first element is assembled. Next, the variable X_1 is eliminated. Finally, the second element is assembled. If the coefficient matrices of the first and second elements are summed up before eliminating X_1, the state of the matrix equation remains the same as equation (5.40). This demonstrates that when the contributions of some variable have been completed, i.e., its corresponding rows and columns have been fully summed, the result is always the same regardless of whether assembly or elimination has been performed first. There are advantages of alternating assembly with elimination. Firstly, matrix A is only used to store those entries affected by the elimination procedure. Thus, the storage is reduced considerably, and the time-consuming operation of scarching a certain entry in a fully formed global matrix can be prevented. Secondly, this alternating process makes it possible to choose an effective pivoting mode without excessive non-zero entry growth. Thirdly, the numerical uncertainty caused by the round-off error is reduced, and the total operation is decreased as the variables are eliminated as early as possible. Lastly, any newly assembled equations will always occupy the initial effective positions at the front, and there is no need to carry out an integrated shift to the equations in the memory. Therefore, the solution efficiency is improved.

It can be seen from above that the symmetric frontal method can solve the problems of storage capacity and operation time. It is beneficial to use this method for solving finite element equations on micro-computers. However, for non-symmetrical coefficient matrices, the most suitable pivot may not be on the leading diagonal. Therefore, it is necessary to improve the conventional pivoting choice mode to meet the requirements of non-symmetrical matrices.

5.5.2.2 Non-symmetrical frontal method

Some obvious differences exist between the two frontal implementation schemes for solving symmetric and non-symmetrical matrix equations. Each scheme uses its own pivoting mode, which has different ways of setting up, handling and storing the frontal information. Comparatively, the frontal routines for solving symmetric matrix equations are simpler as the pivoting search is usually on the diagonal. For solving non-symmetrical matrix equations, other pivoting modes are required, such as the column pivoting or the total pivoting mode. If the order of the frontal coefficient matrix $A_{W\times W}$ is much smaller than that of the global coefficient matrix $A_{N\times N}$, namely $W < N$, and the pivoting choice can only be made among the fully completed rows and columns of the matrix $A_{W\times W}$, a total pivoting mode should be adopted. The non-zero entry growth is not very serious as the local fill produced during variables elimination is limited in the matrix $A_{W\times W}$. Generally, the total pivoting mode hardly increases the memory requirements [15].

It is known from the fundamentals of the frontal method, that when all contributions to a certain node from the other elements around it have been completed, the corresponding variables at the node can be eliminated. The reduced equations will be moved to a buffer storage from the memory. The equations that remain in the memory with their corresponding nodes and variables are called the front. The number of unknowns in the front is termed the front width (w). This width changes continuously since the nodes are constantly being placed and removed from the front. Figure 5.32 shows the progress of the front though a finite element mesh.

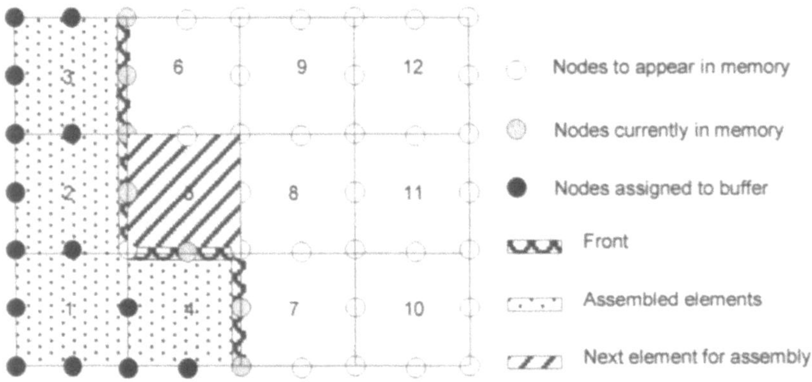

Figure 5.32: Progress of the front though a finite element mesh.

The implementation of the total pivoting mode used in the non-symmetrical frontal method will be explained using the mesh shown in Figure 5.33 as an example. The topology relations in this mesh structure can be represented by a 2-D array NOD as shown in Table 5.1.

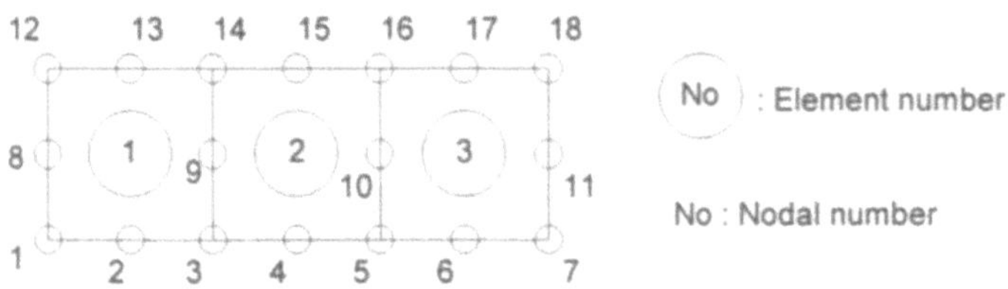

Figure 5.33: Total pivoting mode used in the non-symmetrical frontal method.

Table 5.1: Nodal connections for the example.

Element number	Local nodal number JNOD							
IELE	1	2	3	4	5	6	7	8
1	1	2	3	9	14	13	12	8
2	3	4	5	10	16	15	14	9
3	5	6	7	11	18	17	16	10

Two degrees of freedom are assumed at each corner node while a single degree of freedom is assumed at each middle side node. For every element IELE, using the array NOD(IELE, JNOD), an identifying vector LOC can be constructed. The vector LOC identifies the location, LOC(KVAB) row, of a KVAB variable, within the frontal matrix $A_{W \times W}$. After the pre-frontal operation, if a value in LOC(KVAB) is negative, it means that when the element IELE has been assembled into the matrix A, the row or column corresponding to the variable KVAB is fully formed. For the first element, the values of the identifying vector LOC are tabulated in Table 5.2.

Table 5.2: Values in LOC for element no. 1.

Local nodal number, JNOD	1		2	3		4	5		6	7		8
Local degree of Freedom, KVAB	1	2	3	4	5	6	7	8	9	10	11	12
Identifying vector LOC(KVAB)	-1	-2	-3	4	5	13	18	20	-18	-16	-17	-12

In Table 5.2, the negative values indicate that the nodes corresponding to the variables numbered 1, 2, 3, 9, 10, 11 and 12 in the first element have been fully summed. They will not appear again in the other elements. In addition, using the LOC vector for each element, the destination vectors (KDES and LDES) and the heading vectors (KHED and LHED) can be obtained. After the assembly of the first element, these vectors are of the form shown in Table 5.3.

Table 5.3: Values of destination and heading vectors for element no. 1.

KDES(KVAB) LDES(KVAB)	1	2	3	4	5	6	7	8	9	10	11	12
KHED(NFRN) LHED(NFRN)	-1	-2	-3	4	5	13	19	20	-18	-16	-17	-12

NFRN is the current front width and NFRN < W. When the front moves through the element mesh, the values in these vectors vary.

The destination vectors indicate where the coefficients contributed from each element matrix are to be placed in the front matrix $A_{W \times W}$. The heading vectors label the rows and columns in the $A_{W \times W}$ matrix with the values contained in the LOC vector. Thus, each row and column of the A matrix are associated with a particular nodal variable.

The elimination process is initiated by a search for the values in the heading vector that are coded with a negative sign. The numbers of the fully summed rows and columns in the A matrix are entered into arrays KPIV and LPIV. The pivotal search then continues to the arrays. Based on the total pivoting mode, the largest entry in the summed rows and columns is sought. The pivotal row and column are stored in KPOS and LPOS respectively. Next, the pivotal row is normalised about the chosen pivot. For elimination, the front matrix $A_{W \times W}$ is sub-divided into four distinct parts. Each part is bordered by the pivoting column or row, as shown in Figure 5.34.

For each part, the following total pivoting elimination algorithm is applied.

$$t_{ij} = a_{ij} - a_{i,LPOS}\left(\frac{a_{KPOS,i}}{a_{KPOS,LPOS}}\right) \tag{5.41}$$

1. For entries in part C:
   $i = 1, \ldots,$ KPOS-1. $j = 1, \ldots,$ LPOS - 1. The calculated t_{ij} is assigned to a_{ij}.
2. For part D:
   $i = 1, \ldots,$ KPOS-1. $j =$ LPOS + 1, …, w. t_{ij} is assigned to $a_{i,j-1}$, i.e., entries in D are returned one space to the left.
3. For part E:
   $i =$ KPOS + 1, …, w. $j = 1, \ldots,$ LPOS - 1. The evaluated t_{ij} are assigned to $a_{i-1,j}$, i.e., entries of E are repositioned one space vertically.
4. For part F:
   $i =$ KPOS + 1, …, w. $j =$ LPOS + 1, …, w. t_{ij} is assigned to $a_{i-1,j-1}$, namely, elements in F are moved one space to the left as well as vertically.

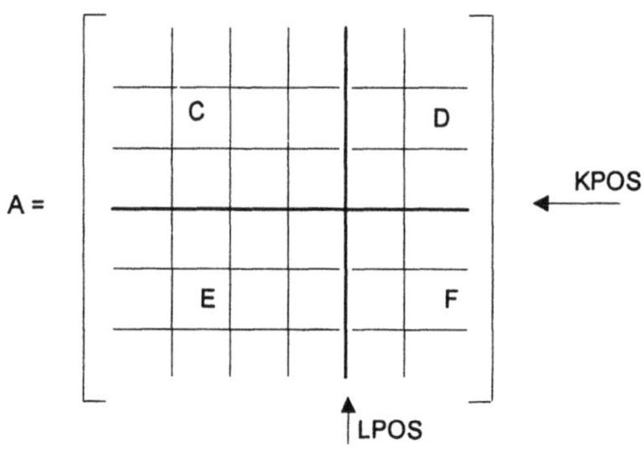

Figure 5.34: A part bordered by the pivoting column or row.

Similarly, the right-hand side vector is divided into two zones, G and H. Each zone is bordered by the pivoting row. In each zone, the Gaussian algorithm is applied.

$$S_i = b_i - a_{i,KPOS}\left(\frac{b_{KPOS}}{a_{KPOS,LPOS}}\right) \tag{5.42}$$

1. For entries in zone G:
   $i = 1, \ldots,$ KPOS - 1. The evaluated S_i is assigned to b_i.

2. For zone H:
   i = KPOS + 1, ..., w. S_i is sent to b_{i-1}.

After the reduction of the non-symmetrical matrix A and the right-hand side vector B according to equations (5.41) and (5.42), the eliminated pivotal row, the heading vectors, the front side and the position of the pivot are recorded for use in the back substitution process. Next, the heading vectors are altered to remove the trace of the last pivotal equation for the elimination of the next variable or the assembly of the next element.

The solution process of the non-symmetrical frontal technique can be stated as follows. The process starts by assembling each of the element coefficient into a matrix in memory until the allocated memory area is filled. Within the assembled part of the matrix, a total pivoting search is then made to determine the largest entry from the fully summed rows and columns. The pivotal row is used to eliminate, using equations (5.41) and (5.42), all the coefficients in the pivotal column. The pivotal row is then stored in a buffer. After sufficient coefficients have been reduced, the next element coefficient matrix can be assembled. Further elimination may be performed next. The two procedures of assembly and elimination are alternately repeated. Finally, when all the unknowns have been reduced, using the data in the buffer in the reverse sequence, the solution can be reached using the back substitution process. The program flowchart for the non-symmetrical frontal solution method is illustrated in Figure 5.35.

5.5.2.3 Examples of fluid finite element calculation

Using both the symmetric and non-symmetrical frontal programs, the finite element simulation of a 2-D incompressible or compressible viscous flow has been carried out on a micro-computer. The two algorithms are compared with respect to memory seizure and execution time.

The first example adopts the basic variables (u, v, p) to solve a 2-D steady incompressible flow problem. An eight-node quadratic and a four-node linear interpolation functions are used to find the velocity and pressure respectively. The calculated domain is divided into 73 quadrilateral elements with 264 velocity nodes and 96 pressure nodes. The total number of equations to be solved is 624. For this example, the results of the comparison are listed in Table 5.4.

For this example, the coefficient matrix is symmetric, and the largest entry in every column of the matrix frequently appears in the leading diagonal. Therefore, if the pivoting search is limited to the diagonal, the operation time will be reduced considerably. It can be seen from Table 5.4 that the memory requirement and computation time of both frontal methods are affected not only by the pivotal mode, but also by the front width. The larger the front width, the wider the pivoting search scope, and the longer the execution time. Thus, it is necessary to determine the optimum value of the front width for each particular problem.

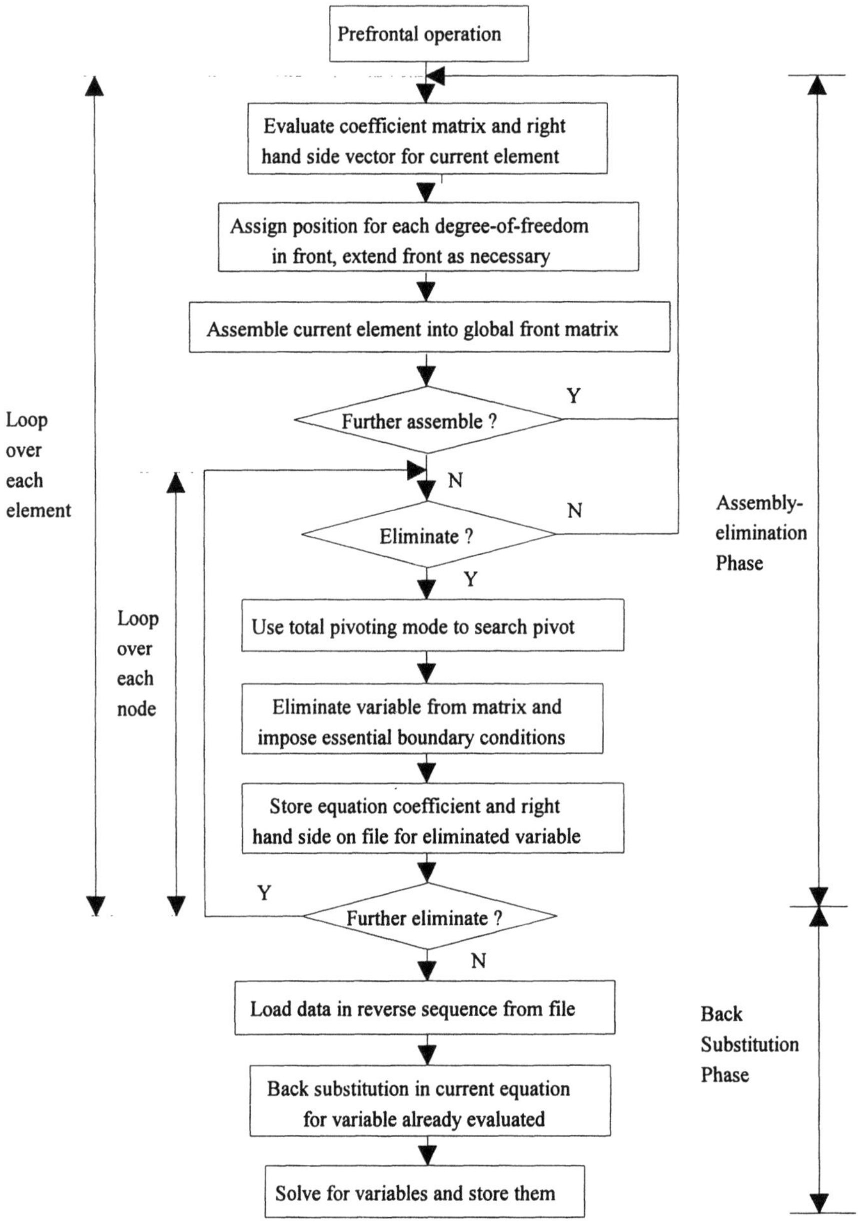

Figure 5.35: Program flowchart.

In the second example, fluid transients in the pipelines of a hydraulic drive system are simulated. The governing equations of this 2-D unsteady compressible flow are solved using the basic variables (u, p). For this example, the coefficient matrix is non-symmetrical. When the symmetric frontal method is employed to evaluate the equations, the solution procedure stopped mid-way due to the occurrence of a very small pivot on the diagonal. The diagonal pivoting mode is changed to the total pivoting mode of the non-symmetrical frontal technique after the break. Since the pivoting choice is now more reasonable, the solution procedure continues until the final answer has been obtained.

Table 5.4: Comparison of two frontal solution methods.

Compared items algorithm	Pivotal mode	Maximum front width	Memory size (KB)	Execution time (minute)
Symmetrical Frontal method	Diagonal pivoting	80	38 880	8.3
		60	38 880	5.7
Non-symmetrical Frontal method	Total pivoting	60	40 096	6.5

This example uses 128 elements and 433 nodes to discretize the flow domain that contains 586 unknown variables. Ten time steps are computed, in which the first eight steps are performed with a time-varying boundary condition that corresponds to the opening process of a control valve downstream. The non-linear finite element equations are treated with an iteration routine. Three to six iterations are needed for each step to converge. It can be concluded from the computed results that the non-symmetrical frontal solution program can provide satisfactory convergence. In addition, the solution speed and the memory requirement are suitable for practical engineering calculations.

References

1. Cavendish J C 1974 Automatic of Arbitrary Planar Domain for the Finite Element Method. *International Journal Numerical Method Engineering* 8:679-696
2. Lawson C C 1977 Software For C^1 Surface Interpolation. In: Rice J (ed) 1977 *Mathematical Software III.* New York, Academic Press, pp 161-194
3. Watson D F 1981 Computing the n-Dimensional Delaunay Tessellation with Applications to Voronoi Polytopes. *Computer Journal* 24(2):167-172
4. Cavendish J C, Field D A 1985 An Approach to Automatic Three-Dimensional Finite Element Mesh Generation. *International Journal Numerical Method Engineering* 21:329-347
5. Frey W H 1987 Selective Refinement: A New Strategy For Automatic Node Placement in Graded Triangular Meshes. *International Journal Numerical Method Engineering* 24:2183-2200
6. Field D A 1988 Laplacian Smoothing and Delaunay Triangulations. *Commun. Appl. Numer. Meth.* 4:709-712
7. Oh S I, Tang J P, Badawy A 1984 Finite Element Mesh Rezoning and its Applications to Metal Forming Analysis. In: *Proceedings of 1st ICTP Conference, Vol 2.* Tokyo, pp 1051-1058
8. Ficke J A, Oh S I, Malas J 1984 FEM Simulation of Closed Die Forging of Titanium Disk using ALPID. In: *Proceedings of NAMRC XII.* Houghton, MI., pp 166
9. Gray W H, Akin J E 1979 An Improved Method for Contouring on Isoparametric Surface. *International Journal of Numerical Method Engineering* 14:451-458
10. Kobayashi S, Oh S I, Altan T 1989 *Metal Forming and the Finite Element Method.* Oxford University Press, New York
11. Ling T, Zhou B K 1984 *The Metalforming Processes* 3:75-83 (In Chinese)
12. Lee C H, Kobayashi S 1973 New Solutions to rigid-plastic Deformation Problems Using a Matrix Method. *Transactions of ASME, Journal of Engineering for Industry* 95:865-873
13. Yang Q Z, Wang Y G 1992 A Finite Element Analysis for Fluid Transients. *Journal of Huazhong University of Science and Technology* 20(5):13-17 (In Chinese)
14. Irons B M 1970 A frontal solution program for finite element analysis. *International Journal of Numerical Methods* 2(1):5-32
15. Yang Q Z , Wang Y G, Huang S H 1992 Application of Unsymmetrical Frontal Solution Method to Fluid Finite Element Calculation. In: *Proceedings of International Conference on Advanced Technology and Machinery in Metal Forming*, pp 609-620

Chapter 6

CAE/CNC of machines for near net-shape operations

Y.G. Wang and Q.Z. Yang

6.1 Introduction

With the development of science and technology, and market competition, the requirement for machines for near net-shape operations is increasing and the design task of these machines is becoming more demanding. The design of machines for near net-shape operations consists of several stages, *viz.*, construction design, kinematics analysis, kinetics analysis, strength analysis, control system design, manufacturing design, etc.

There are several commercial software packages, such as AutoCAD, Unigraphics, Pro/ENGINEER, etc., and other specific software packages such as ADAMS, for the design of machines for near net-shape operations. Early draughting software mainly acts as tools for making 2- and 3-D drawings, and is usually not intelligent. Second generation software uses parametric and feature-based techniques as well as artificial intelligence tools.

CAE software packages are well-suited for performing kinematics, kinetics and strength analyses for the design of machines for near net-shape operations. For each analysis mentioned above, the first step is to establish a mathematical model of the object to be analysed. Solution of the mathematical model can be obtained using either analytical or numerical solution method. The FEM is one of the commonly used tools for the numerical solution method.

CNC is widely used in machines for near net-shape operations. Older control systems are machine-specific. More recent control systems are highly flexible and mainly based on industrial micro-computers.

6.2 Universal CNC systems for near net-shape operations

6.2.1 Composition of a universal CNC control system

Many types of CNC machines are used in near net-shape operations. Although the physical functions of these machines may vary, their control requirements are quite similar to one other. It is highly desirable to produce a universal CNC control system to avoid designing a control system for every machine. A three-stage universal control system is shown in Figure 6.1. The operation desk, programmable logic controller (PLC) and power supply cabinet form the first stage for processing and controlling on-off signals. The second stage consists of an industrial PC and intelligent control module boards. The boards are used to receive instructions and data from the PC. This is combined with different close-loop control sub-systems (e.g., positioning, velocity and pressure control sub-systems) that have feedback elements, amplifiers, hydraulic control valves or driving controllers of motors. The PC is mainly used for man-machine interaction, menu management, data input, displaying and sending action commands to closed-loop control sub-systems for co-ordinating their operations. The auxiliary micro-computer and printer constitute the third stage, which is used for setting, displaying, recording and printing data, off-line programming, simulation and connection with the computer management system in a factory.

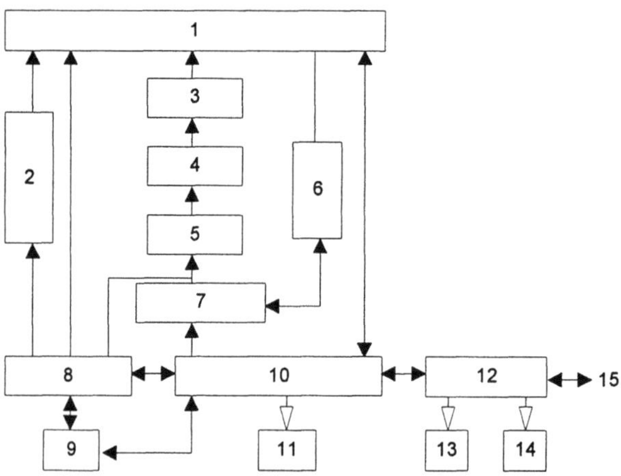

1 - Machine, 2 - Power supply, 3 - Execute components, 4 - Control valves/Driving controller, 5 - Amplifier, 6 - Feedback components, 7 - Intelligent control module, 8 - PLC, 9 - Operation desk, 10 - Industrial PC, 11 & 13 - CRT, 12 - Auxiliary computer, 14 - Printer, 15 - Computer management system of factory

Figure 6.1: Block diagram of a universal CNC control system.

The CNC control system is modular. Problems from the lower stages have no influence on the higher stages. The most important stage (i.e., first stage) uses a high reliability industrial PC.

6.2.2 Intelligent control module board

The execution components for the analogue control of machines in near net-shape operations can either be motors (DC/AC servo motors, stepping motors) or hydraulic cylinders/motors. When servo motors are used, it is necessary to allocate the relevant amplifier and driving controller, while relevant amplifier and control valves (proportional/servo valves) have to be provided when hydraulic cylinders/motors are used. From the control view point, there is no substantial difference between these two types of execution components. Therefore, an intelligent control module board (Figure 6.2) has been developed to amplify and control the different execution components. The core of the board is a 80C196KC 16-bit single-chip micro-computer, an RS-232 serial interface for communicating with the industrial PC, four A/D converters or photo-electronic isolators on-off signals input, one PWM output, twelve photo-electronic isolators for on-off signals output, and a -10V — +10V analogue voltage output. The board, combining with the execution and feedback components, can be conveniently constructed for different multi-axis control systems.

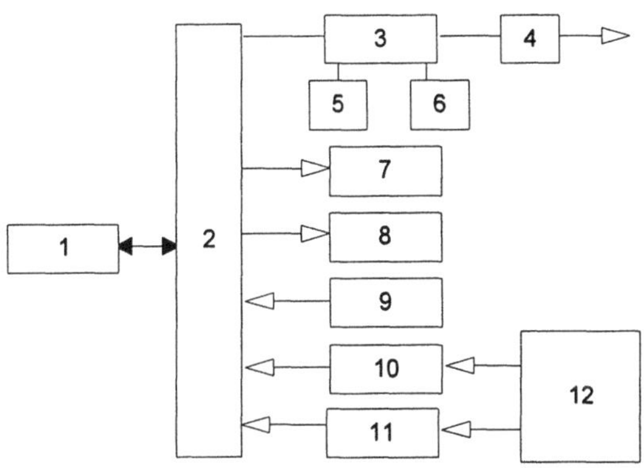

1 - Serial interface, 2 - Single-chip micro-computer, 3 - Decorder, 4 - D/A converter,
5 - EPROM, 6 - RAM, 7 - On-off output, 8 - PWM output, 9 - A/D converter,,
10 - Pulse phase discriminator, 11 - Protector for puls input, 12 - Encorder for pulse input

Figure 6.2: Details of an intelligent control module board.

6.2.3 Module of control software

In the intelligent control module, the control software is modular. It consists of more than ten standard modules, including communication, correction, dead-band compensation, limitation of amplitude, D/A and A/D conversions, reversible count, etc. It is convenient to build modular software based on the allocation of different hardware. The open/close-loop control for position, velocity and pressure can be easily obtained by changing the module of the EPROM control program. Figure 6.3 is an example of a control system for a tube bending FMC. In this system, there are two motor-driven position axes and one hydraulic servo position axis. Therefore, three intelligent control modules are used.

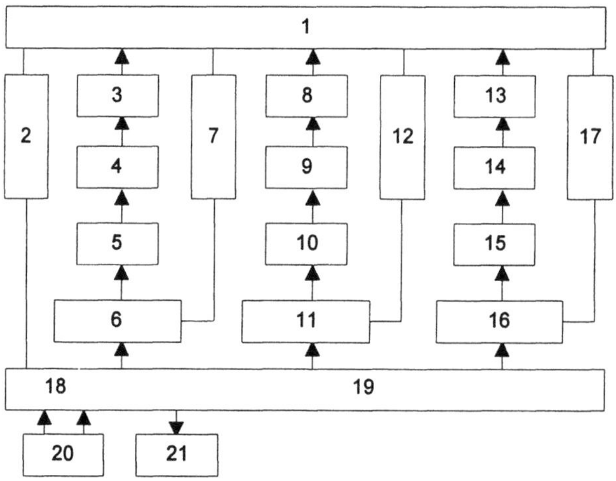

1 - Rube FMC, 2 - Power supply, 3,8 - Servo motor, 4,9 - Servo controller, 5,10,15 - Amplifier, 6,11,16 - Intelligent control module, 7,12,17 - Pulse transducer, 13 - Cylinder, 14 - Servo valve, 18 - Build-in PLC, 19 - Industrial PC, 20 - Operation desk, 21 - CRT

Figure 6.3: A control system for a tube bending FMC.

6.2.4 Communication in an integrated-distributed CNC system

One of the characteristics of an integrated-distributed CNC system is the use of multi-processors. Therefore, it is important to describe the communication between processors.

Figure 6.4 is a communication system for controlling a tube bending FMC. In this system, an industrial PC sends the address signals of the DCC units to call them individually. When all the 80C196-KC single-chip micro-computers have received the call, an interruption occurs and the byte of the address signal in the interruption service program is checked to determine if it is equal to one. The address-equal DCC unit is the required one. It will receive the commands and data from the PC and perform the driving action or send feedback data to the PC. The non-called DCC units continue to work under the original states. When an emergency occurs, the DCC unit sends a signal to the PC for an interruption service. The PC will operate the troubled process program. This communication system has a good open characteristic and can be easily extended.

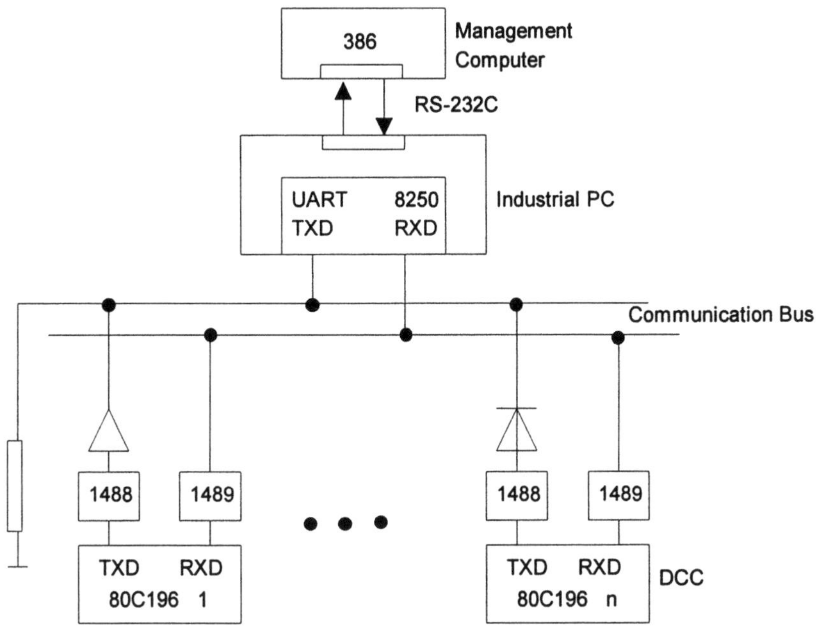

Figure 6.4: Communication system for controlling a tube bending FMC.

Figure 6.5 is a communication system for controlling a high speed hydraulic press. In this system, the communication between the PC and the DCC units is managed by an intelligent communication card with 8088 CPU and eight RS-232C serial communication ports. The PC sends information to the port address of the relevant DCC unit to communicate with the unit. The PC inquires a port address every 8 ms to send commands and read feedback information. The communication system has good reliability and lightens the load of the PC. However, the cost is higher.

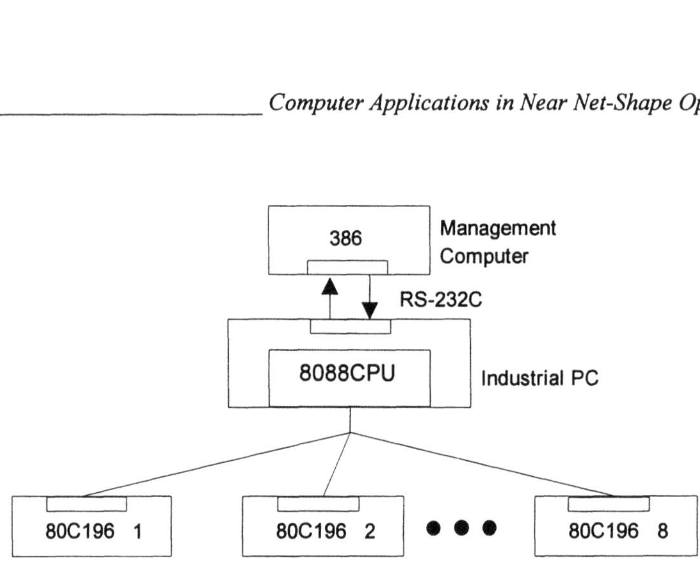

Figure 6.5: A communication system for controlling a high speed hydraulic press.

6.3 CNC system for sheet metal forming machines

6.3.1 CNC system for shearing machines

The CNC system for shearing machines is simpler. Its main function is to control the position of the back gauge. Figure 6.6 is a CNC system for a shearing machine. In this system, the back gauge is driven by an AC servo motor through a ballscrew. The MCS-51 is a single-chip micro-computer. A 2764 EPROM is used for serving the monitoring program. An 8K RAM is used for serving the process program of the product. A 8279 chip is applied for managing thirteen LED digital display tubes and nineteen keys of the keyboard. TXD and RXD are the communication ports between the micro-computer and the AC servo motor. The motor is controlled by a 8096 16-bit single-chip micro-computer.

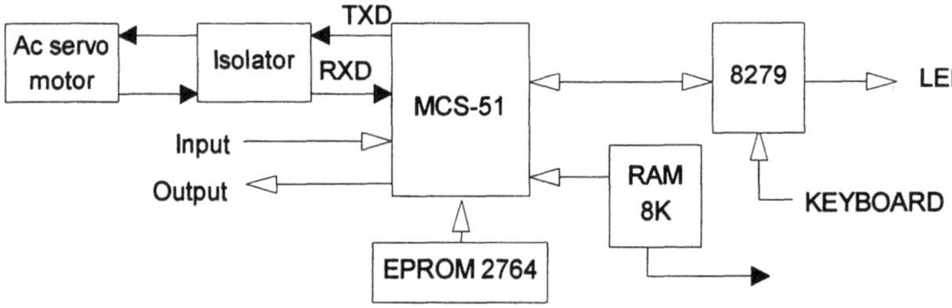

Figure 6.6: CNC system for a shearing machine.

6.3.2 CNC system for press brakes

In a hydraulic press brake, there are usually two movement mechanisms to be controlled, i.e., the back gauge (X-axis) and the slide (Y-axis). Figure 6.7 is a type of CNC systems for a press brake. It is a two-stage system. The first stage is a main controller based on the STD bus industrial micro-computer that performs parameters setting and calculation, function selection and execution, process controlling, and system monitoring and initialising. The second stage is a servo driving unit based on two 8096 single-chip micro-computers that control the AC servo motors for X- and Y-axes respectively. The main controller can send commands to the servo driving units through the communication port, e.g., initialising the positions of the axes, selecting operation modes, positioning points, etc. The driving units can feedback their actions to the main controller, such as the completion of the initialising and controlling operations, information about their troubles, etc.

The CNC system has the following functions:

1. Display
   There is a monitor on the control cabinet that is used for displaying program, input data, trouble information, process parameter, etc., to achieve interaction between the computer and the operator. After every operation, the actual value of the bending for this operation and the setting value for the next operation can be displayed on the monitor.

2. Man-Machine Interaction
   The operator and the computer can communicate through the keyboard and the monitor. For producing a new bending operation, the operator only needs to answer a few questions and set some data. The press can operate automatically.

3. Automatic Calculation of Dead-Point of Slide
   Based on the information of the material, bending and the die, the dead-point of the slide and the bending force can be calculated.

4. Off-Line Automatic Programming
   Besides programming with man-machine interaction, off-line programming can also be achieved using a PC. Once the operator has input the draft of the bending, the CNC system can produce the drawing of the product and the process sequence, select the die set and produce the CNC codes. The CNC codes can be transferred to the main controller through a serial communication port.

5. Parameter Storing
   There is an 8K storage for storing the parameters of 99 bendings. Every bending can have the parameters of 99 bending operations.

6. Self-Diagnosis Functions
   The system can diagnose the following errors.
   (1) Programming Errors
   The system can check all input parameters. If the displacements of the X- and Y-axes are beyond the limiting values, the input parameters have to be modified.

   (2) Operation Control Errors
   If the check points of the axes cannot be found for an operation, the system will send out a warning signal. Other problems in the operation can also be displayed on the monitor.

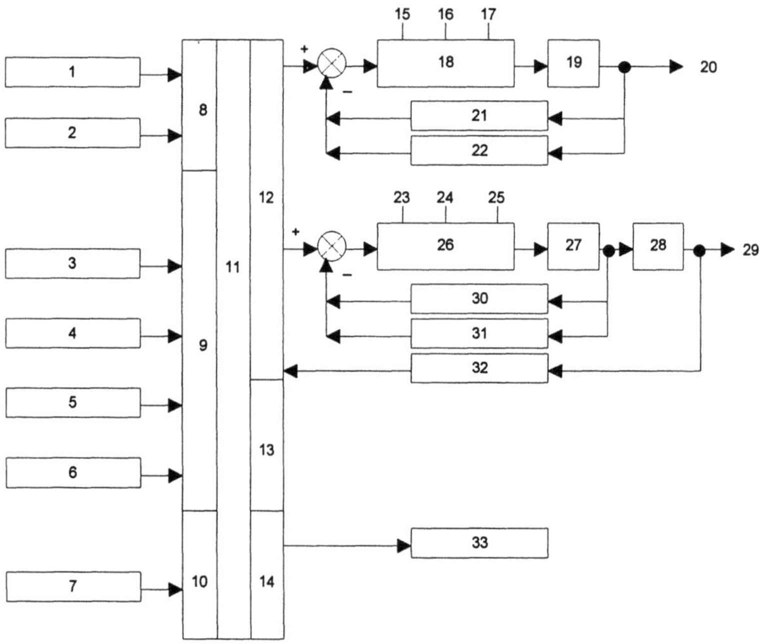

1 - Dector for slide displacement, 2 - Detector for pressure, 3,4,5,6 - Solenoid valve, 7 - Keyboard, 8 - A/D converter, 9 - Power driving card, 10 - Parallel interface card, 11 - STD bus industrial PC, 12 - Serial interface card, 13 - Displacement count card, 14 - CRT control card, 15,16,23,24 - Limit switch, 17,25 - Zero state switch, 18 - Servo controller for X axis, 19,27 - Servo motor, 20 - Back gauge (X axis), 21,30 - Speed feedback, 22,31,32 - Position feedback, 26 - Servo controller for Y axis, 28 - Driving mechanism, 29 - Stopper for Y axis, 33 - CRT

Figure 6.7: A CNC system for a press brake.

6.4 FMC for sheet metal bending

6.4.1 CNC and automatic generation system of software

Figure 6.8 is an outline of a FMC for sheet metal bending. It uses a 9-axis CNC system. It is a versatile sheet bending facility composing of the CNC press brake, loading/unloading robot, positioning table, supporting band, and distributed control system with four-stage micro-computers.

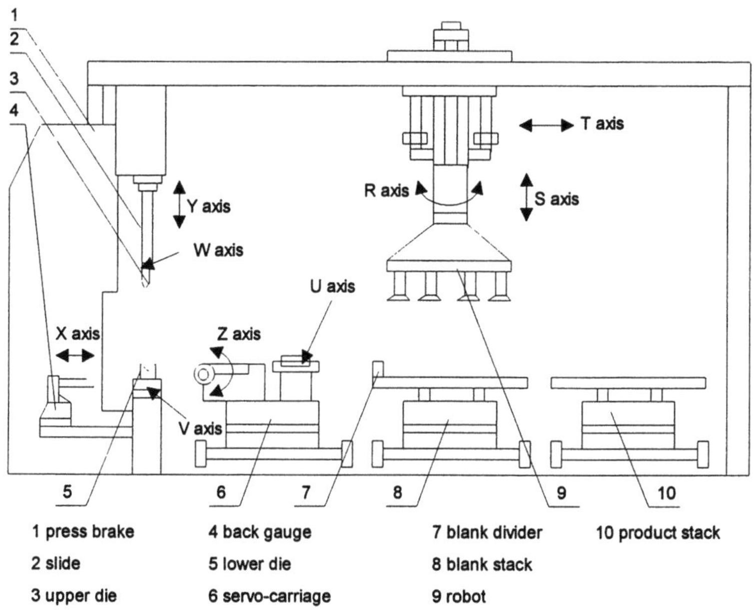

Figure 6.8: Outline of a FMC for sheet metal bending.

6.4.1.1 Computer control of the FMC

In the control system (Figure 6.9), the first stage is a micro-computer (designated as the master computer) that performs functions such as setting the parameters of the FMC, selection of operation modes, auto-programming of the operation sequence, information processing, monitoring display and management. The second stage is communication management, which manages the communication between the master computer and the slave computers to establish a synchronised communication network. The third stage consists of two slave computers to achieve a direct controlling function. The X, Y, W and V axes of the press brake (see Figure

6.9) are controlled by slave computer 1, while the T, S, R, U and Z axes are controlled by slave computer 2. The fourth stage is the servo controlling stage with nine units. Each unit has its own 16-bit single-chip micro-computer to control the operation of nine servo motors for driving the nine axes.

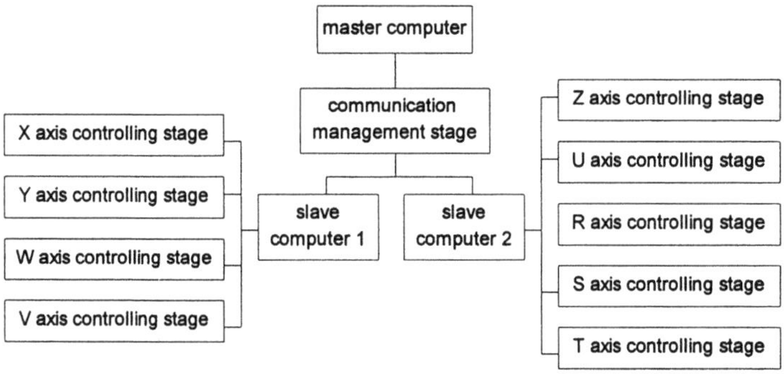

Figure 6.9: The control system of a FMC.

The system uses a computer specially designed for communication with a serial data transfer and a bus type structure. This type of structure has the advantages of simplicity and flexibility for extension. It is suitable for industrial control. The RS-232C interface is used for connection between the master computer and the communication management stage. There is a similar communication control board between the two slave computers and the communication stage. A serial interface on the board is linked to the common data channel via a linear driver and a linear receiver. The access right of the network bus is controlled by the communication management stage.

6.4.1.2 Automatic generation of system software

In a FMC, an integrated CAD/CAPP/CAM software can be generated automatically. The system software is based on feature information modelling. It consists of bending design, process analysis and calculation, process planning, NC coding, and dynamic simulation of the process (Figure 6.10).

With feature-based operations, a designer can use the parametric feature elements in a design feature library to construct a bending. Next, an interface automatically produces the model based on the feature information of the bending. In the design phase, the monitor evaluates the parameters of the input features of the technological constraints and manufacturing rules. When problems have been identified, they will be fed back to the designer immediately to improve the design quality and reduce the development cycle of the products. Since design and manufacturability analysis are performed simultaneously, concurrent design and manufacturing-oriented design of the products can be achieved. The management system of the engineering data library administers and maintains the feature library, knowledge library, product model and data library unitedly. It provides an interactive interface for adding, deleting, modifying and inquiring. The module for feature mapping can extract and transfer the feature information. Based on the background of the application and the knowledge of a certain field, the module can map the feature model to the required product data, and re-organise the application-oriented product model. Therefore, integration and sharing of product information can be realised. The module for graphics display shows the drawing of the designed product. The process planning module calculates the shape and size of the developed blank, spring-back of the bending, and the bending force for every step. Based on the parameters of the dies and the press brake in the data library and the knowledge in the knowledge library, it arranges an optimised bending sequence and calculates the control parameters of every axis to prevent interference between the bending, die and machine. The NC coding module generates the NC programs for the bending sequence and manufacturing parameters produced by the process planning module. Next, the NC program can be input to the control system manually (off-line) or automatically (on-line) by a serial communication interface. Using the simulation module, the bending process can be displayed on the screen of a computer setup-by-setup. The user could check the correctness of the process plan and the axis movements, and test the plan based on the interference and accuracy constraints. As mentioned above, the core of the CAD/CAPP/CAM software system is the feature-based product information model. The system integrates manufacturing, process analysis and checking with design.

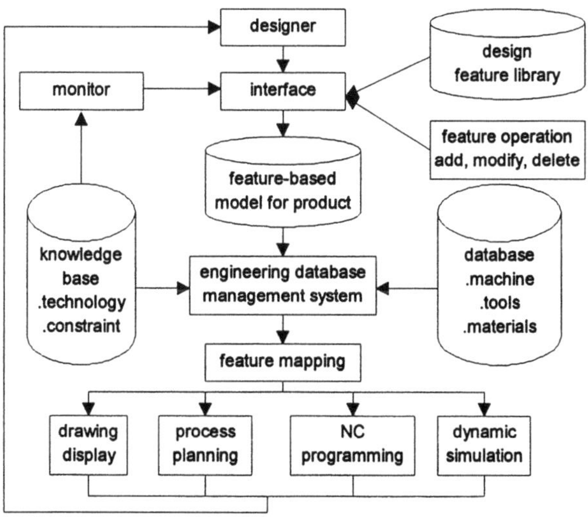

Figure 6.10: An integrated CAD/CAPP/CAM software outline.

6.4.2 Bending design based on features

From the investigation on the shapes and processing of bendings, typical bendings that are commonly used have been found and put into a design feature library in the parametric form. They can be displayed for selection during product design.

The features in the feature library are limited, and cannot express the bendings of all possible shapes. However, using two basic parametric feature elements (board and angle bendings) in the feature library and feature operations, a designer can construct any bendings with complex shapes. In addition, the structure of the feature library is open as a user is allowed to input self-defined features into the library.

Figure 6.11 is such an example. Since the bending (Figure 6.11a) could be considered as a combination of the two feature elements shown in Figures 6.11b and 6.11c, the designer can first select feature element 1 and the relevant parameters to obtain the first part of the bending. Next, the user can deal with feature element 2 using the same procedure to obtain the second part of the bending. Finally, the user can point out the connection state to complete the bending design. As mentioned earlier, bending design based on features can obviously reduce the input data, errors, and design time.

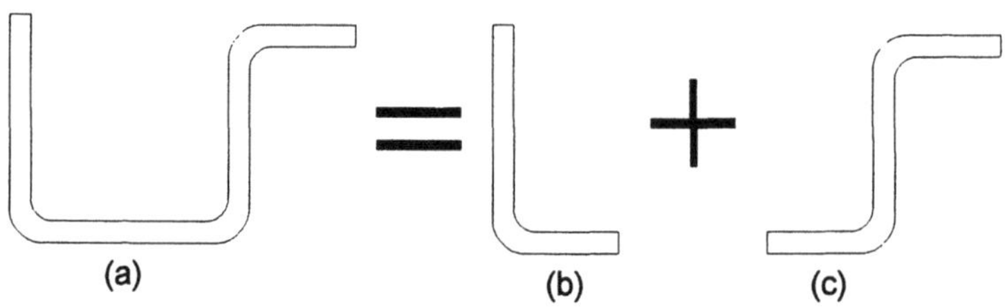

Figure 6.11: Use of features in a smple bending.

6.4.3 Manufacturability criteria for sheet metal bending FMC

Manufacturability criteria for sheet metal bending FMC are used to determine the suitability of the FMC process in generating the bending. The criteria can be summarised as follows:

A. Minimum relative bending radius

$$(r/t)min$$

r is the bending radius and t is sheet thickness. The smaller the r/t ratio, the larger the degree of deformation. If the bending radius is too small, cracks may appear on the surface of the sheet. Therefore, the bending process is limited by the minimum relative bending radius. Usually,

$$(r/t)\min = K$$

K is a coefficient relevant to the material and the rolling direction.

B. Minimum height of straight side of bending (Lmin)

To ensure bending quality, Lmin should be as follows:

$$L\min \geq 0.65V$$

V is the openness of the die. When the bending angle is equal to 90°,

$$L\min = 2t$$

C. Developed size of bending

A FMC can only produce parts in a certain size range. If the size is too small or too large, the FMC would not be able to work well. For the bending FMC, the range of bending size is limited by the vacuum inspiration disk of the loading/unloading robot.

D. Minimum distance from hole to bending line (bmin)

For bendings with holes, if the holes are stamped after bending,

$$b\min = r + 0.5t$$

If the holes are stamped before bending,

$$b\min = r + 1.5t$$

Otherwise, the shape accuracy of the holes cannot be guaranteed.

Manufacturability criteria have been considered concurrently in the design phase to improve the design quality and reduce the design time. When the criteria are violated, the information on these violations would be displayed on the monitor. The designer could then modify the bending design immediately. For example, when a user inputs some parameters of the feature element shown in Figure 6.12a, the monitor evaluates *L*1 and r1 to be too small, which violate criteria 1 and 2. The user can change the feature parameters with the prompt on the monitor to generate the correct feature shown in Figure 6.12b.

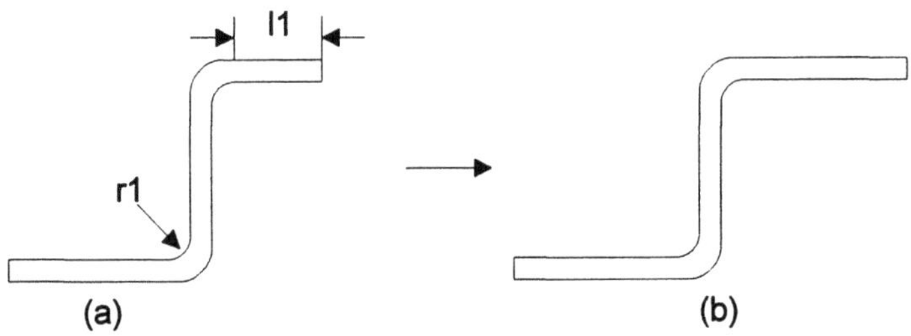

Figure 6.12: Manufacturability criteria evaluation in bending

Bibliography

Hu Y L, Tian Y M, Wang Y G 1993 Study on In-Machine Communication System of Integrated Distributed NC System. *Journal of Machinery & Electronics* 5:26-29 (In Chinese)

Li C X et. al. 1993 Universal NC System Used in Metal-Forming Machinery. *Journal of Metal-forming Machinery* 28(6):48-51 (In Chinese)

Tian Y M, Wang Y G 1992 The Typical Computer NC System. *Journal of Machinery & Electronics* 4:35-37 (In Chinese)

Wang Y G, Ma S, Tai Y M 1994 CNC Automatic Generation System of Software for Sheet Metal Bending FMC. In: *Proceedings of Pacific Conference on Manufacturing*. Jakarta, Indonesia, pp 363-368

Wang Y G, Tian Y M, Yang Q Z 1993 Design Analysis and Control of Flexible Manufacturing Cell for Sheet Metal Bending. In: *Proceedings of 2nd International Conference on Computer Integrated Manufacturing*. Singapore, pp 313-322

Wang Y G, Wang Z W 1991 *System Dynamics*. Huazhong University of Science and Technology Publishers, China (In Chinese)

Wu H X et al. 1991 A NC Shear System Based on Digital AC Servo Driver Unit. *Journal of Machinery & Electronics* 5:26-29 (In Chinese)

Chapter 7

IMOLD®: an intelligent mould design and assembly system

Y.F. Zhang, J.Y.H. Fuh, K.S. Lee and A.Y.C. Nee

7.1 Introduction

In this chapter, IMOLD®, an intelligent system for the design and assembly of injection moulds based on 3-D solid modelling techniques, is described. Firstly, an overview on the current problems and trends in injection mould design practice is given, followed by a brief review of the recent research developments. The key functional modules of IMOLD® are then described, followed by a case study from the solid part model to its complete mould assembly. Finally, discussions on the current approach of IMOLD® and conclusions are given.

7.2 Injection moulding

In the current manufacturing industry, injection moulding is one of the most common processes for plastic parts production. There are generally three steps in the injection moulding process:

1. Plasticising the material which then flows into the mould under pressure.
2. Solidifying molten plastic into the desired objects by the confines of the mould components.
3. Opening the mould to eject the solidified plastic part.

A basic mould consists of two plates that form an impression into which molten material is injected. The surfaces on the two plates (core and cavity) that meet to form a seal when the mould closes are the parting surfaces. The pair of opposite directions along which the two plates of the mould separate are the parting lines. Recesses or projections on the moulded piece that prevent its removal from the mould along the parting directions are called undercuts. Depending on the types of undercut, different mechanical devices such as sliders and lifters are used to help clear the undercut. In general, core, cavity, slider(s), and lifter(s) form the fundamental sub-assembly of a complete injection mould. In addition, an injection mould includes the following components:

- runners and gates
- mould base
- cooling pipes
- ejection pins
- standard parts, such as screws and location rings.

The arrangement of these components is based on the number of cavities and their layout in the injection mould. Figure 7.1 shows a complete injection mould in a 3-D exploded view.

Figure 7.1: An injection mould (courtesy of Manusoft Plastic Pte. Ltd.).

7.3 Computer applications in injection mould design

Recently, the advances of computer applications in design, manufacturing, and engineering analysis have gradually changed the mould design from a completely manual process to a computer-aided process. The more notable changes can be seen in the following areas:

1. More and more plastic part designs given to the mould designer are in the form of 3-D CAD models, instead of 2-D drawings. The mould designer has to use an appropriate CAD software to retrieve the part design information.

2. Computer-aided engineering (CAE) software such as flow analysis programs is used to simulate the plastic injection process to check for possible defects in the moulded parts. Such feedbacks are then used for modifying the moulded part and/or the mould.
3. CAD/CAM software is commonly employed for the generation of NC tool-paths and programs for rough machining of complex 3-D surfaces.
4. CAD/CAM software is used to design electrodes for EDM machining (finishing) of complex 3-D surfaces.
5. CAD/CAM software is used to generate all the 2-D drawings of the parts in the mould assembly.

It is obvious that the conversion from manual to computer-aided processing is far from complete as many phases of mould design are carried out interactively with the basic CAD functions. The effectiveness of using CAD in mould design is largely dependent on the user's skills.

On the other hand, research on how to help automate the geometric design of injection moulds using computer aids has been on-going since the late 1980's. Typically, the object under study is the 3-D model of the moulded part. Research effort is mainly focused on automatic identification of feasible and optimal parting directions, automatic creation of parting lines and parting surfaces, automatic identification of undercuts, and automatic creation of side cores [1], [2], [3], [4], [5], [6], [7], [8], [9], [10]. The developed algorithms, however, still lack the capability and/or consistency to handle moulded parts with complex 3-D surfaces. There is also a report on a proposed object-oriented mould model to support mould design [11]. However, the application of this model has not been reported in any established systems.

In summary, the use of computers in injection mould design has made some progress in both industrial and research communities in the last 10 years. It is clear that the current trend in mould design is migrating from 2-D to 3-D as plastic parts are becoming more complex. On the other hand, mould design lacks comprehensive CAD support as common CAD software provides only basic geometric functions. This results in high demand on CAD skills from the designers and long design time. Although research has made certain headway in automating some of the mould design tasks, it is still far from making any significant impact on industrial applications. It is under such a situation that IMOLD® has emerged as a comprehensive CAD tool to support injection mould design by relieving the designers from routine but tedious geometric manipulation. Furthermore, it provides a powerful and flexible data structure for customisation.

7.4 IMOLD®

Developed by a research team from the Department of Mechanical and Production Engineering, National University of Singapore, IMOLD® is a knowledge-based software designed to capture the designer's intents and apply the expertise in the specific discipline of plastic injection mould design. Built on top of Unigraphics, a schematic diagram of the IMOLD® system structure is shown in Figure 7.2. Starting with a 3-D part model as input, it provides a comprehensive set of functional modules that help the user to complete a mould design process from core and cavity creation to cooling and ejector system design. The output is a complete mould assembly in 3-D which components can be used either collectively or individually for downstream applications. Two notable features of IMOLD® are parametric modelling and automation of routine design tasks. With parametric modelling, the database is concise while a wide range of components with similar shapes can be accommodated easily. Besides, tedious geometric manipulation tasks such as core and cavity creation and assembly are made automatic. Moreover, since many industrial design knowledge and practices are incorporated into the parametric models of the functional parts in an injection mould, the design task is simplified as a user can select from the available alternatives.

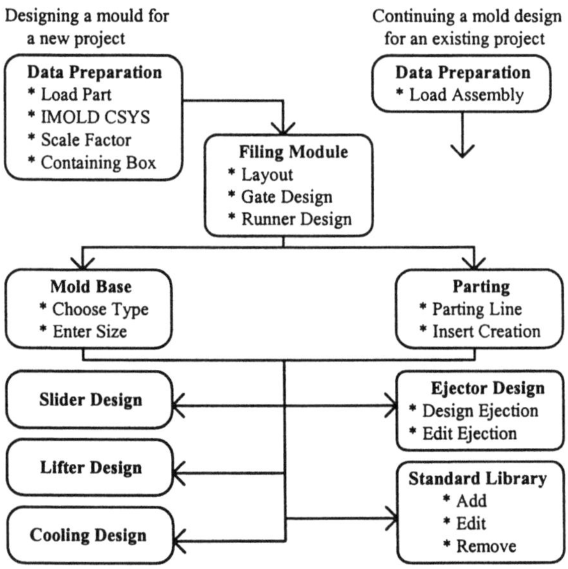

Figure 7.2: Overview of IMOLD®.

7.4.1 IMOLD® functional modules

Under IMOLD® version 3.1, there are a total of 9 functional modules that cover the design tasks for an injection mould. Each module is briefly described in a traditional design sequence as follows:

7.4.1.1 Data preparation and filling module

Upon receiving a 3-D part model, the Data Preparation Module provides options to define part shrinkage and ejection direction. Part model can be imported into IMOLD® from models created using various 3-D CAD software through IGES or STEP. The Filling Module provides various runners, gates, and automatic cavity layout for multi-cavity moulds of up to 128 cavities. A 4-cavity layout created using IMOLD® is shown in Figure 7.3.

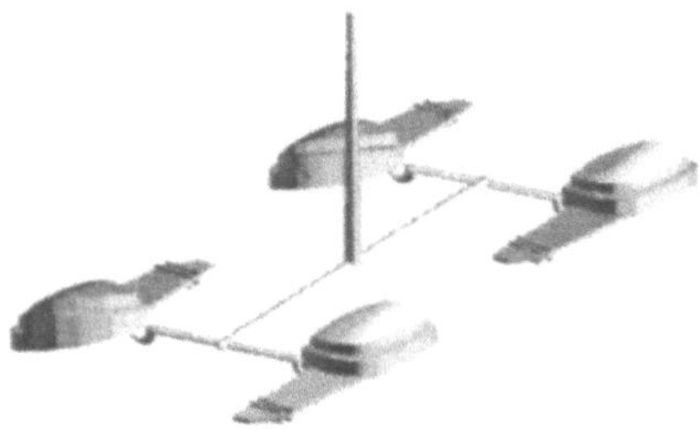

Figure 7.3: A 4-cavity layout.

7.4.1.2 Mould base module

IMOLD® offers a unique 3-D Mould base Module, which automatically creates parametric standard mould bases from mould-makers such as DME, HASCO, and FUTABA. In addition, IMOLD® provides user-friendly tools for easy customisation of mould bases that are unique to a particular company. It also provides other standard components associated with a mould base, such as screws, pillars, and sprue bushings. A mould base created using IMOLD® is shown in Figure 7.4.

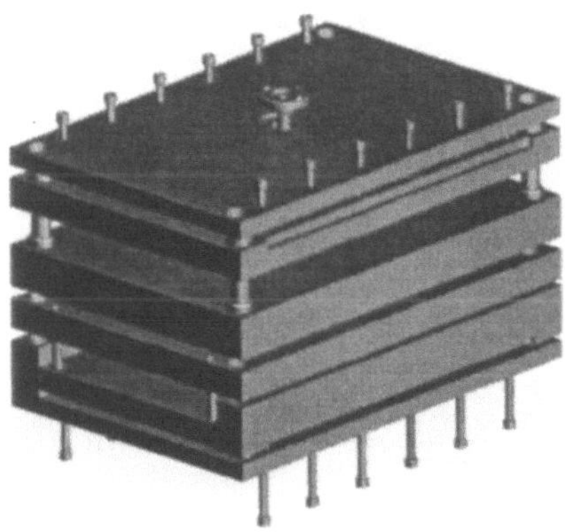

Figure 7.4: A mould base.

7.4.1.3 Parting module

After a containing block is created to enclose the part, the next step is to split the containing block along the parting lines for the creation of core and cavity. The Parting Module provides user-friendly tools for specifying the parting lines interactively. An automatic parting line search algorithm is also available. The parting lines range from simple planar to complex free-form. The module also provides a tool for assisting in patching holes on the part before splitting. Based on the parting lines, the module generates the parting surfaces and carries out the splitting automatically. Figure 7.5 shows the core and cavity for a phone cover after parting.

Figure 7.5: The cavity and core for a phone cover.

7.4.1.4 Slider module

The Slider Module assists in creating sliders for external undercuts. Commonly used slider types and their components are stored in the database as parametric models. Users can simply specify the undercut area and a slider head is generated automatically. Upon selecting a slider type, the slider body and its associated components are created automatically according to the dimensions of the slider head and mould base. Moreover, customised slider types can be easily added to the database for a particular company. A slider created using IMOLD® is shown in Figure 7.6.

Figure 7.6: A slider.

7.4.1.5 Lifter module

The lifter module assists in creating lifters for internal undercuts. Commonly used lifter types and their components are stored in the database as parametric models. Users can simply specify the undercut area and a lifter head is generated automatically. The releasing angle is also calculated automatically. Upon selecting a lifter type, the lifter body and its associated components are created automatically according to the dimensions of the lifter head and mould base. Moreover, customised lifter types can be easily added to the database for a particular company. A lifter created using IMOLD® is shown in Figure 7.7.

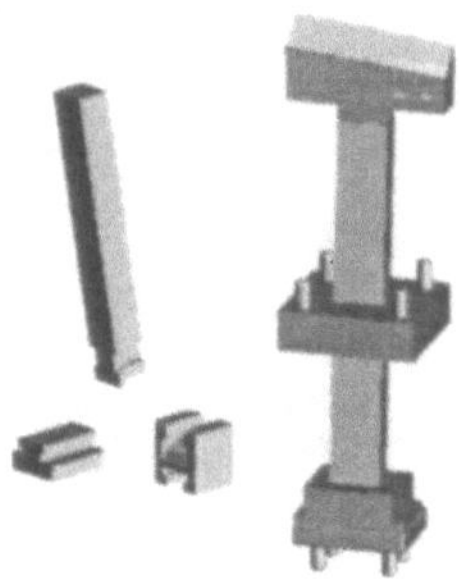

Figure 7.7: A lifter.

7.4.1.6 Cooling module

The cooling module provides extensive options for creating virtually every conceivable type of cooling channel used in the mould design industry today. Besides cooling channels, this module also provides extensive library of standard components such as baffles, O-rings, connectors, and plugs. Figure 7.8 shows some cooling channels created on a mould base.

Figure 7.8: Cooling channels on a mould base.

7.4.1.7 *Ejection module*

The Ejection Module assists in creating ejector pins and adding them into the partially completed mould assembly. It has a library of commonly used ejector pin types based on commercial catalogues. Upon selecting a pin type, diameter, and its location, this module calculates the pin's length and trims the head according to the core's profile automatically. Furthermore, interference check between the ejector pins and cooling channels, and other components is carried out automatically. Figure 7.9 shows a set of ejector pins designed using IMOLD®.

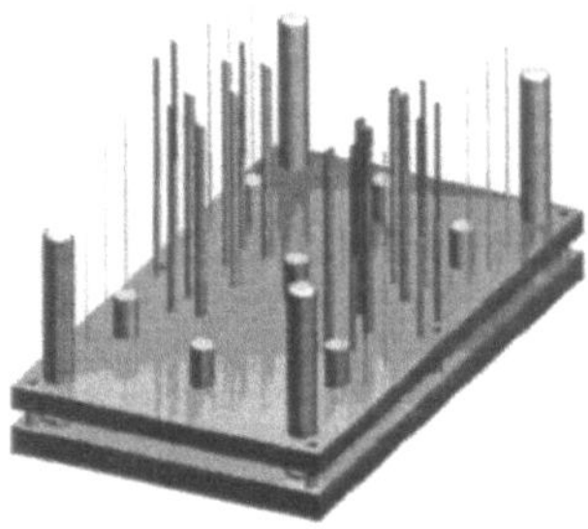

Figure 7.9: Ejector pins.

7.4.1.8 *Standard parts module*

The standard parts module assists in creating the miscellaneous components of a mould, such as springs, screws and nuts, and adding them into the mould assembly. Designers can add their standard components into the database as customisation can be easily carried out. The pockets resulting from the addition of components are created automatically. Some of the standard components available in IMOLD® are shown in Figure 7.10.

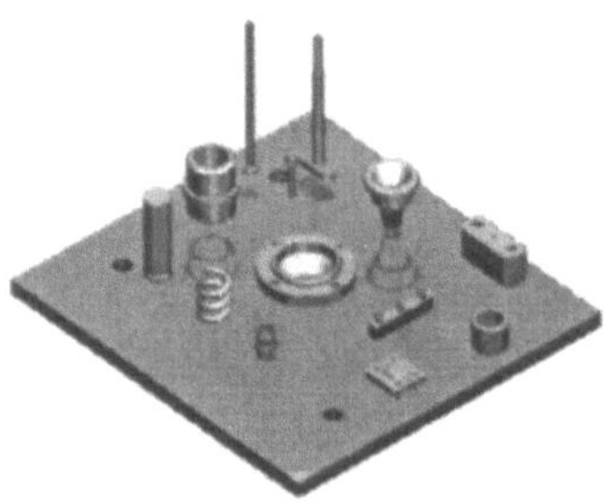

Figure 7.10: Some standard components.

7.4.1.9 IMOLD® tools

IMOLD® tools enhances the practical use of IMOLD®. It provides many user-friendly tools from insert creation to design data management. Some of them are listed as follows:

- Summary Tool provides information such as shrinkage and mould base
- View Manager allows easy access to the specific on-screen part
- Insert Tool allows easy slider and lifter head creation and hole patching
- Version Manager allows easy management of different designs for a mould
- Bill of Material is created automatically

For the design of each functional part in a mould, IMOLD® allows for easy modification. Since the associatively relationships, such as assembly mating constraints, among the parts are stored in the assembly structure, any change on a component will be propagated to those affected.

7.4.2 Design information management in IMOLD®: assembly tree

As there are hundreds of parts in a mould assembly, the efficient management of such a large number of parts becomes an issue in computer-aided injection mould design. By employing the Assembly Navigation Tool in Unigraphics, all the parts in a mould are placed in a hierarchical structure, i.e., assembly tree, according to the assembly relationships among all the parts. The assembly tree has the overall assembly as its root node. Every branch node represents either a sub-assembly or a part. The related nodes, namely the parent-child nodes on the same branch and the child nodes sharing the same parent node, are associated by assembly constraints, such as mating and aligning, and/or referring parameters with each other. This assembly tree structure can be used for easy viewing. More importantly, it can be also used as an editing window. Users can simply browse the branches and click the node to modify. Changes in a node are propagated to other nodes that are associated with it. Figure 7.11 shows a partial assembly tree of a mould.

7.5 An example

An example of using IMOLD® to design an injection mould for a plastic part (a chassis) is given in this section. It comprehensively demonstrates the IMOLD® mould design process. Due to the lengthy process, only the major steps are shown.

(1) Loading the part model
By using the Data Preparation Module, the chassis model is loaded into the IMOLD® system as shown in Figure 7.12. The system also recognises that the

model uses the metric units. The system next asks the user to set up the world co-ordinate system (WCS) where the z axis is the mould parting direction. The user then chooses the type of material of the part and the system automatically scales up the model according to the shrinkage factor of the material.

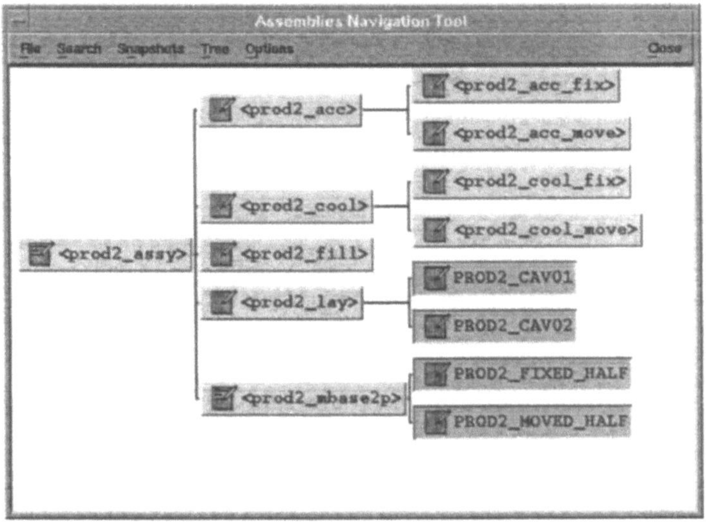

Figure 7.11: A partial mould assembly tree.

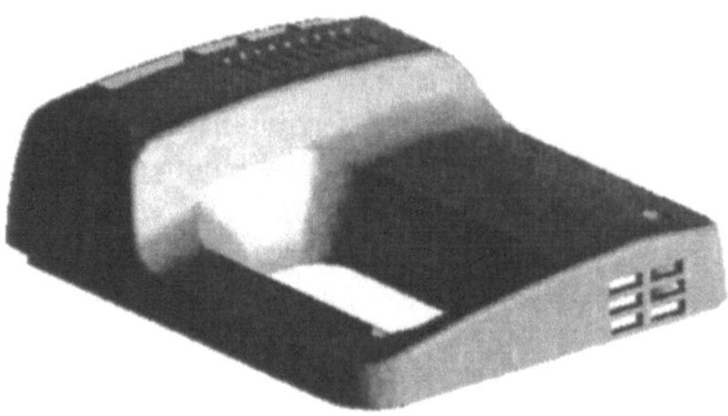

Figure 7.12: The chassis model.

(2) Creating the containing box
A containing block is then created to enclose the part for core and cavity creation. A default offset value is given along each of the x, y, z axis (see Figure 7.13). Users may change the values upon their specific requirements.

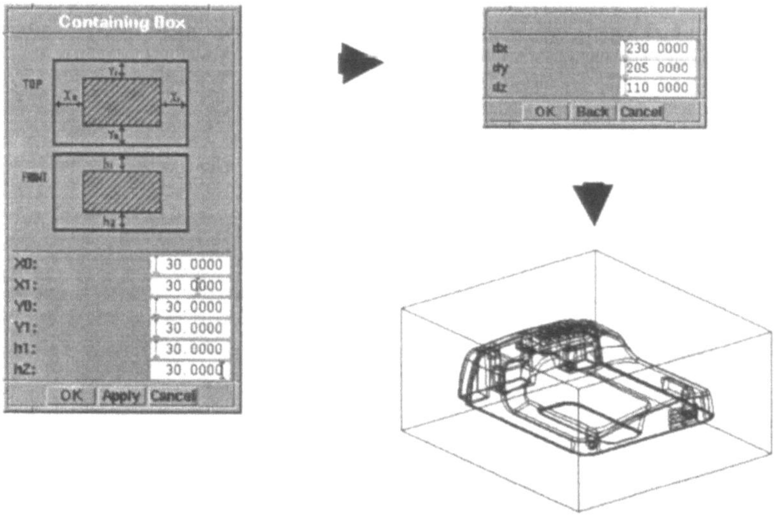

Figure 7.13 Creating containing block.

(3) Creating cavity layout
This includes the selection of the number of cavities in the mould and their pattern of layout. In this case, a 2-cavity layout is selected as shown in Figure 7.14.

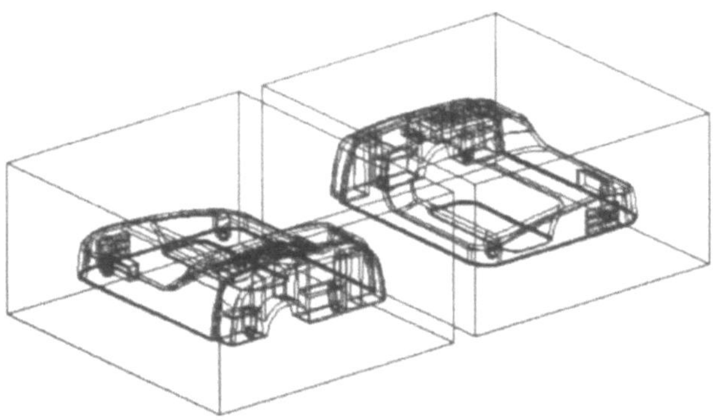

Figure 7.14: A 2-cavity layout.

(4) Creating gates
For gate design, selections are made on gates type, dimension, and position. In this example, a *side gate* is selected and the mid-point on the bottom edge of the chassis is identified as the gating point (see Figure 7.15).

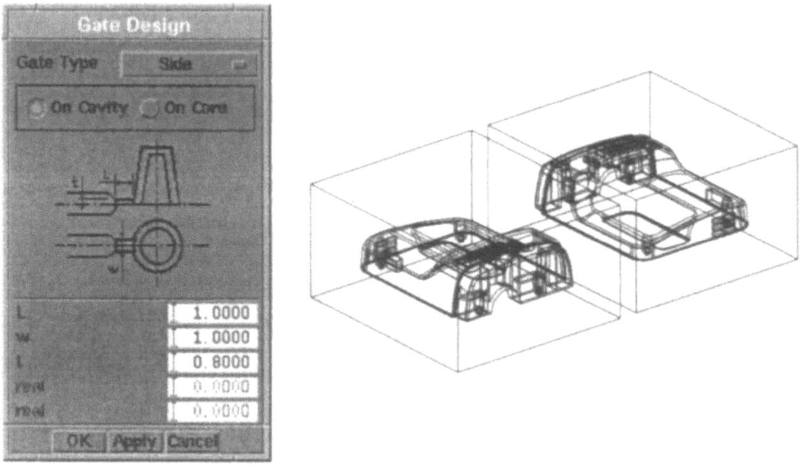

Figure 7.15: Gate design.

(5) Creating runners
For runner design, selections are made on the type of dimensions of the runner. In this example, a *circular runner* is selected. The runner is then linked to the gates automatically (see Figure 7.16).

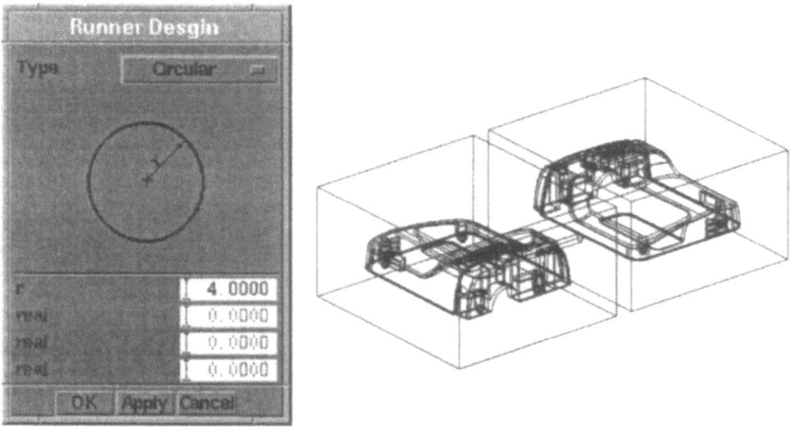

Figure 7.16: Runner design.

(6) Creating mould base
For mould base design, selections are made on the mould base type and its dimensions (according to the series in the catalogue). In this example, LKM is the mould base type selected with a dimension of 40×55. The mould base is then loaded and assembled with the mould layout (see Figure 7.17).

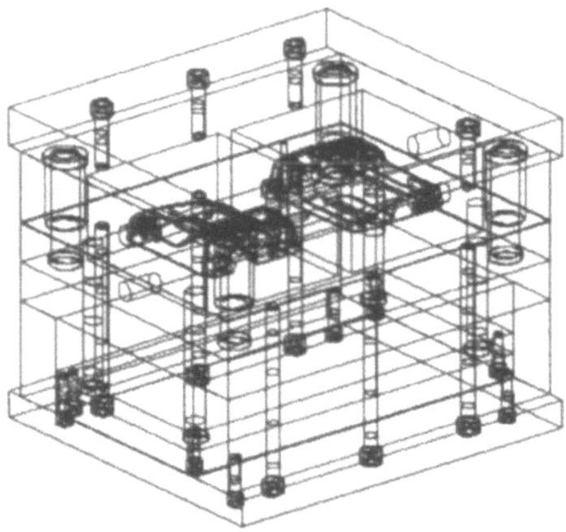

Figure 7.17: Mould layout within a mould base.

(7) Patching holes
All the through holes on the part need to be patched before the core and cavity can be created through splitting the containing block. For through holes in the parting direction, the system will simply patch them using faces. While for through holes along other direction, inserts will be created to block them. These inserts are also stored for downstream applications, such as slider and lifter design. The chassis has both holes along the parting direction and others. The part model in the containing block after patching is shown in Figure 7.18.

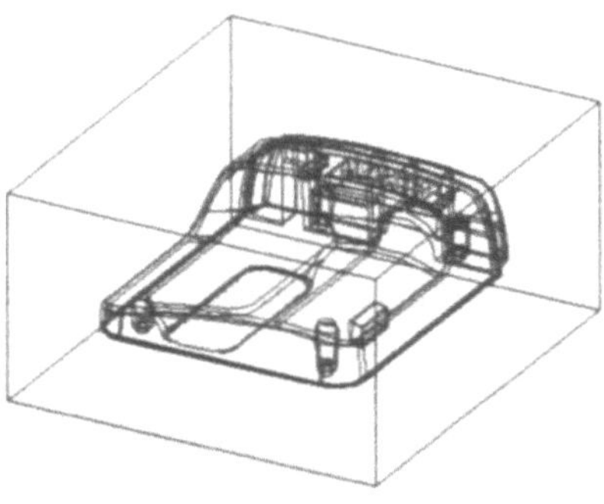

Figure 7.18: Part in containing block after patching.

(8) Creating core and cavity
The patched part is first subtracted from the containing block. The user next specifies the parting lines one by one, the extruding directions for creating parting faces are created automatically but would need to be confirmed by the user (see Figure 7.19a). The containing block is finally split into two solids, i.e., cavity and core, along the parting faces automatically (see Figure 7.19b).

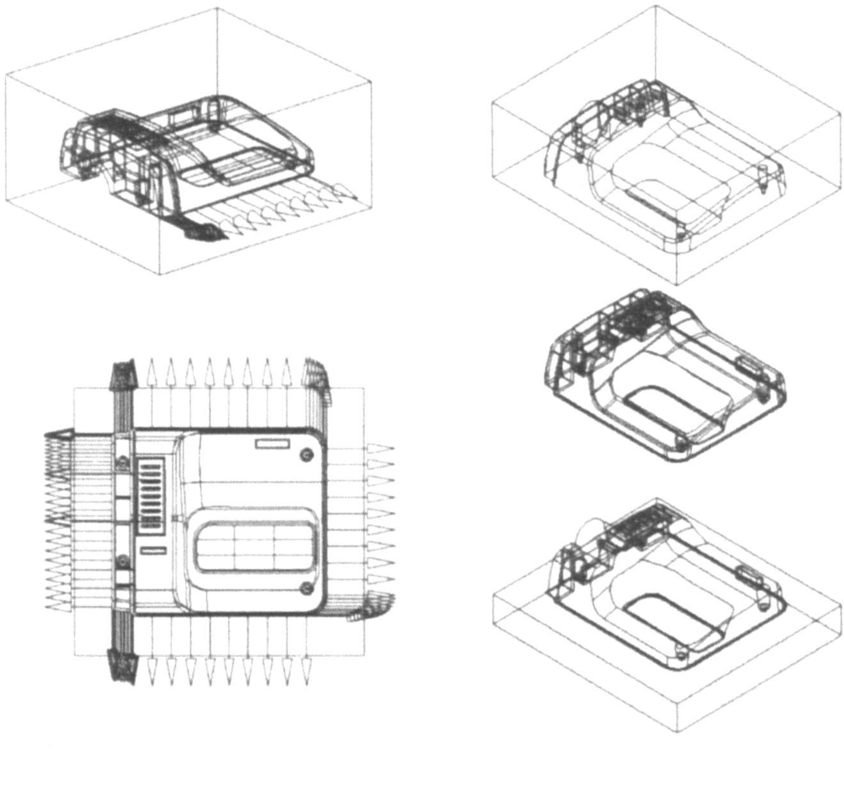

(a) (b)

Figure 7.19: Core and cavity creation.

(9) Creating heads for sliders and/or lifters
In case where there are undercuts which need sliders and/or lifters, the user need to use the Insert Tool module to create the heads before the slider and/or lifter assemblies can be attached to them. In this example, there is only one side which needs a slider. The user first specifies the boundary of the undercut area and a default block is automatically created that covers the undercut. Next, the geometry (size, draft angle, etc.) of the block is modified by the user. This modified block is then used as the head for this undercut (see Figure 7.20).

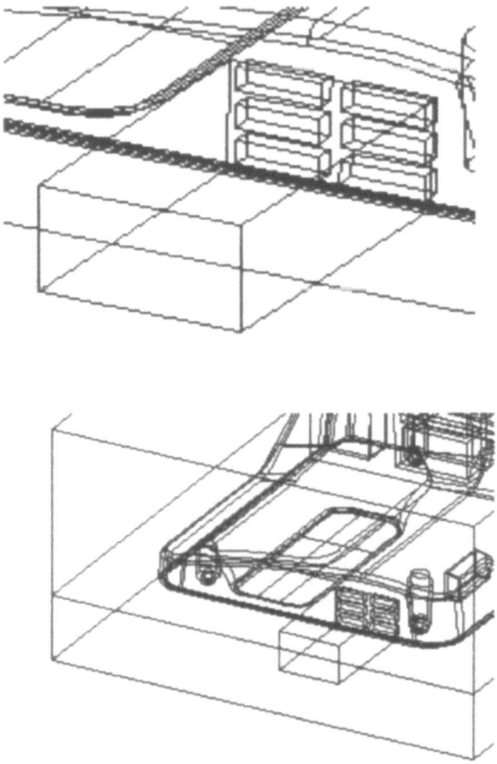

Figure 7.20: Creating slider head.

(10) Creating sliders

To create a slider, the user first selects the slider head and the type of slider. Upon this selection (*finger-cam* selected), two windows containing the sketches of the slider body and the finger-cam and their default dimensions will appear on the screen. After confirming with the parameters, the user selects the face on the head where the slider body is attached to and the slider body together with the finger-cam are incorporated into the mould assembly (see Figure 7.21). Next, the user creates the accessory components for the slider, namely guides, wear plates, heel block, and stop block. The selection of these component types is done automatically with their parameters. The user simply confirms or changes the values of the parameters and the components are assembled to the slider body automatically. Pockets on the cavity and or core are also created automatically. For this example, the slider assembly is created as shown in Figure 7.22.

(11) Creating cooling system

Generally, cooling holes are needed on both cavity and core. The user selects the cross-section type and dimensions. The cooling channel position and parameters are next selected. Cooling channels are then created automatically (see Figure 7.23).

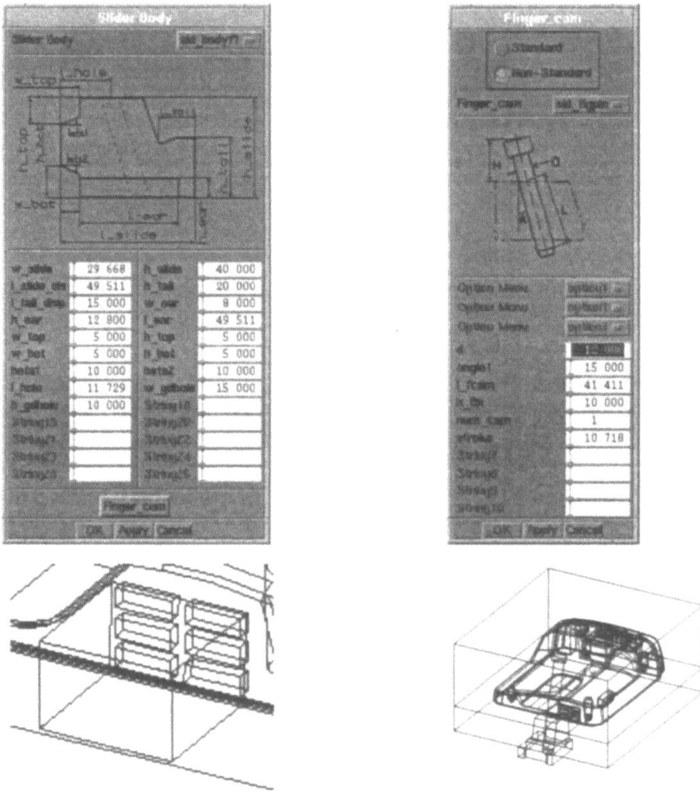

Figure 7.21: Slider main body creation.

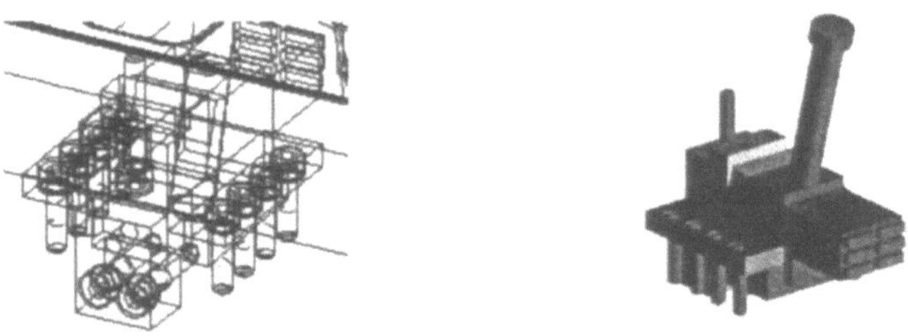

Figure 7.22 A complete slider assembly.

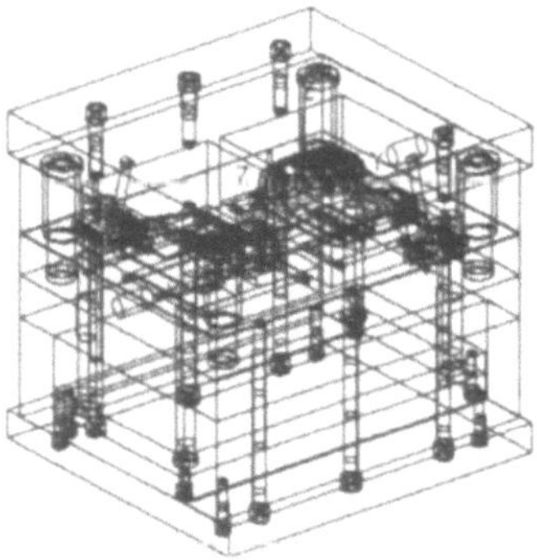

(a) on core

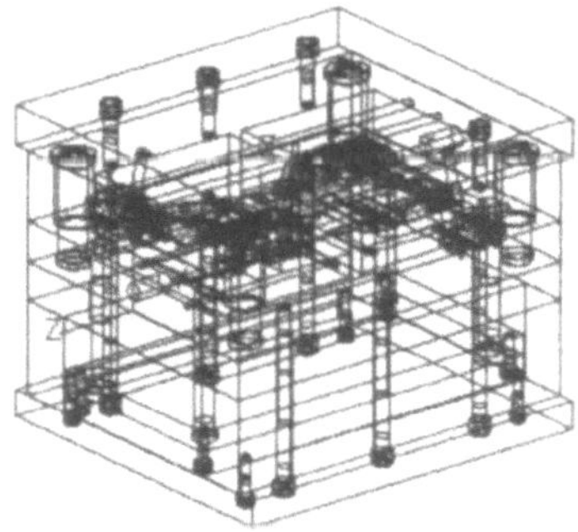

(b) on cavity

Figure 7.23: Cooling holes.

(12) Creating ejecting system

The user selects the type of ejectors and the their parameters. Next, the ejection positions on the core surfaces are selected (Figure 7.24a) and ejector pins created (Figure 7.24b). Upon confirmation, the top of each ejector pin is trimmed to conform to the profile of the core (Figure 7.24c).

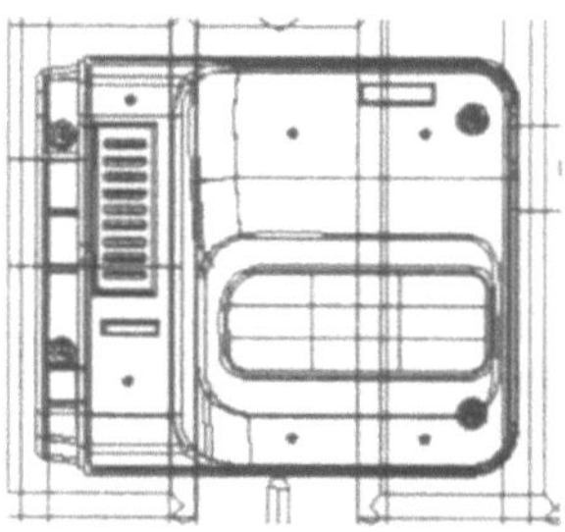

(a) select positions

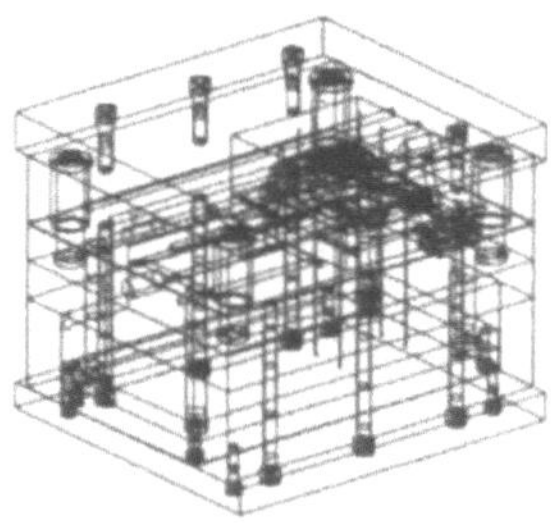

(b) before trimming

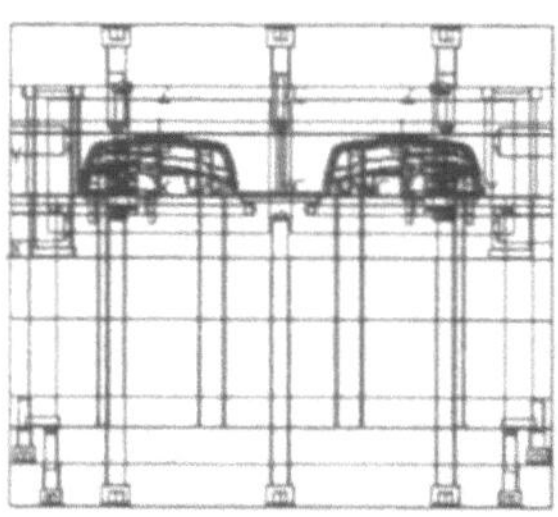

(c) after trimming

Figure 7.24: Creating ejecting system.

(13) Creating standard components

For each standard component, the user opens the Standard Parts Library and selects the type of component and its serial number (dimension). Upon specifying the assembly reference, the standard component is attached to the main assembly. For this example, the standard components include connectors (Figure 7.25a) and O-rings (Figure 7.25b) for cooling pipes, locating ring (Figure 7.26a), sprue bush (Figure 7.26b), and side locks (Figure 7.26c).

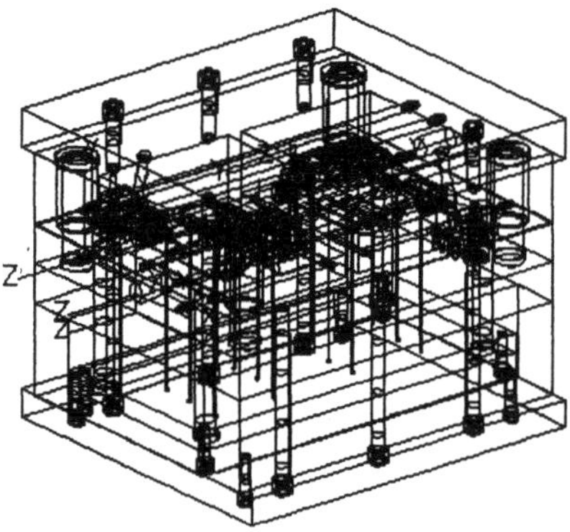

(a) Connectors

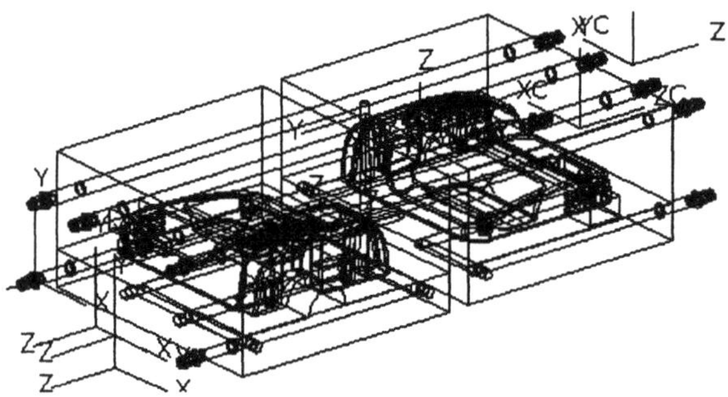

(b) O-rings

Figure 7.25: Creating standard components for cooling pipes.

So far, all the components of the mould have been created. The final mould assembly in an exploded view is shown in Figure 7.1. All the solid part files of the components can be retrieved and used individually for down-stream applications, such as NC program generation and part ordering.

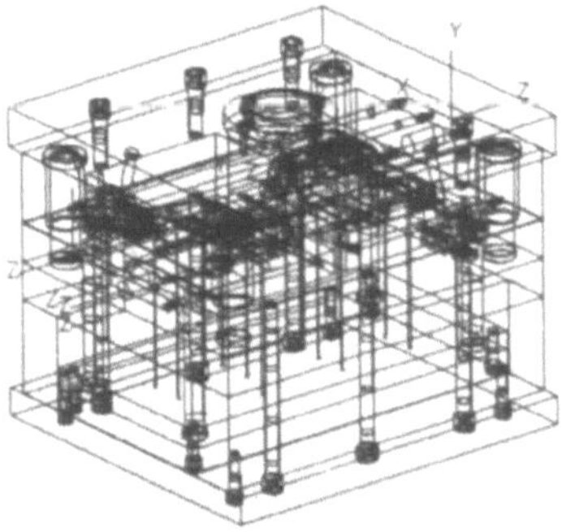

(a) Locating ring

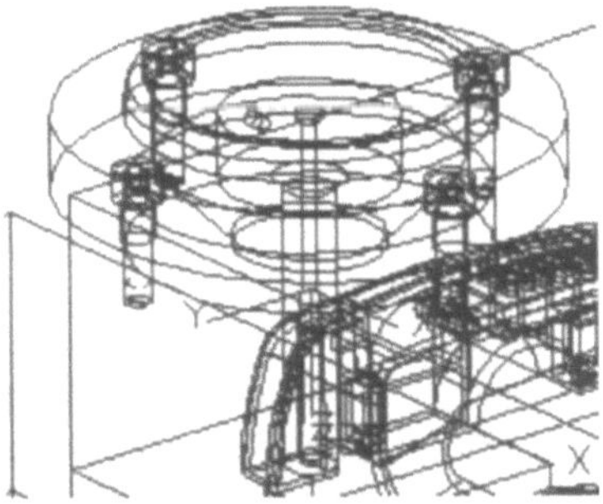

(b) Sprue bush

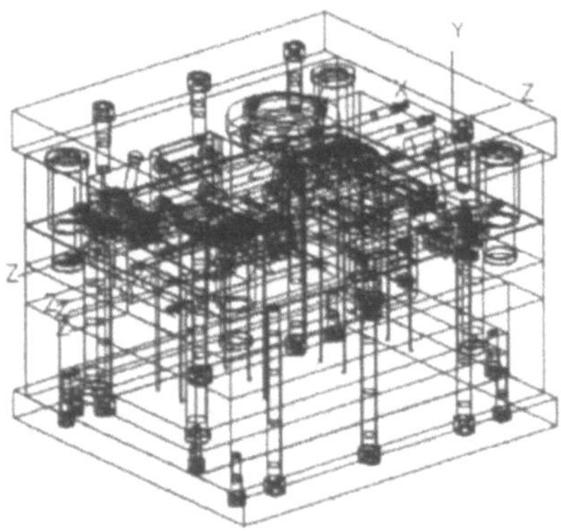

(c) Side locks

Figure 7.26: Creating other standard components.

7.6 Conclusions

In this chapter, the application of CAD in injection mould design has been discussed. An intelligent mould design and assembly system, IMOLD®, based on 3-D techniques, has been introduced comprehensively. An example of using IMOLD® to design a mould for a plastic part from industry is given step by step. It successfully demonstrates that designing a mould totally in 3-D is realisable and highly recommended. The use of CAD tools can cut the design time of a traditional method by up to 70%.

On the other hand, the development of computer-aided mould design technique has just started. There are certainly more areas in the mould design process which will benefit from this technology. For example, functional modules for designing hot runners and electrodes for EDM are the useful features to be considered.

References

1. Ravi B, Srinivasan M N 1990 Decision criteria for computer-aided parting surface design. *Computer-Aided Design* 22(1):11-18
2. Tan S T, Yuen M F, Sze W S, Kwong K W 1990 Parting lines and parting surfaces of injection moulded parts. *Proceedings of IMechE, Part B: Journal of Engineering Manufacture* 204:211-221
3. Hui K C, Tan S T 1992 Mould design with sweep operations – a heuristic search approach. *Computer-Aided Design* 24(2):81-91
4. Chen L L, Chou S Y Chou, Woo T C 1993 Parting directions for mould and die design. *Computer-Aided Design* 25(12):762-768
5. Shin K H, Lee K 1993 Design of side cores of injection moulds from automatic detection of interference faces. *Journal of Design and Manufacturing* 3:225-236
6. Rosen D W 1994 Towards automated construction of moulds and dies. In: *Proceedings of ASME Annual Meeting – Computer in Engineering, Vol. 1.* pp 317-326
7. Sawai S, Kakazu Y 1994 A study on automatic generation of the mould cavity and core geometry. In: *Proceedings of Japan-USA Symposium of Flexible Automation – A Pacific Rim Conference.* Kobe, Japan , pp 729-732
8. Nee A Y C, Fu M W, Fuh J Y H, Lee K S, Zhang Y F 1997 Determination of optimal parting directions in plastic injection mould design. *CIRP Annals: manufacturing Technology* 46(1):429-432
9. Zhang Y F, Lee K S Lee, Wang Y, Fuh J Y H, Nee A Y C 1997 Automatic side core creation for designing slider/lifter of injection moulds. In: *Proceedings of the International Conference and Exhibition on Design and Production of Dies and Moulds*. Istanbul, Turkey, pp 45-50
10. Nee A Y C, Fu M W, Fuh J Y H, Lee K S, Zhang Y F 1998 Automatic determination of 3-D parting lines and surfaces in plastic injection mould design. *CIRP Annals: manufacturing Technology* 47(1):95-98
11. Kruth J P, Willems R 1994 Intelligent support system for designing injection moulds. *Journal of Engineering design* 5(4):339-351

Chapter 8

Computer applications in intelligent progressive dies design (IPD)

B.T. Cheok and A.Y.C. Nee

8.1 Introduction

Progressive dies are tools used to mass produce metal stampings. Metal stampings are important structural components of most modern day electrical and electronics appliances. In a progressive die, the workpieces are advanced from one station to another. At each station, one or more die operations, such as piercing, notching, blanking, lancing, shaving, drawing, embossing, coining and forming are performed on the sheet metal strip. The die design process can be divided into three discrete stages: the input of the product description (features modelling), the design of the strip layout (stamping process planning) and the construction of the progressive die (die configuration).

This chapter discusses a variety of intelligent computer-based techniques and presents the methodology to unify them to develop a progressive die design system.

8.2 Overview of the computer architecture

An overview of the computer architecture for the die design software is shown in Figure 8.1. It is based on a client-server concept. The overall architecture can be symbolically represented by two broad layers: an Applications Layer (clients) and a Design Knowledge Layer (server). The Applications Layer provides the front end for the design system. It consists of the four main modules for die design, i.e., Features Modeller, Unfolder, Strip Layout Planner and the Die Configurator. These design modules provide the programs to control the man-machine communications during the iterative feedback design cycle and the dynamic data exchange (DDE) conversations and transactions between the clients and the server applications. The AutoCAD Husk provides the functionalities for the 2-D and 3-D wire-frame graphical representation of design objects in AutoCAD. The ACIS husk provides the functionalities for the construction of 3-D solid models to represent die objects for display in AutoCAD or ACIS viewer. The Design Knowledge Layer is a DDE server application that provides information to the client applications during the design process. It also provides a storage mechanism for the objects and rules required to support die design. The Design Knowledge Layer is supported by a suite of "intelligent" libraries that provides the class and rule-based templates for the construction and manipulation of the various information structures (design trees) representing the product, the flat pattern, the process plan (strip layout) and the progressive die.

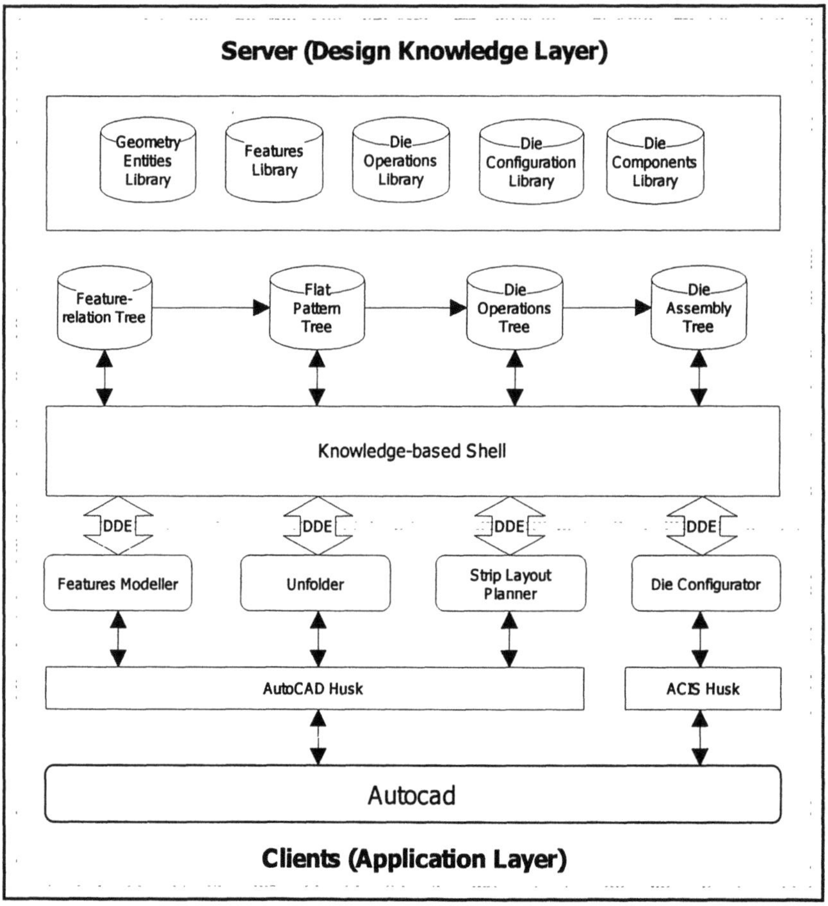

Figure 8.1: Overall systems architecture for the die design system.

This modular architecture helps to ensure that the design system can be ported from one CAD/CAM system or solid modelling kernel to another. It is noted that a large portion of the codes for the design system is developed solely using the C++ programming language and the Microsoft Foundation Class Library (MFC) provided by Microsoft Visual C++. The only modules that are dependent on the CAD/CAM system and the solid modelling kernel are the husks. This means that the core design modules can be readily integrated with other CAD/CAM systems or solid modelling kernels that run under the Microsoft Windows NT operating environment. This can be achieved by implementing the appropriate CAD/CAM or solid modelling kernel husks.

8.3 Advanced knowledge-based techniques for the modelling and generation of progressive dies

A variety of knowledge-based implementation techniques has been used to develop the die design system. The object-oriented concepts and rule-based representation used to model the engineering knowledge have been described at the appropriate sections of this chapter. However, the framework for the development of overall design system is based on two advanced knowledge-based methodologies. They are a model-based reasoning knowledge approach for the synthesis of the intermediate and final design models and some special shape representation and spatial reasoning techniques to assist decision making.

8.3.1 A model-based reasoning (MBR) approach for die design

Dym and Levitt described an architecture for building knowledge-based engineering systems using the model-based reasoning (MBR) approach. The following sections will describe how their architecture and methodology have been adopted for the development of the die design system.

8.3.1.1 Obtaining component descriptions from CADD databases

Selection, refinement and topological mapping of system components are the principal tasks of design synthesis of most engineering disciplines. The descriptive information process by the die design system includes:

1. Component geometry. This is described using the primitive geometrical entities.
2. Topological information. In die design, the topological information of the initial, intermediate and final states of the design is modelled by the Feature-relationship Tree, the Flat Pattern Tree, the Die Operations Tree

and the Die Assembly Tree. These trees are referred to as the Information Trees.
3. The user enters physical properties, technical specifications and administrative information via dialogue boxes and stored as attributes in the appropriate objects in the knowledge base.

8.3.1.2 How components can inherit behaviour

Dym and Levitt had stated that the key to inheriting component behaviour is that objects retrieved from the CADD system must be correctly associated with the component abstraction hierarchy. This has been achieved by dynamically linking CADD entities to objects in the knowledge base so that this inheritance is automatic and instantaneous. For example, a line drawn in AutoCAD modelled as a Bend feature will automatically and instantaneously inherit the transformation behaviour to unfold its connecting walls.

8.3.1.3 How components can deduce their functions

The MBR approach specifies that the knowledge-based system can deduce a substantial amount of knowledge about the roles and functions of individual components by reasoning about the part-of hierarchy for the product and the topological links among its components. For example, the system can deduce the function of a die opening (represented in CADD system as a circle or segloop) which is part of a plate. This opening can either be a structural support (to hold a punch), or a locating hole (for a dowel) or a fastening hole (for a fastener). Using the topological links in the Die Assembly Tree, the type and topological information of the male parts for insertion in the opening can be further deduced.

8.3.1.4 Generating system behaviour in MBR

In the die design system, MBR is used to predict the behaviour of the modelled system by simulating the design process. The components in the respective information trees inherit the desired behaviour from their respective class descriptions. The topological information stored with each component allows the system to deal with the interactions among its members. A "cascading" approach is used to deduce the intermediate design models until the final model of the die assembly is obtained, that is, design information cascades downwards from one tree to another until the Die Assembly Tree is constructed.

1. The MBR approach for the synthesis of the Flat Pattern Tree from the Feature Tree is shown in Figure 8.2. The individual components inherit their behaviour from the class descriptions of the various Features (i.e. Walls, Bends, etc.) using the "Is-a" relationship. Their topological

relationships are defined by their respective locations in the information tree. Note that the Dym and Levitt's Product part-of hierarchy for the representation of information trees for die design has been modified. In the Feature-relationship tree, the topological relationship is expressed by "connected-to" (wall-bend relationship) and "part-of" (internal features of a wall) relationship. The components in the Flat Pattern Tree are deduced from the Feature-relationship Tree as follows:

- The external profile is deduced from the transformation and summation of the Walls and Bends.
- The components under the "internal profile" and "bend" branches are deduced from the transformation of the respective components in the Feature-relationship.

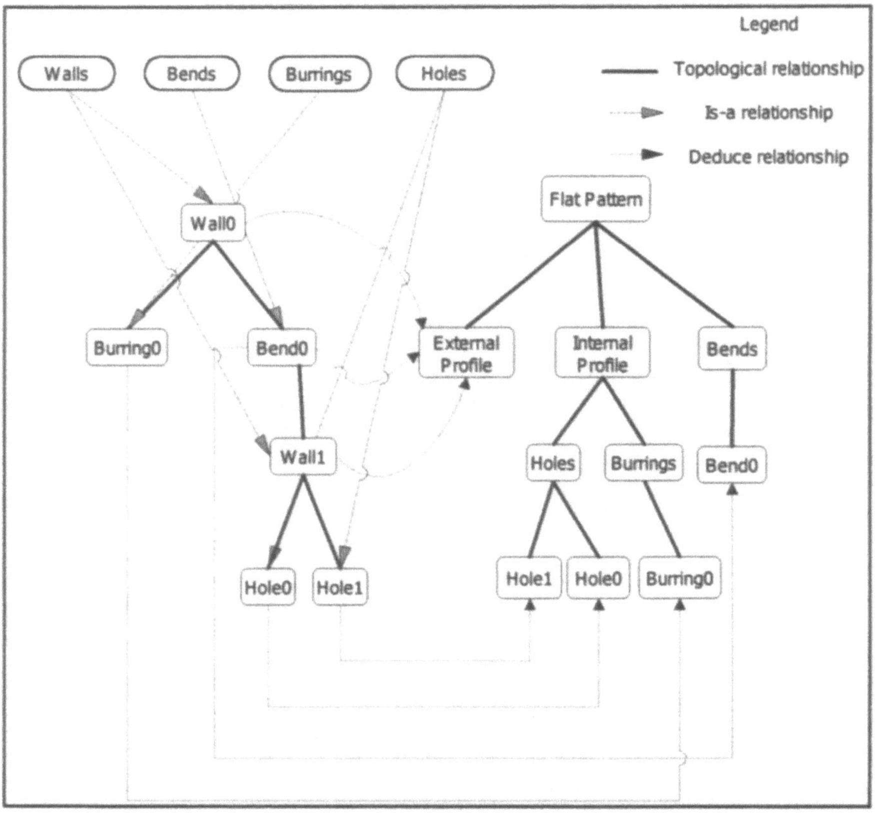

Figure 8.2: MBR approach to the synthesis of the flat pattern.

2. The MBR approach to the synthesis of the Die Operations Tree is shown in Figure 8.3. The generation of information for the piercing operations for

the holes and the piercing and extruding operations for the burring is rather straight forward and are deduced directly. Pilots may either be selected by the user on the flat pattern or selected automatically. Thereafter the associated piercing and piloting operations can be deduced automatically (the symbolical deduction relationship for the synthesis of piercing and piloting operation to guide the workpiece in the progressive is not shown in Figure 8.3). Similarly, the user may either design the notching profiles using interactive tools or use the strip approach to generate (and modify) the notching profiles on the flat pattern. Once this is done, the system will automatically deduce the notching punches in the Die Operations Tree.

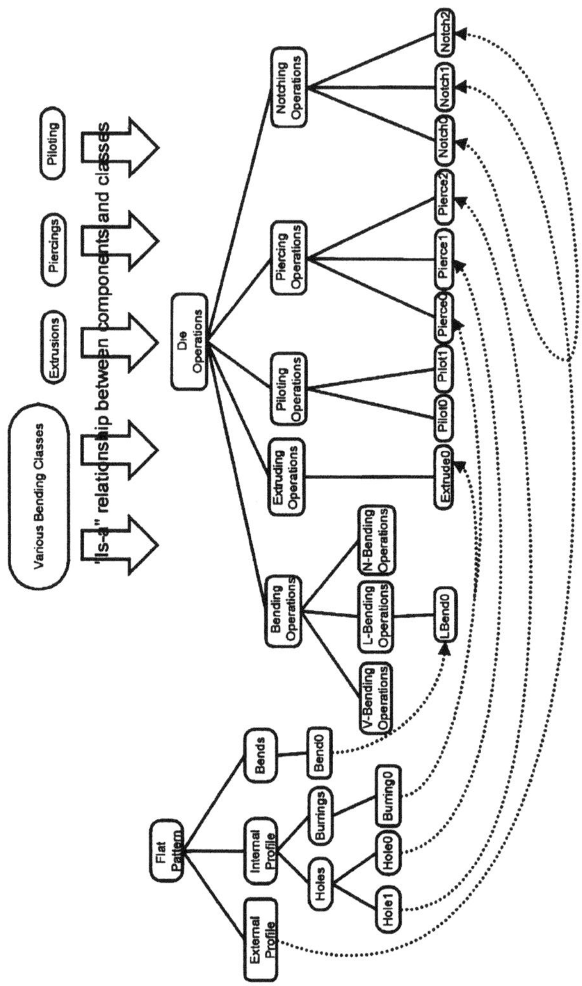

Figure 8.3: MBR approach for the synthesis of the strip layout.

3. Finally, the MBR approach for the synthesis of the Die Assembly Tree is shown in Figure 8.4. Here, the Die Operations Tree provides the 2-D spatial information for the deduction of the 3-D geometrical description of the die components. As noted earlier, the structure of the template Die Assembly Tree is not fixed, but constructed from the Die Assembly Tree Library according to the user's selection. The topological arrangement of the ancillary components not directly provided in the strip layout (such as springs, dowels, fasteners, etc.) are deduced using spatial reasoning techniques based on heuristics described in the previous chapter. The final topological arrangement of all the components in the assembly is deduced from the function-spatial topological description statements built in the respective objects to be described later.

8.3.2 Shape representation and spatial reasoning techniques

One of the main tasks of the progressive die designer is to use the vastly superior human image processing ability to make decisions based on shapes and spaces he sees. Shape representation and spatial reasoning techniques are used to mimic this human ability. A formal treatment of the shape representation and spatial reasoning techniques used at the various stages of die design is briefly given below:

1. Two-dimensional boundary description techniques are used to characterise objects in terms of the shape of their boundary. The information can be used for calculation of distances, areas, centroids to check whether certain design (such as spatial and strength) requirements are fulfilled. This technique is primarily used to represent the shape information of objects on the strip layout and used to locate the positions of ancillary components (i.e. fasteners, dowels, pins, etc.) of the die on a plate. Boundaries can be described by the actual profiles of the objects, or by their circumscribing circles/rectangles/polygons. The spatial reasoning of these boundaries is supported by a set of routines to perform the following functions:
   - Calculation of intersection point(s) of two boundaries
   - Combining two profiles into one (Union operation)
   - Determination of whether a shape is inside, outside or overlaps another shape
   - Enlargement of a shape by offsetting its edges
   - Calculation of geometrical properties of the boundaries (area, edge length, centre of pressure of cutting edges, centroid, etc.)
   - Transformation of boundary

2. 1-D skeletal simplification of 2-D shapes for punch shape recognition.

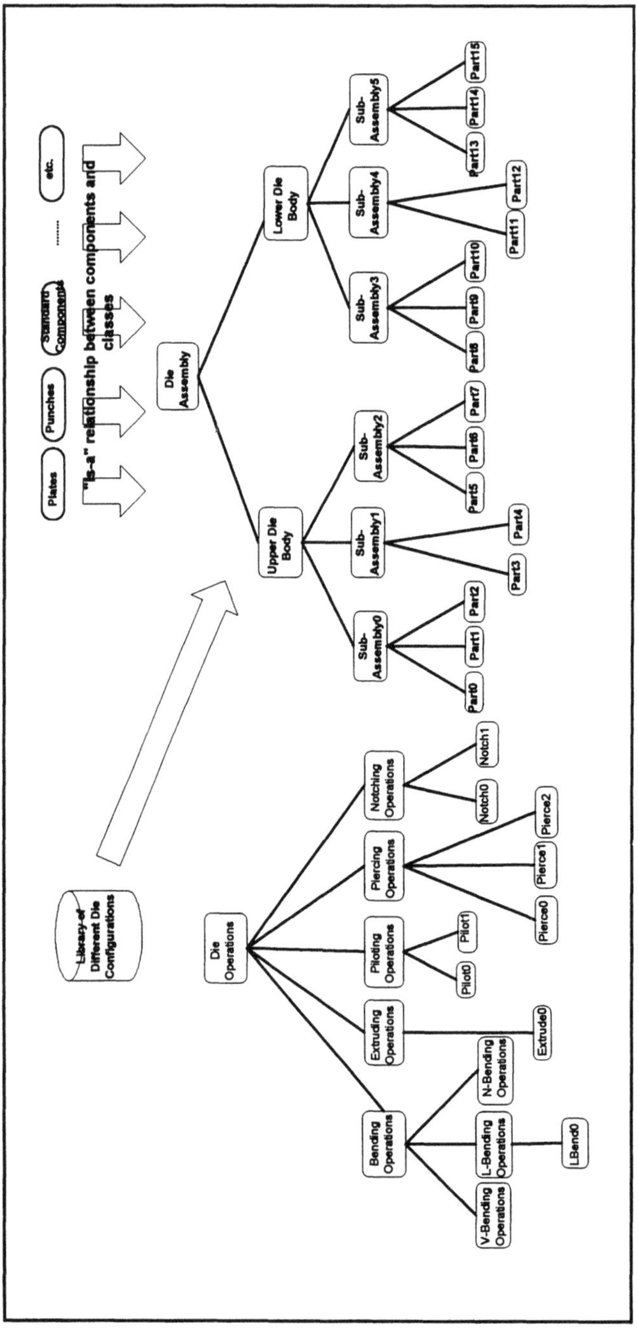

Figure 8.4: MBR approach for the synthesis of the die assembly.

3. Volumetric (2-D area projected by material thickness) description techniques have been used to describe the metal stamping features. The part is then assembled by identifying the dimensions, locations and orientation for each feature.
4. Part-whole representation is used to represent the die components. Here, objects are described in terms of the parts of which they are made up of when assembled. The parts are arranged in an array of complex inheritance network. An intermingling of spatial and functional reasoning is used to configure the die.

8.3.3 A function-spatial language for the configuration of a die

A simple function-spatial language can be used for the spatial reasoning leading to the location of some of the prismatically-shaped and orthogonally-oriented parts in the die assembly. The basic syntax of this topological description statement of a part is as follows:

> *{function} {relative position descriptor} {neighbouring part list} {additional description}*
>
> where:
> *{function}* is a word describing the function of the owner part, for example: *Secured, Placed, Inserted, ...*
>
> *{relative position descriptor}* is a word describing the relative position of the owner part to the *{neighbouring part list}*. For example, *AtTop, AtFront, FromTop, Through, ...*
>
> *{neighbouring part list}* is a list identifying the neighbouring parts that are used to provide information to calculate the relative location or dimension(s) of the owner part.
> *{additional description}* is an optional word used to provide additional information for the topological description statement. For example, *Blind, Through, ...*

Topological description statements constructed using the above syntax can be used to select the positions and certain dimensions of prismatic parts that are orthogonal in orientation such as plates, screws, dowels, springs, pins, etc. Examples of the use of topological description statements to deduce the relationship between die components are shown in Table 8.1. The arrangements of other components that cannot be handled by this spatial reasoning technique are "hard-coded" into the respective objects.

The topological description statements are pre-defined for each relevant class of objects in the die component library and stored as slots in the respective class. Each class of object may have more than one topological description statements. Hence, objects created in the Die Assembly Tree will inherit these statements. All parts in the upper die body take reference from the upper die shoe. In other words, spatial reasoning starts from the plate just below the upper die shoe. Similarly, parts in the lower die body are assembled with respect to the lower die shoe. The spatial reasoning to assemble the parts is implemented as a goal-driven forward chaining system. For a plate which dimensions are derived from design consideration (and not from topological links), the goal is to find the location of the part. For ancillary components such as screws and dowels, the goal is to find the location and to calculate certain missing dimensions (e.g. the length of a screw).

The function-spatial language can also be used to deduce the configuration of the openings on die plates.

Table 8.1: The use of some topological description statements.

Parts	Statements	Illustrations
Screw1	1. Secured FromTop Plate1,Plate2,Plate3 Blind 2. Flushed AtTop Plate1	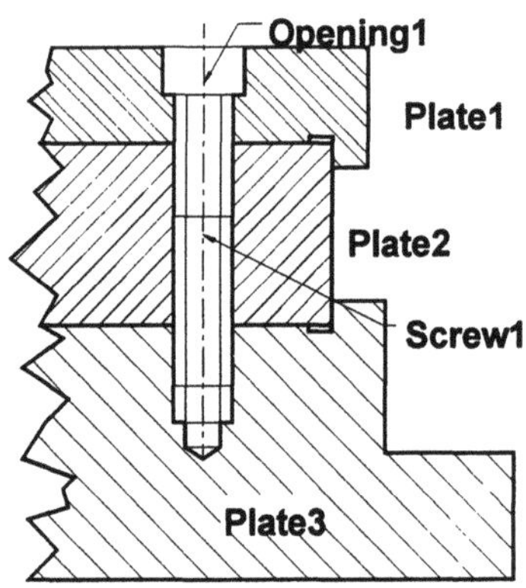
Plate1	Placed OnTop Plate2	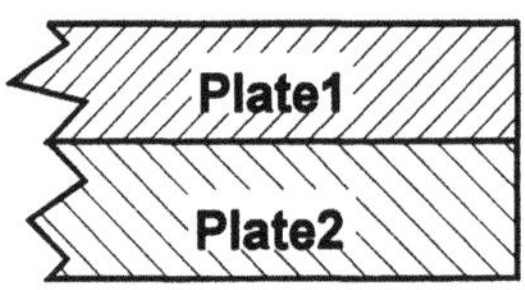

8.4 An industrial case study

8.4.1 Introduction

An industrial case study is used to illustrate the concepts described in this chapter and to explain in greater details, the steps required to design a die. It also illustrates some of the user interfaces of the IPD software.

8.4.2 Description of stamping

The metal stamping to be manufactured is shown in Figure 8.5

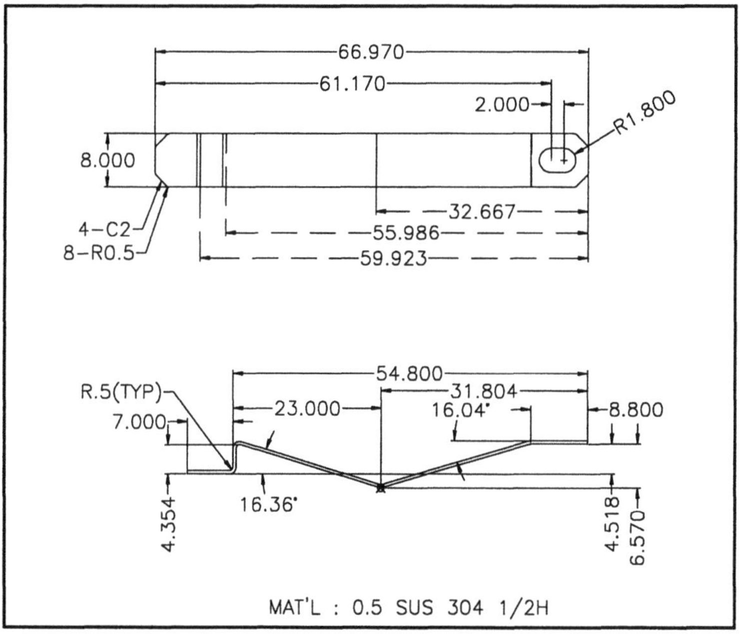

Figure 8.5: Stamping to be manufactured.

8.4.3 Features of the part

There are two ways to represent the features of a part: using "internal" dimensions or "external" dimensions. This is to allow the thickness of the part to be considered. Using "external" dimensions, the walls required to represent the part are shown in Figure 8.6.

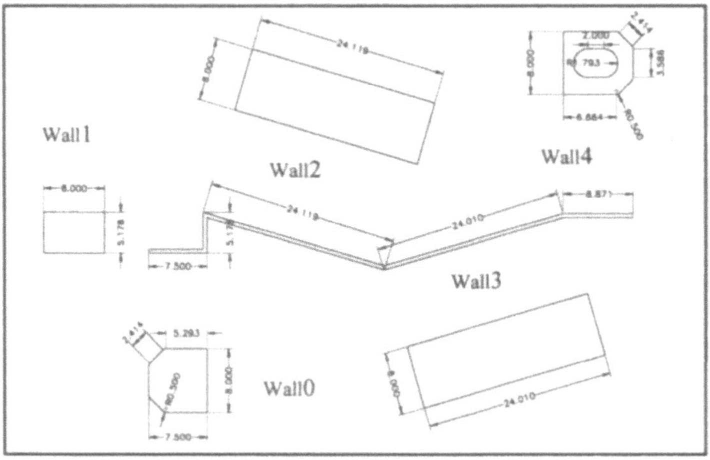

Figure 8.6: Walls required to model the stamping.

8.4.4 Modelling the part

The initial interface for the input of a stamping into the knowledge base is shown in Figure 8.7.

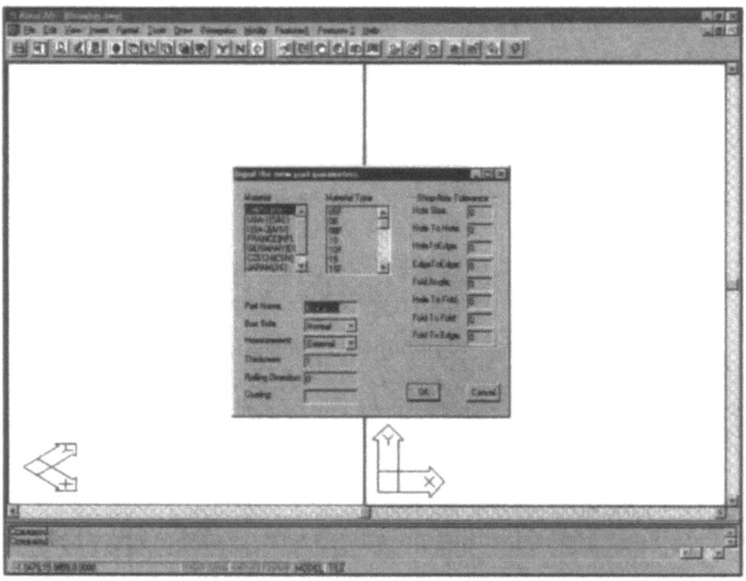

Figure 8.7: Initial interface for the modelling of metal stampings.

In Figure 8.7, the AutoCAD window is divided into two viewports. The left viewport is the 3-D Modelview and the right viewport is the 2-D Sketcher. The user draws a wall and its associated features (such as holes and burrings) in the 2-D sketcher, and then using "*Features*Make" commands, attach the wall to the actual 3-D model which is displayed in the Modelview.

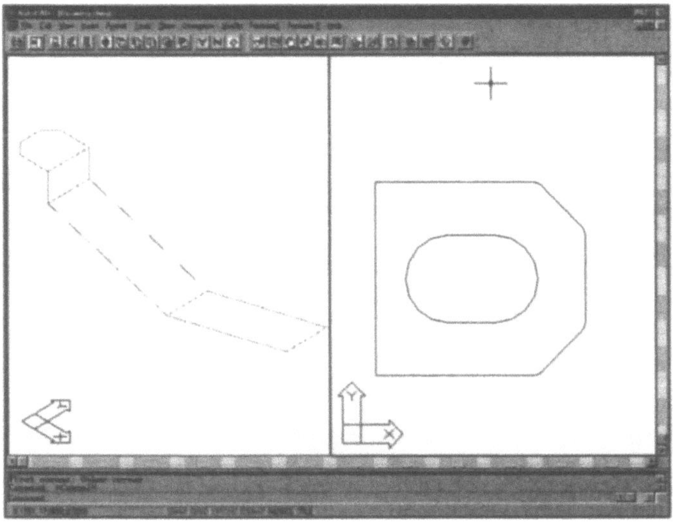

Figure 8.8: Graphical representation of features in Sketcher and Modelview.

For example, Figure 8.8 shows the profile of Wall4 and Hole0 as drawn in the Sketcher, ready for attachment to the other walls modelled earlier and displayed in the Modelview.

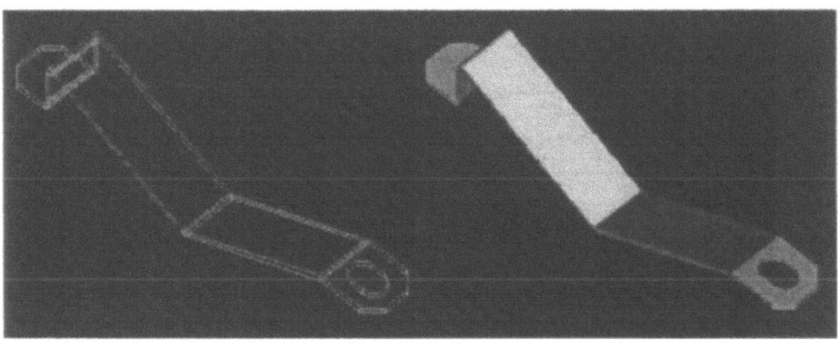

Figure 8.9: Solid model representation of input part produced by the ACIS husk for the Feature-relationship Tree.

An ACIS husk for the generation and display of ACIS solid models from the Feature-relationship Tree can be used to verify the accuracy of the input part

(Figure 8.9). As a feature is created graphically in AutoCAD, its geometrical and topological information is captured by the system to build the respective objects in the Feature-relationship Tree in the knowledge base. The Feature-relationship Tree for the part is shown in Figure 8.10.

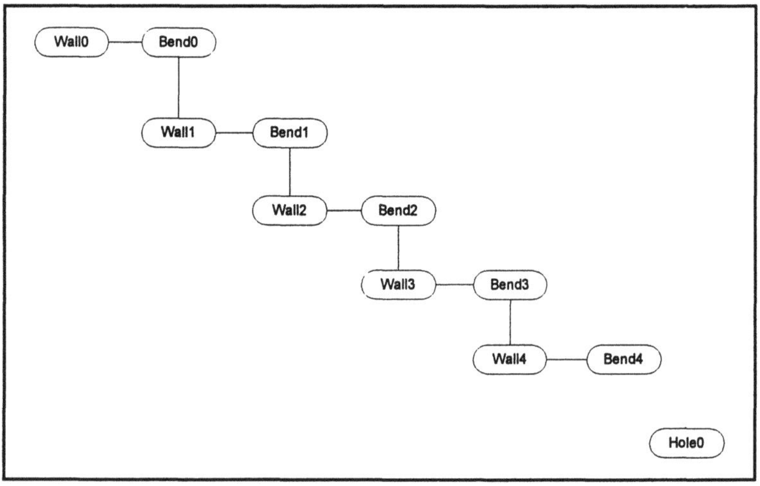

Figure 8.10: Feature-relationship Tree of the part.

8.4.5 The flat pattern

After the part is successfully modelled in the knowledge-base, the unfolder can be used to generate the its flat pattern (Figure 8.11). As the Flat Pattern Tree is also stored in the knowledge-base, the geometrical information of the external and internal profiles can be used to generate the dimensions of the flat pattern. This provides a fast and accurate means for the user to check the flat pattern.

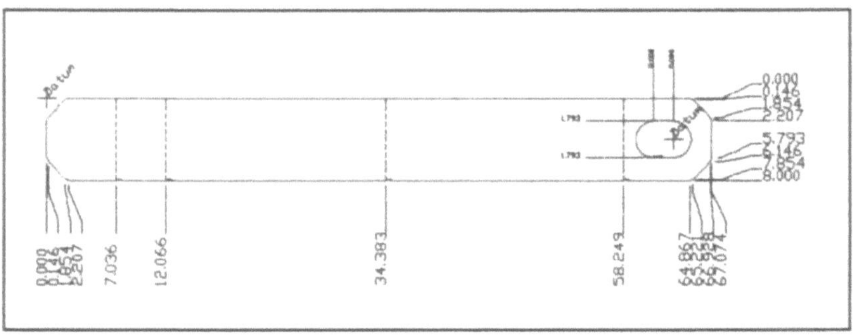

Figure 8.11: Flat pattern (with dimensions) generated automatically by the system.

8.4.6 The strip layout

As the flat pattern is nearly rectangular in shape, the nested arrangement of the strip layout is rather straight forward (Figure 8.12).

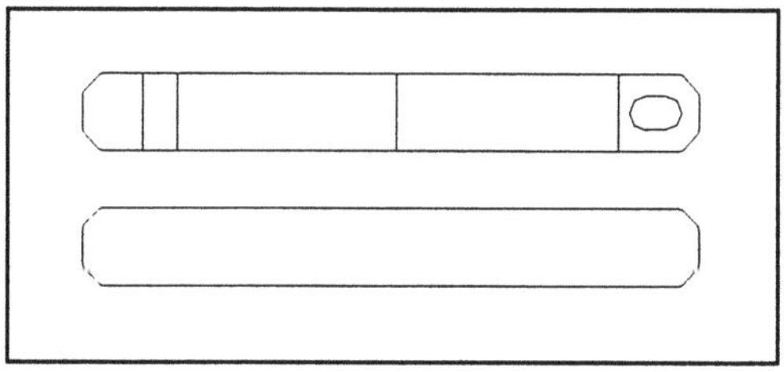

Figure 8.12: Nested arrangement of flat pattern.

The profiles of the notching punches required to stamp out the external profile of the part are drawn using AutoCAD drafting commands. Thereafter, the punch shape creation command can be used to convert these AutoCAD geometrical entities into punch profiles in the Die Operations Tree. Design heuristics is used to generate the bending areas and sequence of the bend automatically. Note that Wall3 has been selected as the "Base Wall". The pilots, punches and bend areas selected are shown in Figure 8.13.

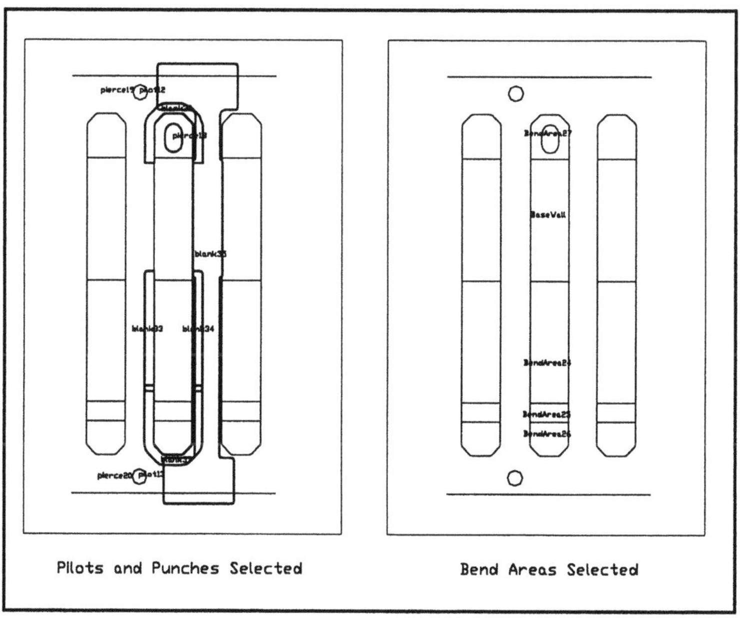

Figure 8.13: Pilots, punches, and bend areas selected to manufacture the test part.

Next, design heuristics is used to automatically generate the strip layout.

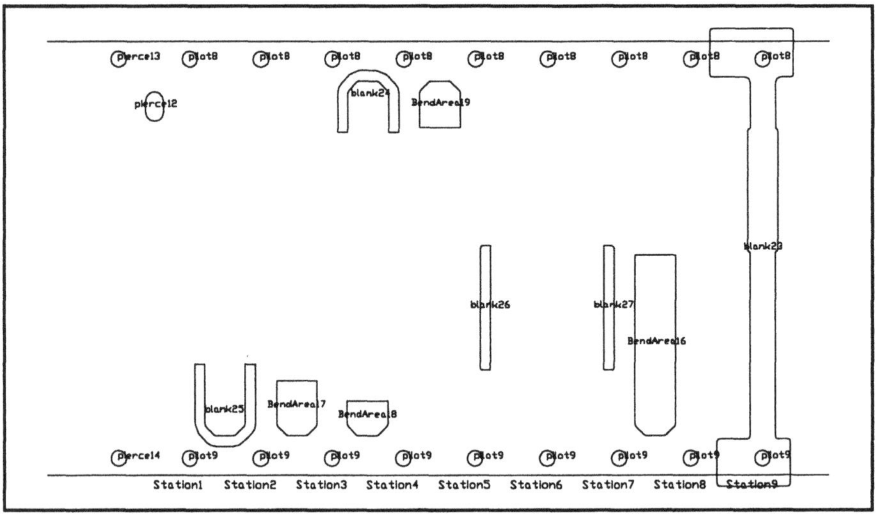

Figure 8.14: Strip layout automatically generated by the system.

However, this layout is not acceptable, hence, the finally strip layout shown in Figure 8.15 is obtained by interactively moving the stamping operations and inserting idle stations in the following manner.

8.4.7 3-D strip layout

The 3-D strip layout can be generated using information from the Die Operations Tree and Feature-relationship Tree. Figure 8.16 shows the 3-D strip layout generated using the ACIS husk.

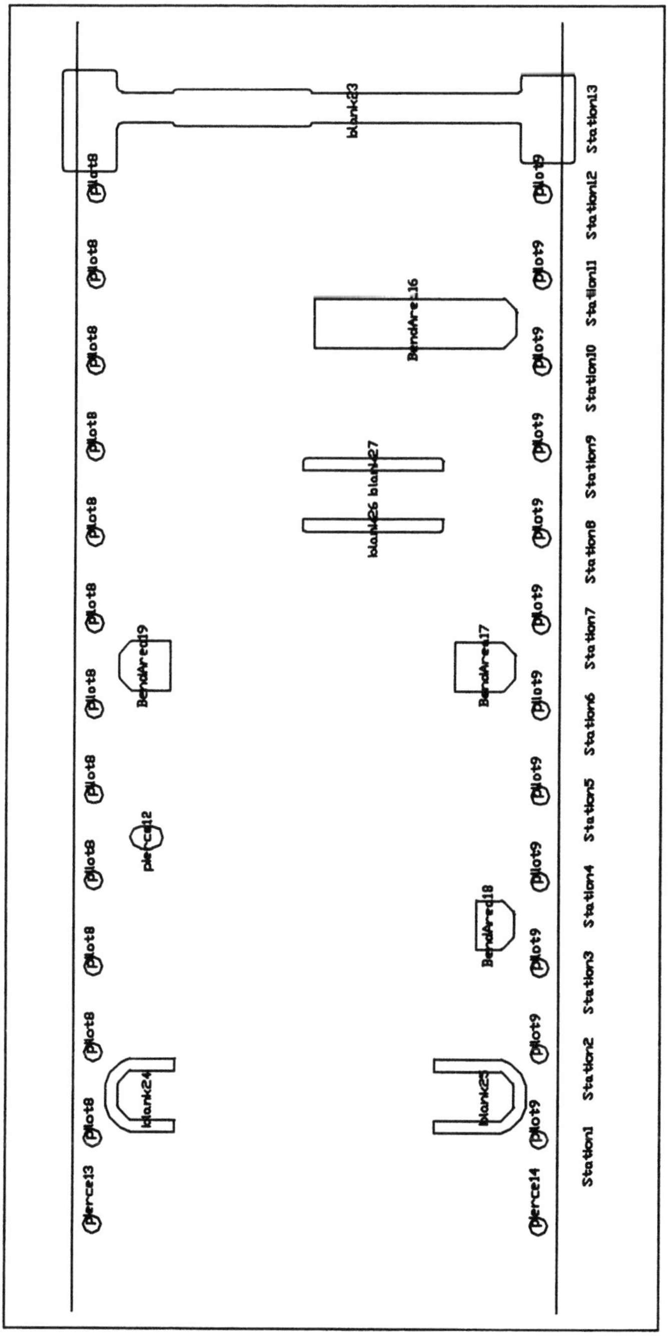

Figure 8.15: Final strip layout selected by the user.

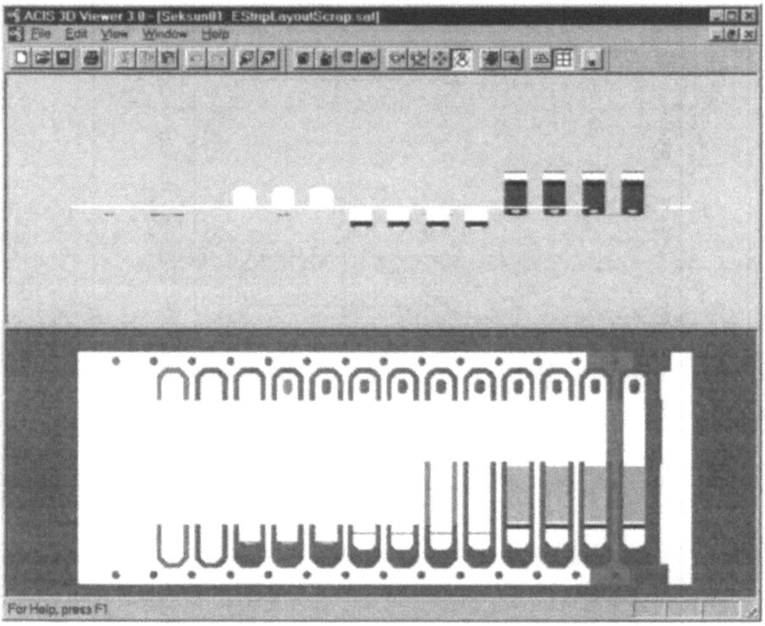

Side and top view of 3D strip layout as shown in AutoCAD.

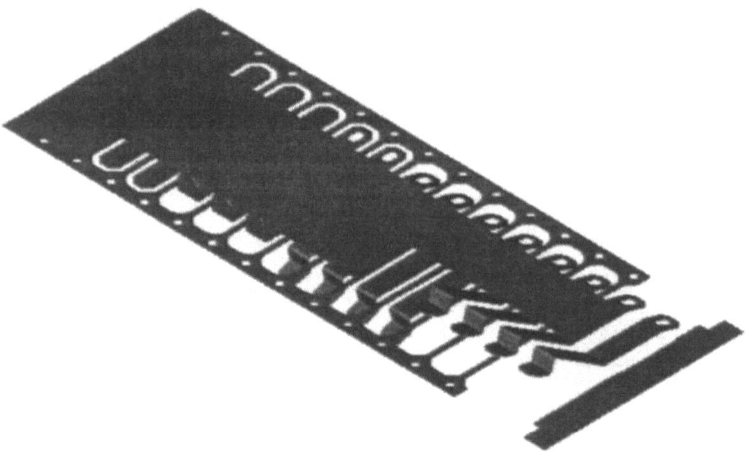

Isometric view of 3D strip layout.

Figure 8.16: 3-D strip layout generated for visual inspection.

8.4.8 Configuration of the progressive die

Once the strip layout is developed. The system has sufficient information to develop the Die Assembly Tree. Once the Die Assembly Tree is developed, the solid model of the progressive die can be built using the ACIS husk.

8.4.8.1 The die assembly

The various views of the die assembly tree are shown in Figure 8.17.

	Isometric View of Die Assembly.
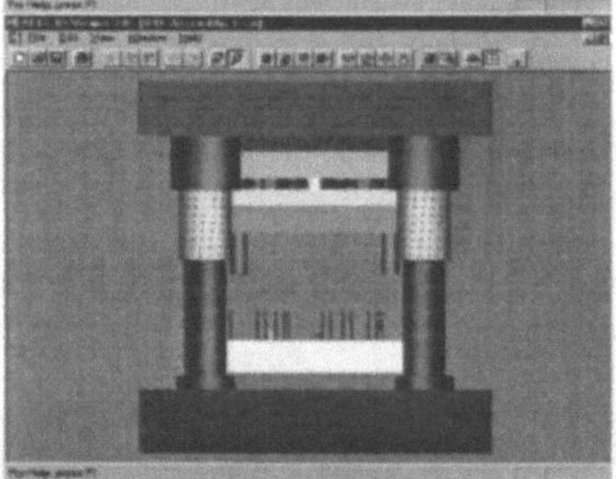	Side View of Die Assembly
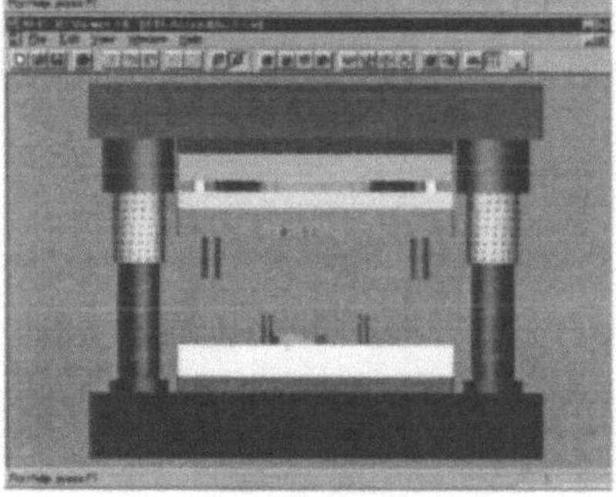	Front View of Die Assembly.

Figure 8.17: Various views of the die assembly.

The details of the upper die sub-assembly and lower die sub-assembly are shown in Figure 8.18.

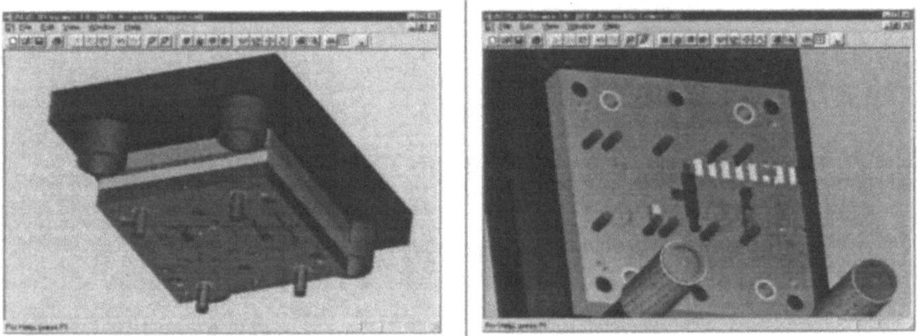

Upper die sub assembly	Lower die sub assembly

Figure 8.18: Close-up views of the upper and lower die sub-assemblies.

8.4.8.2 The die plates

The die plates generated by the system are shown in Figure 8.19.

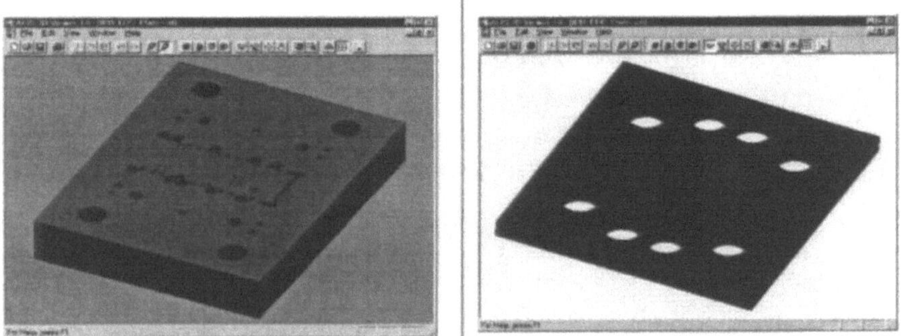

Lower Die Shoe (LDS)	Die Backing Plate (DBP)

Figure 8.19: Plates generated by the system.

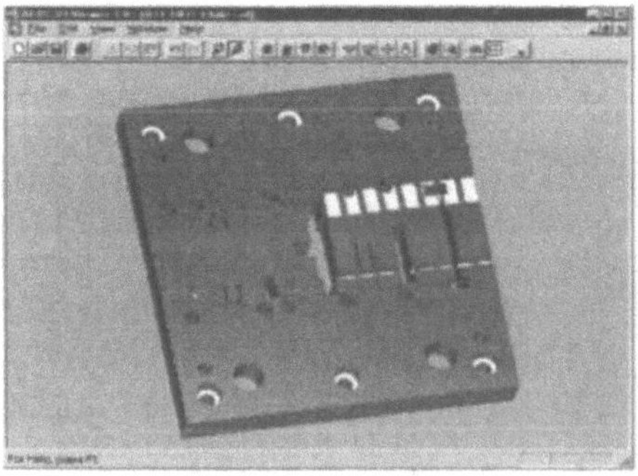

Die Plate (DLP)

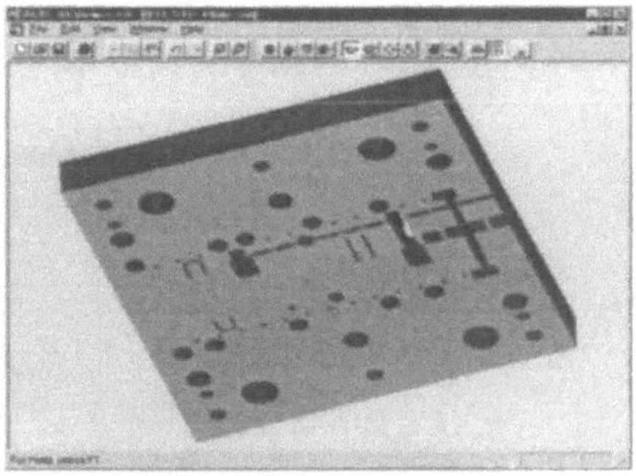

Stripper Plate (STP)

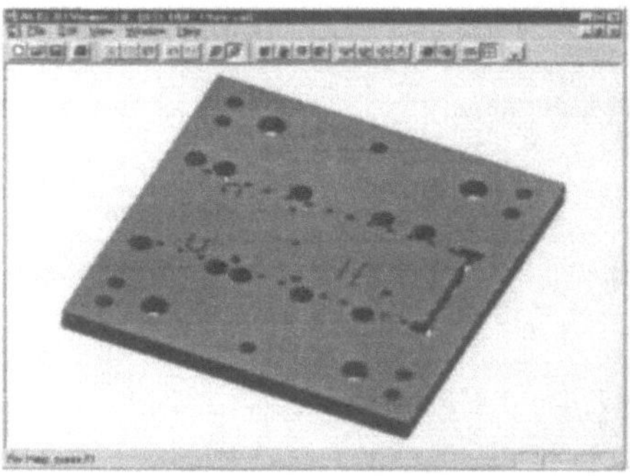

Punch Backing Plate (PBP)

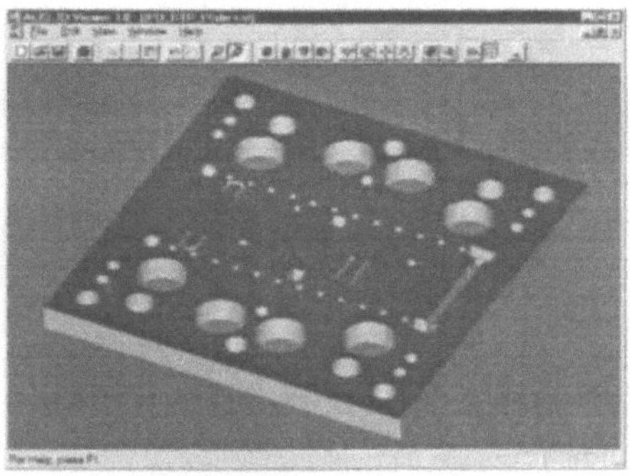

Top Backing Plate (TBP)

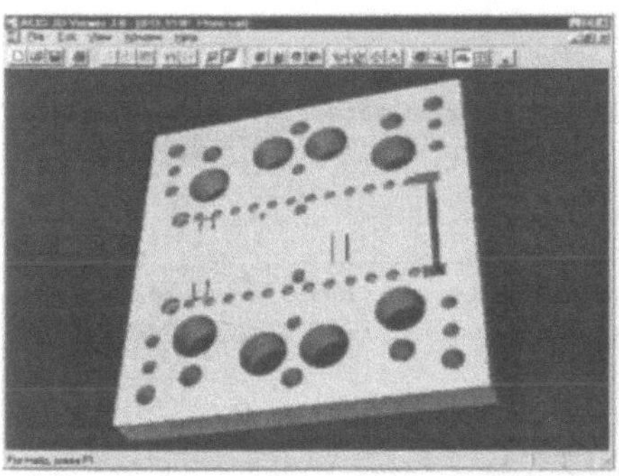

Punch Plate (PHP)

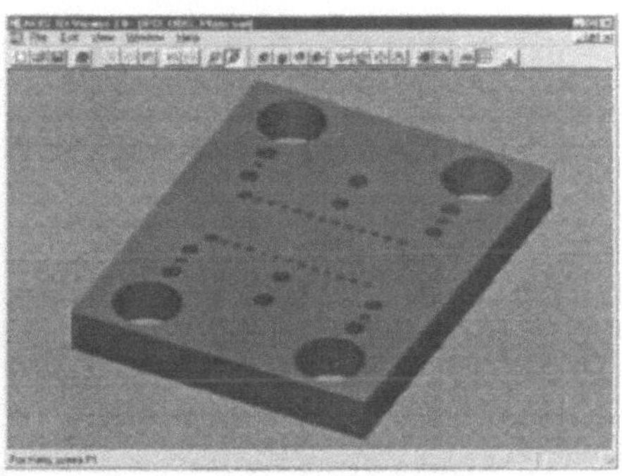

Upper Die Shoe (UDS)

Figure 8.20: Plates generated by the system.

8.4.8.3 Punches

Some of the punches generated by the system are illustrated in Figure 8.21

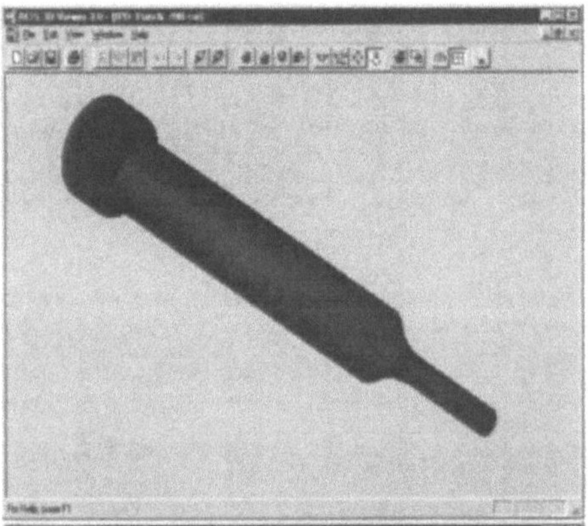	Standard Punch for Pierce13.
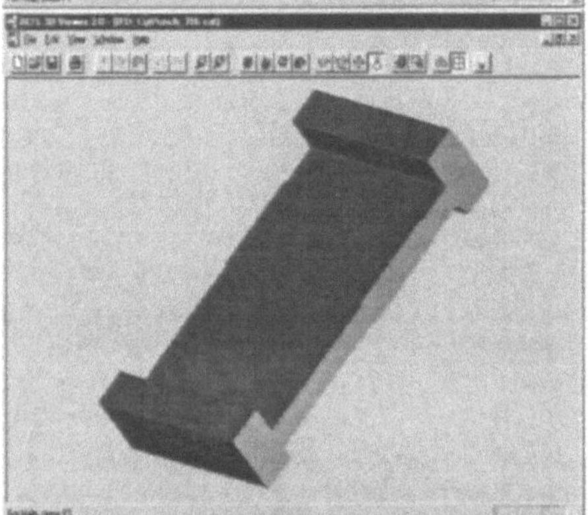	Straight Punch for Blank23.
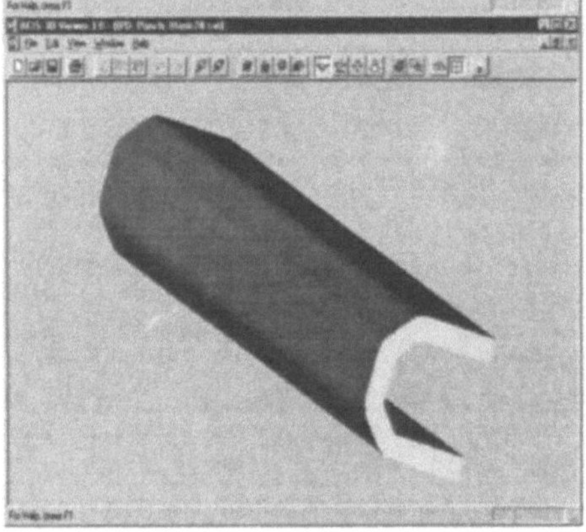	Straight Punch for Blank24. The User had ignored the need for a punch holder.

Figure 8.21: Punches generated by the system.

8.4.8.4 Bending punches and blocks

The L-bend configuration to form Wall0 and Wall4 (i.e., BendArea19 and BendArea17) are shown in Figure 8.22.

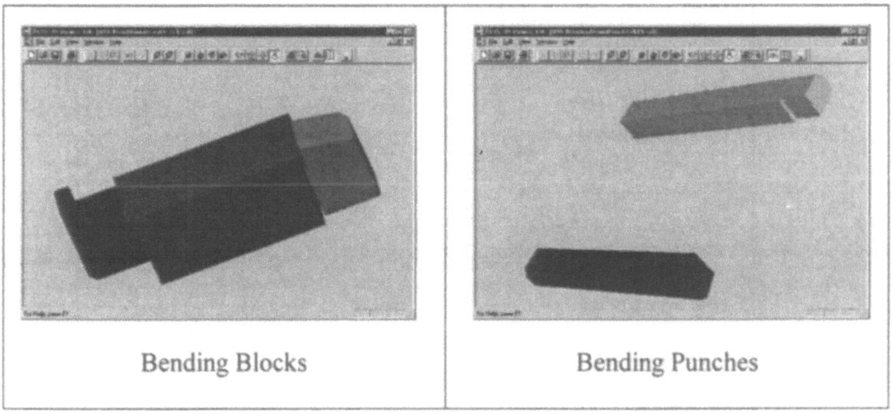

Figure 8.22: L-bend configuration to form Wall0 and Wall4.

The L-bend configuration to form Wall3 (i.e., BendArea16) is shown below.

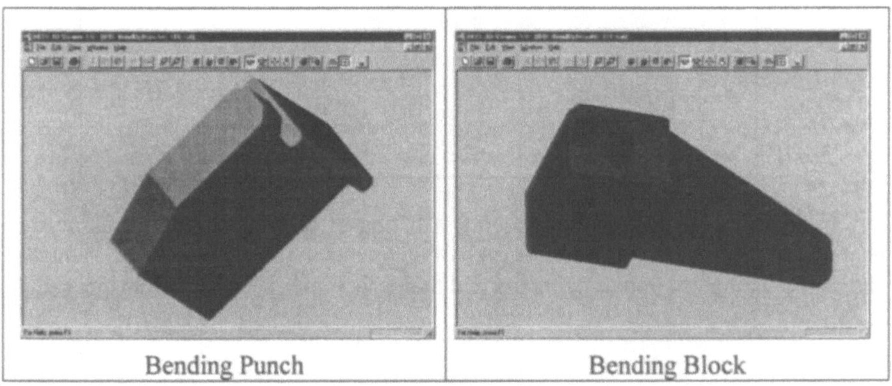

Figure 8.23: L-bend configuration to form Wall3.

8.4.9 Technical data generated by the system

The information stored in the Die Assembly Tree can be used to generate technical data that are useful to the shop-floor. A listing showing the standard components in the die assembly is shown in Table 8.2.

Table 8.2: Listing of standard components in the die assembly.

Standard Parts List:

No.	Name	Code	Qty
===	====	====	===
Fixed on the UDS plate:			
1	IPD_AdjustPin_200	PIN-N 6 - 55.0	24
2	IPD_AdjustPin_200	PIN-N 10 - 53.0	2
3	IPD_ScrewPlug_200	MSW 8	24
4	IPD_ScrewPlug_200	MSW 14	2
Fixed on the PBP plate:			
Fixed on the PHP plate:			
1	IPD_CoilSpring_200	SWH 30 - 70	8
2	IPD_Dowel_200	MSTM 10 - 70	4
3	IPD_Fastener_200	CB 10 - 55	6
4	IPD_Pilot_200	STAS 3 - 72 - P2.99	24
5	IPD_Punch_200	SPAS 8 - 70 - P3.00	2
6	IPD_StripperBolt_200	MSB 13 - 65	6
7	IPD_SubGuidePin_200	SGPH 16 - 120	4
Fixed on the BTP plate:			
Fixed on the STP plate:			
1	IPD_Dowel_200	MSTM 8 - 40	4
2	IPD_Fastener_200	CB 8 - 30	6
3	IPD_InsFastener_200	CB 5 - 20	2
4	IPD_SubGuideBushing_200	SGBH 16 - 25	4
Fixed on the DLP plate:			
1	IPD_Dowel_200	MSTM 10 - 70	4
2	IPD_Fastener_200	CB 10 - 55	6
3	IPD_GuideLifter_200	GLP 10 - 50.0- 2.5	10
4	IPD_InsFastener_200	CB 5 - 25	1
5	IPD_ScrewPlug_200	MSW 16	10
6	IPD_SubGuideBushing_200	SGBH 16 - 30	4
7	IPD_WireSpring_200	WH 14 - 65	10
Fixed on the DBP plate:			
Fixed on the LDS plate:			
1	IPD_GuidePost_200	RS 38 - 170 - M100 - 120	4

In addition, the system can generate a listing of openings and their associated geometrical and machining data for each of the plates in the die assembly. Table 8.3 shows the listing of circular openings and their machining data for the stripper plate (STP).

8.5 Conclusions

This case study illustrates the capability of the system developed. It shows the steps required to generate a progressive die from the part description. It also shows the intermediate interactions between the user and the system required to build up the respective knowledge-based trees described in this chapter. The outputs generated from the respective knowledge-based trees are also shown. The system is able to produce results that are able to meet most of the requirements of the industrial die designer. Therefore, it is believed that the intelligent techniques and the computer methodology proposed provide an excellent framework for the development of an industrial progressive die design system.

Table 8.3: List of circular opening and machining data for the stripper plate.

[STP_PLATE_01]: Circular Opening's Positions List:

1: Dia3.00 W/C: +0.005/s; Dia10.0 Drill Thru'; 8.0 F/F;
(1): x= -7.695 y= -30.690
(2): x= -7.695 y= 44.662

2: Dia2.99 W/C: +0.005/s; Dia4.0 Drill Thru'; 8.0 F/F;
(1): x= 6.305 y= -30.690
(2): x= 6.305 y= 44.662
(3): x= 20.305 y= -30.690
(4): x= 20.305 y= 44.662
(5): x= 34.305 y= -30.690
(6): x= 34.305 y= 44.662
(7): x= 48.305 y= -30.690
(8): x= 48.305 y= 44.662
(9): x= 62.305 y= -30.690
(10): x= 62.305 y= 44.662
(11): x= 76.305 y= -30.690
(12): x= 76.305 y= 44.662
(13): x= 90.305 y= -30.690
(14): x= 90.305 y= 44.662
(15): x= 104.305 y= -30.690
(16): x= 104.305 y= 44.662
(17): x= 118.305 y= -30.690
(18): x= 118.305 y= 44.662
(19): x= 132.305 y= -30.690
(20): x= 132.305 y= 44.662
(21): x= 146.305 y= -30.690
(22): x= 146.305 y= 44.662
(23): x= 160.305 y= -30.690
(24): x= 160.305 y= 44.662

3: Dia12.0 Drill Depth 4.8 F/B;
(1): x= -19.000 y= -35.799
(2): x= -19.000 y= 52.280
(3): x= 1.000 y= 52.280
(4): x= 33.000 y= -35.799
(5): x= 50.000 y= -35.799
(6): x= 54.000 y= 52.280
(7): x= 92.000 y= -35.799
(8): x= 110.500 y= 52.280
(9): x= 134.000 y= -35.799
(10): x= 138.000 y= 52.280

4: Dia22.010, W/C: Actual Size; Dia27.0, Drill Depth 4.0 F/B;
(1): x= 12.000 y= -86.760
(2): x= 12.000 y= 103.240
(3): x= 150.000 y= -86.760
(4): x= 150.000 y= 103.240

5: M10 Tapping Thru';
(1): x= -18.000 y= -96.760
(2): x= -18.000 y= 113.240
(3): x= 81.000 y= -96.760
(4): x= 81.000 y= 113.240
(5): x= 180.000 y= -96.760
(6): x= 180.000 y= 113.240

6: Dia8.00 W/C: +0.004/s;
(1): x= -18.000 y= -76.760
(2): x= -18.000 y= 93.240
(3): x= 180.000 y= -76.760
(4): x= 180.000 y= 93.240

7: Dia15.0 Drill Depth 9.0 F/B; Dia9.0 Drill Thru';
(1): x= -18.000 y= -61.760
(2): x= -18.000 y= 78.240
(3): x= 81.000 y= -61.760
(4): x= 81.000 y= 78.240
(5): x= 180.000 y= -61.760
(6): x= 180.000 y= 78.240

Bibliography

Adachi M, Inoue K, Funayama T 1983 Integrated CAD System for Progressive Dies. *Fujitsu Scientific and Technical Journal* 19(2):133-148

Bergstorm K, Kivivuori S, Osenius S, Korhonen A 1988 Computer aided design of a progressive die. In: Chersot J L, Ohate E. (eds) 1988 *Modelling of Metal Forming Processes*. Kluwer Academic Publishers, pp 155-162

Cheok B T, Foong K Y, Nee A Y C 1996 An intelligent planning aid for the design of progressive dies. *Proceedings, Institution of Mechanical Engineers, Part B: Journal of Engineering Manufacture* 210:25-35

Cheok B T, Foong K Y, Nee A Y C, Teng C H 1994 Some aspects of a knowledge-based approach for automating progressive metal stamping die design. *Computers in Industry* 24:81-96

Cheok B T, Nee A Y C 1996 A Knowledge-based approach for the planning and design of progressive dies. In: *Proceedings of the 12th International Conference on CAD/CAM Robotics and Factories of the Future*, London (14-16 August 1996), pp 265-271

Cheok B T, Nee A Y C 1997 A computer-based tutoring aid for progressive die design. *International Journal of Mechanical Engineering Education* 25(4):263-278

Cheok B T, Zhang Y F, Leow L F 1997 A skeleton-retrieving approach for the recognition of punch shapes. *Computers in Industry* 32:249-259

Choi S H, Wong K W 1992 A CAD/CAM Package for sheet metal blanking dies. In: *Proceedings of International Conference on Manufacturing Automation*, Hong Kong (10-12 August 1992), pp 674-679

Chow W W 1979 Nesting of a single shape on a strip. *International Journal of Production Research* 17:305-322

Dighe R, Jakiela M J 1996 Solving pattern nesting problems with genetic algorithms employing task decomposition and contact detection. *Evolutionary Computation* 3(3):239-226

Duffey M R, Sun Q 1991 Knowledge-based design of progressive stamping dies. In: Venkatesh V C, McGeough J A (eds) 1991 *Computer-aided Production Engineering*. Elsevier Science Publishers B.V., pp 241-247

Dym C L, Levitt R E 1991 An architecture for model-based reasoning. In: *Knowledge-based systems in engineering*. McGraw-Hill, New York, pp 179-190

Esche S K, Khamitkar S, Kinzel G, Altan T 1996 Process and Die design for multi-step forming of round parts from sheet metal. *Journal of Materials Processing* 59:24-33

Face Standard Components for Press Dies 1996.5 → 1998.4 1996 Misumi Corporation, 2-Chome, Toyo, Koto-Ku, Tokyo, 135 Japan

Fukuoka K, Horiuchi K 1986 Integrated mold and die design and manufacturing system. *Fujitsu Scientific and Technical Journal* 22(5):438-450

Hoffman E G 1991 *Fundamentals of Tool design*, Third Edition. Society of Manufacturing Engineers, USA

Huang K, Ismail H S, Hon K K B 1996 Automated design of progressive dies. *Proceedings Institution of Mechanical Engineers, Part B: Journal of Engineering Manufacture* 210:367-376

Ismail H S, Chen S T, Hon K K B 1996 Feature-based design of progressive press tools. *International Journal of Machine Tools and Manufacture* 36(3)367-378

Lazaro A S, Engquist D T, Edwards D B 1993 An intelligent design for manufacturability system for sheet-metal. *Concurrent Engineering: Research and Applications* 1(2):117-123

Lee I B H, Lim B S, Nee A Y C 1993 Knowledge-based process planning system for the manufacture of progressive dies. *International Journal of Production Research* 31(2):251-278

Lee R S, Chuang L C, Yu, T T, Wu M T 1995 Development of an assessment system for sheet metal forming. In: *Proceedings of the International Conference on Precision Enginering '95 (2nd ICMT) "Trends and Innovations in Precision Manufacturing Technologies*, Singapore, 1995, pp 515-518

Li Z 1993 Intelligent strip layout progressive design. *Zhongguo Jixie Gongcheng/China Mechanical Engineering* 4(3):1-3

Li Z, Chen Z, Hong Y, Xiao X, Xiao J 1990 Computer aided design of progressive die construction. In: *Proceedings of International Conference on Die & Mould Technology '90*, Shanghai, China, May 18-21, 1990, pp 357-363

Li Z, Li J, Li Z, Wang G, Xiao X, Xiao, J 1992 CAD/CAM of progressive dies for household electronic appliances. In: *Proceedings of International Conference on Advanced Technology and Machinery in Metal Forming*, Wuhan, 1992, China, pp 71-77

Lin Z C, Chang H 1994 An investigation of expert systems for die design. *Journal of Intelligent Manufacturing* 5:177-192

Lin Z C, Hsu C Y 1996 An Investigation of an expert system for shearing cut progressive die design. *International Journal of Advanced Manufacturing Technology* 11:1-11

Lin Z C, Hsu C Y, Yao K H 1989 Planning and building an integrated design software of blanking and piercing dies (ESSCP) with a micro expert system as a main structure. *Journal of the Chinese Society of Mechanical Engineers* 10(2):101-120

Mahajan P V, Poli C, Rosen D W, Wozny M J 1993 Design for stamping: A feature-based approach. In: *Design for Manufacturability* – 1993 ASME Design Engineering Divison, pp 29-49

Mantripragada M, Kinzel G, Altan T 1996 A computer-aided engineering system for feature-based design of box-type sheet metal parts. *Journal of Materials Processing* 59:241-248

Murakami H, Shirai K 1984 Cad system for press tools. In: *Proceedings of International Symposium on Design and Synthesis*, Tokyo, 1984, pp 86-91

Murakami H, Shirai K, Yamada O, Isoda K 1980 A CAD system for progressive dies. In: *Proceedings of 21st Machine Tool Design and Research Conference*, pp 687-592

Nakahara S, Toshio K, Tamura K, Asuke F, Soda C, Nakamura T 1978 Computer aided progressive die diesign. In: *Proceedings of 19th Machine Tool Design and Research Conference*, pp 171-176

Nee A Y C 1984 A heuristic algorithm for optimum layout of metal stamping blanks. *Annals of CIRP* 33:317-320

Nee A Y C 1989 PC-based computer aids in sheet-metal working. *Journal of Mechanical Working Technology* 19:11 -21

Nee A Y C, Cheok B T, Foong K Y 1992 A system for automated punch selection and staging design for progressive dies. In: *Proceedings of the International Conference on Manufacturing Automation*, Hong Kong, 1992, pp 10-12

Nee A Y C, Foong K Y 1992 Some considerations in the design and automatic staging of progressive dies. *Journal of Materials Processing Technology* 29:147-158

Okada M 1980 Computer aided manufacturing system for sheet metal parts. In: *Proceedings of 21st Machine Tool Design and Research Conference*, pp 603-609

Ostergaard D E 1963 *Basic Diemaking*. McGraw-Hill Book Company, New York

Ostergaard D E 1967 *Advanced Diemaking*. McGraw-Hill Book Company, New York

Poli C, Dastidar P, Graves R J 1993 Design for stamping-analysis of part attributes that impact die construction costs for metal stamping. *Journal of Mechanical Design, Transaction of ASME* 115(4):735-743

Prasad Y K D V 1994 A set of heuristic algorithms for optimal nesting of two-dimensional irregularly shaped sheet-metal blanks. *Computers in Industry* 24(1):55-70

Prasad Y K D V, Somasundaram S 1992 CADDS: An automated die design system for sheet-metal blanking. *Computing and Control Engineering Journal* 3:185-91

Ruan X Y, Zeng X Z, Wang L L, Li Y T 1987 A Blanking Die CAD System on Microcomputer. *Advanced Technology of Plasticity, Germany, Vol. 1*. pp 139-143

Schubert P B 1966 *Die methods: Design, fabrication, maintenance, and application, Book Two*. Industrial Press Inc., New York

Schubert P B 1966 *Die methods: Design, fabrication, maintenance, and application, Book One*. Industrial Press Inc., New York

See Toh K H, Loh H T, Nee A Y C 1995 A feature-based flat pattern development system for sheet metal parts. *Journal of Materials Processing* 48(1-4):89-95

Shirai K, Murakami H 1985 Development of a CAD/CAM system for progressive dies. *Annals of the CIRP* 34(1):187-190

Shirai K, Murakami H 1989 A compact and practical CAD/CAM system for progressive die. *Bulletin of the Japan Society of Precision Engineering* 23(1):25-30

Song X, Ying D, Wang E 1996 Feature representation product metal for progressive die CAD/CAM system. In: *Proceedings of the SPIE – The International Society for Optical Engineering, Vol. 2644*. pp 381-385

Strasser F 1971 *Functional design of metal stampings*. Society of Manufacturing Engineers

Takashi J, Fumitaro S 1990 On high speed precision blanking of IC lead-frames using a progressive die. Journal of Materials Processing 22:291-305

Tiza M 1995 Expert systems for metal forming. *Journal of Materials Processing* 53:423-432

Waller J A 1978 *Press tools and Presswork*. Portcullis Press, Bristol

Wang E, Pan G, Ying D 1993 Knowledge representation and uncertainty reasoning model in intelligent mould and die design system. In: *Proceedings of the Third International Conference on CAD and Computer Graphics*, Beijing, 1993, pp 678-681

Xiao X, Li Z, Li J, Li Z, Ma J, Xiao J 1993 An intelligent approach to the strip layout of progressive die design. In: *Proceedings of the Fourth International Conference on Technology of Plasticity*, pp 1817-1821

Yuan B W, Wang L, Zhang W D, Zhu J 1990 Graph processing and strip layout in a CAD system on mirco-computer for the multistage progressive die with bending and forming. In: *Proceedings of International Conference on Die & Mould Technology '90*, Shanghai, China, 1990, pp 434-441

Index

D

E

F

G

H

I

J

K

L

M

Q

R

S

MIX
Papier aus verantwortungsvollen Quellen
Paper from responsible sources
FSC® C105338
FSC
www.fsc.org

If you have any concerns about our products,
you can contact us on
ProductSafety@springernature.com

In case Publisher is established outside the EU,
the EU authorized representative is:
Springer Nature Customer Service Center GmbH
Europaplatz 3, 69115 Heidelberg, Germany

Printed by Libri Plureos GmbH
in Hamburg, Germany